Building the American Highway System

Technology and Urban Growth, a series edited by

Building the American Highway System

Engineers as Policy Makers

Bruce E. Seely

Temple University Press
Philadelphia

Temple University Press, Philadelphia 19122

Published 1987
Printed in the United States of America

The paper used in this publication meets the minimum requirements of American National Standard for Information Sciences—Permanence of Paper for Printed Library Materials, ANSI Z39.48-1984.

——CIP TO COME——

Contents

Series Preface

Historian Thomas L. Haskell observes that "all of us defer to experts." Bruce E. Seely explores a facet of that deference in this study of the successes enjoyed by engineers at the U.S. Bureau of Public Roads in shaping American highway policy. The key to the ability of leaders at the bureau to influence highway development at the federal and even the state levels, argues Seely, was invocation of the notion of technical service without political interest. This process, as Seely labels it, was one of apolitical expertise. Engineers at the bureau were fond of reporting that they simply collected statistics, checked on standards, and distributed funds in conformance with law. But it is Seely's contention that officials at the bureau actually shaped policy, in part through making their expertise available and in part through lobbying, and lots of it. In effect, the Bureau of Public Roads served as a mediating agency. Government officials had created the bureau in order to service the demand for roads, which was indiscriminate and insatiable. The roads that were constructed, however, were based on esoteric knowledge in civil engineering regarding highway location, design, and finance. Expertise and political savvy were merged at the bureau without disturbing the image of non-politics. That stance enhanced the authority of federal road engineers in a democratic society.

Seely locates the success of the bureau in the political styles of its directors between 1893 and 1956. In this scheme, Thomas H. MacDonald, chief of the bureau from 1918 to 1953, emerges as one of the principal figures in American road politics. During the course of negotiations leading up to the Federal Highway Act of 1921 (which determined the course of federal sponsorship of highway building for decades), MacDonald worked with road engineers and politicians to fix funding levels and allocation formulas, removing crucial decisions from the congressional arena. During the next two decades, he presided over construction of a national system of rural roads, the U.S. highway system. Although engineers at the bureau only checked on standards and reimbursed the states for no more than 50 percent of costs, MacDonald and his staff actually determined standards and helped guide route location, an area of particular contention as officials struggled to provide every American with a paved road. In 1944, MacDonald and his associates helped persuade President Franklin D. Roosevelt and members of the U.S. Congress to authorize construction of the Interstate Highway System.

Historian Stuart W. Leslie identifies MacDonald as an entrepreneur-bureaucrat. In retrospect, argues Seely, the inter-war decades under MacDonald's stewardship represented a "golden age" in American highway programming.

By the late 1930s, the consensus about highway building that the bureau had shaped began to dissolve into frantic competition for funds and disagreements regarding technical standards and construction priorities. The confusion intensified after the war, as the number of vehicles on the nation's highways doubled between 1946 and 1955. Furthermore, post-war inflation reduced real spending by 50 percent; and road engineers in many states had enhanced their own technical competence as well as their political know-how and clout. Limited funds, traffic jams, and a less deferential group of colleagues at the state level reduced the effectiveness of bureau engineers in the proverbial corridors of power. By 1955, despite a widespread demand for accelerated construction of the Interstate Highway System, members of Congress refused to follow the recommendations of President Dwight D. Eisenhower and bureau officials, but could not coalesce around a bill of their own. Not until 1956 could members of Congress agree to the terms of a bill that satisfied the conflicting goals of modest increases in gasoline taxes and rapid construction of the costly Interstate system along with thousands of miles of rural and urban highways. Although the new road legislation failed to satisfy several of the more ambitious schemes submitted by the bureau as well as President Eisenhower, still intact were bureau determination of technical standards and the system of state-federal cooperation in the administration of highway programs. "As they had always done . . . ," argues Seely, "the leading engineers at the Bureau of Public Roads resorted to politics, but without disturbing their reputation as objective experts." One must add that the Federal Aid Highway Act of 1956, by including substantial sums of money for construction of the Interstate system during the next decade and beyond, also held out the prospect of a technical fix for the inability of engineers at the bureau to secure relief for motorists through routine lobbying on a biennial basis. The authority of experts in the highway field was becoming attenuated.

Although events after World War II reduced the bureau's influence, the overwhelming impression up to that point is one of an agency possessing a remarkable skill at adapting to broad political currents. During the era of enthusiam for progressive causes, expertise and a non-political stance won wide favor. From about 1920 to 1940, the bureau's system of cooperation with state engineers in testing materials and fixing standards squared neatly with the ascendance in national politics of the cooperative ideal and notions of the associative state. After the war, the bureau's influence declined in the face of state agencies possessing enhanced skills and the rhetoric and methods of apolitical expertise. By the 1980s, as Seely observes, the hierarchy of prestige had been reversed, and younger engineers assumed that state transportation departments were the more logical loci for the application of advanced skills.

Seely's book is thus about politics. Curiously, that topic has fallen largely out of favor in a number of scholarly arenas. Particularly his analysis of politics proved uninviting in the history of technology, where scholars have generally failed to highlight the role of government or public policy in the initiation, financing, construction, operation, and regulation of technological systems. The omission is a remarkable one, because large engineering projects, especially public works such as bridges and roads, often relied on action taken by political leaders. Nonetheless, a minority tradition has matured, even flourished, in the area of the history of government and technologically oriented enterprises. Scholars such as David A. Hounshell have identified government as the initiator of massive changes in manufacturing technology and business; other such as Paul Barrett and Albro Martin find government exercising a negative influence on railroad and trolley operations; and still others including Christopher Armstrong and H. Viv Nelles as well as Thomas P. Hughes, K. Austin Kerr, and Harold L. Platt contend that the outcomes of negotiations between government officials and technologists were contingent on shifting alliances, court decisions, local circumstances, and, in Chicago, the presence of an always-venal politician. Studies by Mark S. Foster and Kenneth T. Jackson assess the political and ideological settings within which federal highway politics and its swarm of experts operated. Seely's book, with its exploration of an engineering agency that was created by politics and sensitive to it, joins this literature. Finally, Seely's analysis informs the work of younger scholars such as Guy Alchon and Donald T. Critchlow, who are examining the role of professionals in directing organizational change.

The implications of Seely's work also extend to transcendent questions in American historical scholarship. For instance, Seely's analysis permits a partial sense of closure regarding a fundamental paradox in the American political experience. In brief, leaders at the bureau routinely staffed professional committees, wrote legislation, and lobbied with favorite legislators as well as with state engineers and industrial representatives. Rarely were they silent, especially regarding funding levels. On the other hand, MacDonald and his colleagues spoke of expertise without political interest; they served technique, or so ran their voluminous rhetoric. Seely's book suggests that the paradox lies in the minds of observers rather than with engineers at the bureau. They operated within the framework of a common set of assumptions: gasoline tax funds should be dedicated to highway building; roads ought to serve traffic rather than local economic or urban development; and the Interstate Highway System, by virtue of its geometry of design and route coordinates, was a superb investment as well as exquisite engineering. Ideology guided design and technique. Even so, these ideological components of civil engineering had appeared in an inchoate form in their introductory courses in college, and had matured through several decades at an overburdened federal agency during the first several decades of massive highway building and spending. Anthropologist Clifford Geertz' contention that crafts and professions often "work by the light of local knowledge" helps bring this analysis into sharper focus. Civil engineering, one of the many engineering professions about which scholars have developed surprisingly little disciplined knowledge, operated by the light of its own set of assumptions. Engineers at the Bureau of Public Roads had been charged to perform immense tasks of national import within the confines of an imprecise federalism. The rhetoric of apolitical expertise served (reasonably well) both the ideals and ideology of their profession and the tumultuous politics of highway construction over the course of more than a half-century. Because historians have often overlooked the habits and symbols associated with engineers' work, we have frequently confused ourselves, a state Seely's book begins to rectify.

Books in the Technology and Urban Growth Series have developed several lines of analysis. Each has informed our recognition of technology as a system of patterned relationships that was guided by decisions made in political and social contexts. In turn, we have come to understand that choices made regarding a technology often have exercised broad and long-term consequences, especially in urban arenas. Few will doubt that the highway system, like its predecessor the trolley, has influenced form and function in city and countryside and the relationships between them. Seely's focus is not on the local scene but on the agency that helped initiate and guide American highway building. This is the sixth book in the series.

The Editors, Technology and Urban Growth Series

Acknowledgments

The road by which I came to study the United States Bureau of Public Roads and the development of the American highway system is not a straight one. Perhaps it had roots in a childhood fascination with the huge machines that I saw building the Interstate system during family vacations in the 1960s. Later a graduate school search for a dissertation topic led me to consider writing a history of what Eugene S. Ferguson called "the earthmoving revolution." But as often happens in historical research, new questions emerged from my digging. When the activities of federal highway engineers captured my attention, I found myself moving from examining the impact of earthmoving machinery to questioning American highway policy and the role of engineers in forming that policy. Of central importance in helping me make this transition were the patient advice and consistently helpful suggestions of my professors at the University of Delaware in Newark: George Basalla, Glenn Porter, David Hounshell, and, most importantly, Eugene Ferguson. I was privileged to study with and learn from this group of scholars who have provided me with a model of original intellectual inquiry that I hope to emulate always. I also owe a particular debt to the Hagley Graduate Program of the Hagley Museum and Library in Wilmington, Delaware, for its support. Hagley is a special place that provides an atmosphere especially conducive to graduate instruction by virtue of its staff, students, and visiting scholars such as Lynwood Bryant and Merritt Roe Smith. Hagley and the people there started my research moving in the right direction.

The list of people who have read part or all of this manuscript since I left Hagley is long, and I extend thanks to numerous colleagues, including Martin Melosi, Gary Anderson, James Rosenheim, Sally Hacker, John Smith, James Schneider, Bill Leslie, and Joel Tarr. I save special thanks for two scholars whose work I greatly respect and who have provided far more assistance than a simple acknowledgment can indicate. From the beginning of my work, I have crossed paths with Mark Rose, and without fail his comments have been cogent and his advice supportive and enthusiastic. I also owe special thanks to Paul Barrett, whose helpful, detailed comments as reader for Temple University Press kept me on the right track at crucial points in the revision process. I also wish to express my appreciation to my editor, Janet Francendese, the entire staff of Temple Press, and copy-editor Sherry Bobbitt for their patience and help as they turned the manuscript into a book.

I received consistently excellent assistance from a number of archivists and librarians. At the Washington National Records Center in Suitland, Maryland, where I camped for weeks on end, Wilbert Mahoney, Ellie Mellaman, and Stanley Brown answered a steady stream of questions as they helped me plow through the Bureau of Public Roads records. Mary Jo Pugh at the Bentley Library at the University of Michigan in Ann Arbor not only answered requests for information and records but also brought a number of important collections to my attention. Leroy Barnett at the Michigan State Archives in Lansing worked patiently with my requests for material, as did Alice Dalligan, Chief of the Burton Historical Collection in the Detroit Public Library, and Richard A. Shrader of the Southern Historical Collection at the University of North Carolina at Chapel Hill. I also was cordially received and assisted at the Library of Congress, Washington, D.C.; the University of Illinois at Champaign-Urbana Archives; the Folger Library at the University of Maine, Orono; the Texas A&M University Archives, College Station; the Iowa State University Archives, Ames; the Iowa State Department of Transportation Library, also in Ames; the Memphis Public Library; the Memphis/Shelby County Archives; the Carl Albert Congressional Research and Studies Center at the University of Oklahoma, Norman; the Franklin D. Roosevelt Library, Hyde Park, New York; the American Society for Testing and Materials Library, Philadelphia; the Transportation Research Board Library and the National Academy of Sciences—National Research Council Archives, both in Washington, D.C.; the South Dakota Historical Resource Center, Pierre; and the Barker Texas History Center at the University of Texas at Austin. I also owe special thanks to G. Donald Kennedy, an engineer for the Michigan Highway Department and the Automotive Safety Foundation and head of the Portland Cement Association, for opening his papers and spending several hours with me talking about highway developments in Michigan and the origin of the Interstate highway system. I also have to thank David Cook, Staff Archaeologist at the Iowa State Department of Transportation, and Dr. Chris Hendrickson, Professor of Civil Engineering at Carnegie-Mellon University, Pittsburgh, for sharing their thoughts with me.

Most of this research would have been impossible without the generous financial assistance of several institutions. The Hagley Museum and Library merits a second mention for the graduate fellowship they awarded me. The travel funds that enabled me to visit manuscript and other repositories across the country were provided by Grant SES-8007899 from the National Science Foundation's Doctoral Dissertation Research Grant Program. Further assistance came from the National Endowment for the Humanities Travel-to-Collections Program, Grant

RY-20079-84, as well as from Texas A&M University, which provided me with not only the encouragement to continue my research and writing but also reduced teaching assignments, use of a computer, and two minigrants for additional travel. I also received truly valuable help from a postdoctoral Fellowship in Technology, Science, and Society at Carnegie-Mellon University for the academic year 1984–85. This award allowed the time for uninterrupted writing, a luxury that proved vital to the completion of this study.

Despite all the wonderful help and assistance I have received, writing and research can at times become a lonely business. During the preparation of this manuscript I have been fortunate to have always had the support and love of my wife, Nancy, and my son, Michael. No words can suffice to thank them for tolerating my absent-mindedness and my inattention as I worked at "Daddy's 'puter." At times they saw more of my back than my face. Yet their support and encouragement were the most important help throughout the years of my work. To them I dedicate this book, with gratitude and love.

Building the American Highway System

INTRODUCTION
Experts in a Democratic Society

Between 1890 and 1960, the American highway system was transformed from two million miles of poorly built, disconnected roads into a coordinated system of four million miles of paved roads. During these years the focus of American highway policy also changed from farm-to-market roads to a federally aided system of roads between cities. In 1956 this culminated in the creation of the Interstate system, a 41,000-mile network of limited access expressways. Although urban aspects of road building have attracted much attention, there are few general studies of roads, and the basic assumption has been that highways are an expression of the vast interest in the automobile.[1] In my judgment, however, the origin and administration of highway policy were not so deterministic but rather the result of human decisions, and the central actors behind these choices were engineers in the federal Bureau of Public Roads (BPR), which was formed, as the Office of Road Inquiry, in 1893. They emerged as the arbiters of not only the technical but also the political and financial aspects of American highway policy because they established and maintained a reputation for apolitical expertise for more than sixty years.

The role of BPR engineers in highway development will be developed in three sections. The first examines the origin of the bureau and with it a Progressive pattern of promotional and technical efforts by which two of its directors shaped federal highway legislation between 1912 and 1921. The policies they fashioned also produced a consensus among the highway community that guided road building through the mid-1930s, which is the subject of the second section. These chapters trace a pattern of state and federal cooperation in building federal-aid roads, setting construction and materials standards, and conducting research in which BPR engineers were dominant despite a federalist framework intended to balance state and national authority. Superior expertise accounted for the influence of the federal engineers, especially because the BPR operated within the ideals of the associative state that emerged in the 1920s. The final section focuses on the turbulent two decades after 1936, when this consensus among road builders collapsed and was only laboriously restored in the 1950s. After 1940 the BPR's influence was not as overpowering as it had been, and the pattern of policy making shifted toward a pluralism in which elites guided the outcomes. Yet even though the context of their efforts had shifted, federal highway engineers remained the most important participants, relying on their image as apolitical experts to shape crucial elements of the policy that unraveled the political deadlock. Thus the fundamental concern of this book is the shaping of federal highway policy.

But the central position of engineers in making public policy raises a more general question about the interaction of experts and society. Experts have been variously described as the ideal leaders of society, docile and powerless providers of public wants, servants of corporate capitalism, and technocratic usurpers of power.[2] At issue here is the place of experts in an American society that has faced the occasionally contradictory demands of democratic traditions and technological change. The emergence of greater social equality after 1820 produced the first tensions, which were visible in a distrust of learned experts. Only the Gilded Age's "search for order,"[3] led by railroads and manufacturing firms confronting the problems of scale and complexity posed by technical change and national markets, challenged this outlook. To control dispersed operations, they pioneered new management systems and also turned hesitantly to scientific research. By the 1890s, engineers such as Frederick Taylor were developing the concept of scientific management as a means of solving both production and labor problems.[4] At the same time, urban engineers demonstrated they could mitigate sanitation, power, and transportation problems, while social scientists attacked a political system they saw flawed by corruption.[5]

After 1900, the rising status of experts was visible in Progressive attempts to use specialists to solve the social and technical problems of an emerging urban industrial society. Several factors underlay the reformers' choice of problem solvers.[6] First, because their optimistic belief in the ability to build a better society was buttressed by the technical wonders displayed by industry and engineering after 1870, the experts responsible for these advances seemed the logical managers of reform. Second, Progressives assumed people were rational and would do the right thing if it were shown to them. Set apart by their university degrees, experts epitomized the scientific rationality increasingly favored by the times. Third, most reformers demanded that public behavior correspond to a morality committed to such principles as the "general welfare" and the "common good."[7] Engineering and other professions that embraced codes of ethics embodied such altruistic ideals of public service. These assumptions combined to produce calls for nonpartisan government. Many Progressives loathed what they considered an inefficient political system; others simply doubted the wisdom of majority rule. In either case, scientifically trained experts presented an alternative: Rational experts devoted to efficient public service could be given decision-making or regulatory authority and thus replace politicians prone to machine corruption.[8]

At the outset, one should recognize that this portrayal of experts as apolitical automatons was unrealistic. Yet after 1905 these assumptions about expertise appeared in a number of reform initiatives. Municipal bureaus of research, staffed by young social scientists, furnished statistical data supporting proposed changes. Local governments headed by city managers removed daily operation from the hands of corrupt officials and gave it to experts, many of them engineers. Schools became the province of professional educators concerned with uniform standards. The high point of efficiency reforms may have been Gifford Pinchot's conservation movement, which sought to give federal experts control of resource allocation and management.[9] The same emphasis on apolitical expertise, however, was very evident in the reform effort to improve the American highway system, which had the initial goal of creating local and state highway agencies directed by engineers and thus divorced from politics. Although talking only about roads, civil engineer T. F. Hickerson described the rationale for *all* reforms that relied on expertise in his 1919 paper to the Road Institute at the University of South Carolina:

> There can be no guarantee of a wise expenditure of [highway funds] . . . without the freest application of scientific non-political methods. Engineers must have ample time and opportunity to investigate and plan thoroughly all the important features of roads that are to be built and kept up, so that whatever is done, is done in a thoroughly scientific manner.[10]

The Progressives' reforming zeal collapsed in the aftermath of World War I, and historians have long debated to what extent their influence survived. This study contends that the Progressives firmly affixed an ideal of apolitical technical experts into the American conscience that has continued to define popular assumptions about such expertise. This does not mean that this image correctly described experts nor that the definition of expertise remained constant. Yet the image has retained credibility in part because experts and others found it congenial. Engineers, especially those from humble backgrounds, used it to raise their social status. Politicians appealed to apolitical expertise as a political strategy to limit debate. In other words, the rhetoric of expertise could survive without an appreciation for the historical meaning of the ideal.[11] Yet one need not assume that only cynical or self-serving factors explain the survival of this ideology of expertise. As one observer has suggested, the appeal of this apolitical image of experts may lie in the willingness of technical experts to tackle sticky problems that resist resolution by democratic institutions.[12] In addition, many technical experts sincerely believe they can make objective decisions for the benefit of society, and on occasion they have been able to live up to this claim. Thus for many reasons the idea gained a life of its own after the Progressive period.

This study of the policy-making role of highway experts will demonstrate that the appeal of apolitical experts as problem solvers certainly was evident in the way America built its road system. Nor is this the only instance that attests to the durability of this image. Herbert Hoover's popularity in the 1920s rested on his organization of food relief after 1916 and his promotion of industrial efficiency after the war. Trained as a mining engineer, he seemed to embody all the best reasons for relying on experts, and the cooperative philosophy of the associative state that he championed relied on the ideal of apolitical expertise. The emergence in the 1920s of the Brookings Institution as a respected source of nonpartisan policy analysis also showed the survival of this image. Donald Critchlow has argued that Progressive ideals were the basis for the founding of this organization and that "its elitist origins, its continuing belief in nonpartisan expertise, and its faith that public policy could be depoliticized by separating administration from politics" guided the body through the 1950s.[13]

The New Deal also relied on experts, with many of its ideas coming from men who had retained the Progressive assumptions about expertise. One who embraced those beliefs was Joseph Eastman, Federal Coordinator of Transportation and author of the first plan for federal regulation of motor trucks. Other Depression-era evidence of the image of experts as ideal problem solvers was the sudden but brief fascination with Howard Scott's technocracy movement, which claimed that the Depression proved that engineers should control society. The rapid decline of his movement followed the discovery that Scott himself had no technical credentials.[14] Yet no one seemed to protest the motto of the Chicago Century of Progress Exposition in 1933—"Science Finds, Industry Applies, Man Conforms."[15]

This image of expertise has not faltered since World War II. A central issue in the debate over the creation of the National Science Foundation in the late 1940s was whether political or scientific criteria should guide the allocation of its funds.[16] The willingness of society and Congress to entrust the development of nuclear technology to the physicists, engineers, and military and corporate officials who controlled it also assumed that experts are objective. Alvin M. Weinberg based his call for creation of a priesthood that would assume responsibility for nuclear power on a similar image of expertise.[17] It is also instructive of the strength of this image of experts that many contemporary technologies have encountered difficulties of popular acceptance or support only after confrontations between experts who had reached opposite conclusions from the same data. Only dramatic public disagreements and controversies, such as have been raised by nuclear power, the environment, and many individual technological projects, including the American supersonic transport plane and the Alaska pipeline, have caused the public to question the image of experts as apolitical. In other words, it was a long time before the public, in an analogy used by historian of technology Eugene S. Ferguson, began to learn the consequences of asking the barber if it needed a haircut.[18] Even so, the image continues to survive, as evidenced by recent proposals to adjudicate technical disputes in a Science Court where experts will render objective decisions.[19]

The development of American highway policy in the twentieth century illustrates not only the emergence but also the continuation of this ideology of expertise. This does not mean that the activities of federal engineers always matched the rhetoric surrounding this image. Nor does it mean that highway engineers alone determined this country's highway program. Policy decisions emerged from a complex, pluralist political process that strained the engineers' belief that policy could be made on narrowly technical grounds. Moreover, highways were administered through federalist procedures, which meant that only states and local governments built roads. But to a surprising degree, the government efforts and policies were the product of the federal highway engineers' vision of what roads should be and do. What follows is an examination of American highway policy from 1890 to 1956 that illustrates how the engineers at the BPR relied on the image of apolitical expertise to influence a democratic society.

PART I

Building the Constituency and Setting the Patterns, 1890–1921

In late nineteenth-century America, the railroad was king—the basis of transportation and the most profitable and powerful sector of the country's economy. American roads, on the other hand, were in a horrible state. During the early national period, Robert Morris, Albert Gallatin, and Henry Clay expressed considerable interest in roads, yet by 1840 enthusiasm for railroads had pushed highways into the background. State bankruptcies in the late 1830s inspired numerous state constitutional amendments forbidding expenditures for internal improvements, including roads, and cemented the tendency of governments to ignore all but western military roads and make the railroad the national transportation system.[1] Thus few people in 1890 would have predicted that by 1920 the railroad's unrivaled position in transportation would be challenged and later overtaken by a system of highways. Even the first statistical compilation in 1904 showed only 154,000 of the more than two million miles of American roads were improved in any way. In 1916, however, Congress authorized federal funds for road building and in 1921 focused these funds on a federally-aided system of primary highways. The creation of popular and political support for a road-building program, the determination of the goals highways should serve, and the establishment of the procedures for reaching these ends took place between 1890 and 1921. Significantly, highway engineers in a small office in the United States Department of Agriculture emerged as the central figures in the good roads movement that orchestrated these promotional, legislative, and technical developments.

Chapter One

The Early Good Roads Movement, 1890–1905

This department is to furnish information, not to direct and formulate any system of organization, however efficient or desirable it may be.

Secretary of Agriculture
J. Sterling Morton, 1893[1]

The effort to improve American highways dates to the 1880s, with the inspiration coming from the appearance of the high-wheeled "ordinary" bicycle in 1876 and the "safety" version from England after 1887. Although bicyclists spearheaded the first demands for better roads, by 1895 the newly formed Office of Road Inquiry (ORI) in the Department of Agriculture was leading an amorphous movement at the local level designed to convince rural residents that good roads were their key to a better life. Central to both the ORI's position in the movement and the acceptance of the good roads theme was the office's almost instant reputation for technical expertise. To this it joined a series of promotional ventures, usually run in cooperation with other good roads supporters. The two men who headed the office from its creation in 1893 until 1905, Roy Stone and Martin Dodge, established these approaches, which long remained the basis of the federal highway agency's operations. In some respects, the office moved toward the pattern of technical professionalism that had begun to emerge in this country by 1880, but it also retained the traditional political orientation of a federal bureaucracy. This chapter examines this combination of technical and political activity in the divergent styles of the ORI's directors. Although their efforts strengthened the good roads movement, their activities also suggest why turn-of-the-century Americans showed a growing willingness to employ apparently impartial experts as problem solvers.

Creating Public Support: Roy Stone and the LAW

During the 1880s, interest in highways revived in the United States for the first time since the plank-road craze of the 1840s. The demands came from bicyclists who encountered miserable roads, and the

League of American Wheelmen (LAW), a bicycle-riders group formed in 1880, led the first campaign for state funds for local road construction after 1888. It had little success, however, because of opposition from farmers, who feared that improvements for "idle-rich" urban cyclists meant higher property taxes.[2] In response to this impasse, the Wheelmen began an educational program on the benefits of better roads, which established the basis for the good roads movement for the next quarter century. In pamphlets, weekly bulletins, and, after 1892, the magazine *Good Roads*, the LAW claimed that good highways would raise land values, open new markets, provide access to manufactured goods, end rural poverty, increase political participation by farmers, and improve education. In short, the LAW argued that better roads would end rural isolation, and added as a final touch that road building cost less in the long run than existing maintenance programs.

By 1892, Albert A. Pope, the leading bicycle manufacturer and organizer of the LAW, decided that there was sufficient support for a campaign for federal funds for highway improvement. A New York civil engineer and Wheelman named Roy Stone prepared a bill creating a National Highway Commission, while the Wheelmen organized the National League for Good Roads (NLGR) at a national conference on roads in late 1892 to promote the bill. Senator Charles Manderson, the sponsor of Stone's bill, was elected president of the NLGR, Stone was made vice-president and secretary, and the conference participants endorsed his plan. The bill failed, but Congress earmarked $10,000 in the Department of Agriculture's 1894 budget for investigations of road construction and management. The secretary formed the ORI on October 3, 1893, appointing the LAW's Roy Stone as special agent and engineer for road inquiry.[3]

One Wheelman later boasted that the NLGR "accomplished the adoption of the Government Good Roads' Bureau."[4] But as important a reason for the creation of the ORI was the lack of basic information regarding highways. Roads were primitively built by untrained local officials and maintained by statute labor performed in lieu of tax payments. That even the total road mileage in the United States was unknown strengthened the argument for such an inquiry. In addition, placing the office under Stone in the Agriculture Department signified federal acceptance of the Wheelmen's vision of rural road improvement. Finally, the ORI provided a focus for what was a locally oriented good roads movement. Highways were strictly a province of local government. New Jersey passed the Wheelmen's bill giving state aid for roads (probably drafted by Stone) in 1892, and Massachusetts created the first highway department a year later, but no state built roads directly. The LAW, with its poor image among rural residents, had only begun to build support for good roads. Thus the ORI became the leader of the good roads movement the day it was founded.

The office that Roy Stone headed was the product of traditional political lobbying, directed by Stone himself and another Wheelman. But the secretary of agriculture limited its goals in important ways, directing Stone to focus on the technical task of gathering information and to avoid politics, which he feared would cause problems.[5] This pattern prevailed in other bureaus of the Agriculture Department as it became one of the country's most respected scientific institutions in the 1880s, with a scientific staff that solved a variety of practical plant, animal, and pest problems. It also matched the efforts of engineers in cities, who were assuming control of sanitation projects.[6] Stone, a civil engineer, certainly understood the arguments about professionalism that underlay these trends. But even though he had a yearly budget of only $10,000 until 1896, after which it dropped to $8,000, he never broke completely free of the earlier desire for political involvement. In the end, he pioneered three enduring patterns of activity for the ORI: build a reputation for technical knowledge, promote the gospel of good roads, and utilize cooperation to reach those goals. The first fulfilled the office's mandate from Congress, and the second grew from the promotional goals of the Wheelmen, but the third was Stone's hallmark, even if it was necessitated by a small budget.

At the outset, Stone focused on collecting and publishing general information about road conditions and construction, relying on cooperation from several sources. The ORI had only a single clerical assistant, so Stone appealed to governors, secretaries of state, congressmen, state geologists, and railroad presidents for information on state road laws, supplies of road materials, and rates for rail haulage. As he explained to the governors, "Such knowledge can be practically reached and disseminated only through a central agency, but that agency will need the assistance of all State and local officials concerned in order to bring its work within the means of Congress, and within a proper time limit."[7] Stone thus saw the ORI as a clearinghouse for information on roads, a role it assumed almost by default. Not one university taught highway engineering until Harvard's Nathaniel Shaler launched a program in the 1890s. And even after engineering texts appeared, local road builders still needed basic information, which they found in the bulletins and circulars that emerged from Stone's compilation efforts. Among the early subjects were road materials, railroad freight rates, and speeches on road laws; a few were written by experts on specific technical topics. In its first two years, the ORI released eighteen bulletins and twenty-three circulars, and their wide circulation quickly gained the office the reputation as the source for answers to road construction and administration questions.[8]

While data collection always remained important to the ORI and was even expanded to include the preparation of county road maps by 1895, the office also inaugurated more formal technical tests and investigations. The first concerned an inquiry into the traction resistance of different road surfaces, which grew into a study of steel-track roads in 1897.[9] But the ORI's most successful and popular technical activity through World War I was the object-lesson construction program. The idea came from Massachusetts, where the highway commission used its meager funds to construct a short sample of good road in every county. The concept was warmly received in an Agriculture Department in which the problem-solving style stressed practical demonstrations as much as scientific research. Stone built the first quarter-mile demonstration section at the National Good Roads Parliament in Atlanta in 1896, and quickly launched a regular object-lesson program with goals that paralleled those of the agriculture experiment stations created in 1887 to introduce farmers to new procedures and crops.[10] Federal engineers visited communities willing to furnish materials and labor and supervised construction using machinery donated by manufacturers, thereby teaching otherwise untrained local officials the best procedures for building simple but durable roads. The program was instantly popular. The office's part-time engineers built two object-lesson roads in 1897 and twenty-one in nine states by 1899, yet failed to meet the demand. A 1904 survey by the office found that the program worked: "Once the object-lesson road is established in a community with all the advantages to travel over its smooth and hard surface, the people are forced to a realization of the economy of good roads and the result is generally an increased expenditure along that line." Each $500 for object-lesson roads stimulated the building of an average of ten miles of new roads at a cost of $30,000.[11]

These demonstrations cemented the ORI's reputation as the central source of technical information about roads, and in 1897 ORI replies to information requests totaled two thousand pages. But this growing popular enthusiasm was not translated into increased federal spending for roads, which led Stone to develop the second component to ORI efforts—promotion of the gospel of good roads. In 1893, the secretary of agriculture had warned the ORI to avoid an advocacy role that might lead to criticism. The object-lesson work, however, was one way of circumventing this limitation. Although a technical venture, these projects naturally spread the gospel of good roads through the excitement they always created. Nor was this the only way in which Stone continued what he had started for the LAW, namely building public support for good roads. He called this work "educating the public" and gave it as much attention as the technical projects. Stone remained a leading speaker at numerous conventions of good roads supporters as the ORI became the information leader. Furthermore, along with technical pamphlets, the office printed the proceedings of five road conventions, and fully two-thirds of its pamphlets were promotional.[12]

Although Stone maintained rigid standards of propriety, he never sacrificed his promotional agenda. By the mid-1890s Stone was devoting more time to advocacy than to technical questions, thanks again to cooperation from outside sources. He relied heavily on the LAW, using his position as secretary of the NLGR to mask his direct political activities. In fact, Stone made little distinction between the work of the ORI and that of the NLGR. While the LAW prepared mailing lists and wrote bulletins for the ORI, paid the salary of an ORI engineer, and met the expenses of office personnel at LAW conventions, Stone extended his franking privilege to the Wheelmen, formed chapters of the NLGR, and frequently consulted LAW leaders. The message of both groups was the same. In more than five million pamphlets and on an extensive lecture circuit from 1889 through 1900, the Wheelmen promulgated the gospel of rural road improvement, and ORI bulletins, circulars, and *Yearbook of Agriculture* articles echoed that theme.[13] This message became even more persuasive with the inauguration of experimental Rural Free Delivery (RFD) mail service in 1896. Mail routes had to be passable in all weather, and as RFD spread, farmers began to clamor for good roads. By the late 1890s, the gospel of good roads was finally being accepted in rural America.[14]

Such cooperative promotion, however, offered more than a chance to build public support for good roads. Stone also found that the LAW could pursue activities off-limits to the ORI, especially the advocacy of state legislation. As an LAW lobbyist, Stone had authored the league's model state-aid law, and he continued to prepare and encourage similar bills after 1893. He fielded requests for model legislation from Kansas, Iowa, Connecticut, New York, Rhode Island, and Michigan, and drafted "a complete system of State-aid and supervision . . . [that] was speedily adopted by the legislature" in California.[15] Cooperation with the LAW expanded Stone's ability to shape state legislation, and one Wheelman who grasped the possibilities of such an arrangement proposed an "understanding" of official cooperation between the LAW and the ORI: "We can certainly be of great assistance to each other. . . . In the matter of agitation and methods of working local politics to obtain appropriations and the election of men who would vote for appropriations, that I presume, could not properly come from your office, still it should be one of the most important parts of *our* work."[16]

Stone instantly rejected this plan, but his position within the NLGR achieved a similar result. In 1896, for example, the LAW pushed unsuccessfully for a larger ORI appropriation. And in Minnesota in 1897–98, the clearest cooperative project began as the LAW headed a drive to repeal a ban on state-funded internal improvements. Wheelmen prepared two pamphlets that the ORI published and mailed as part of this successful campaign.[17] In the final analysis, cooperation between the LAW and the ORI created significant enthusiasm for road improvement and state legislation. The ORI noted in 1898 that "the press of the country seems to have become thoroughly aroused in behalf of the road movement, and without doubt the papers have printed as much, if not more, in advocacy of good roads last year as they have on any other one subject," sometimes copying ORI bulletins. Moreover, popular magazines began to echo the ORI's arguments.[18]

In 1899, Roy Stone left the ORI to become president of the NLGR; however, the bicycle craze was already collapsing, and the NLGR disappeared within a year. Although Stone played little role in the good roads movement after 1899, he had, by assisting state and local road activities as a compiler and organizer more than as technical expert, helped define the goals of the movement. In addition, he established the ORI as both a source of technical information and a cheerleader for good roads, introducing but not always adhering to a pattern of technical professionalism. Little physical progress had been made, and while six states had created highway commissions, none actually built highways and only Massachusetts set minimum engineering standards for state-aided local projects. But the rationale for better roads was planted firmly, as the secretary of agriculture explained: "Public interest in the whole subject of road improvement has become thoroughly aroused . . . and great hopefulness has . . . developed. The usefulness of a central good-roads propaganda such as this office affords has been amply illustrated."[19] Indeed, "by the end of the century," observed Phillip Mason, the ORI had emerged as "the recognized leader of the good-roads movement."[20]

Promotion and Politics: Martin Dodge and the OPRI, 1899–1905

If most rural Americans remained bogged in the mud of poor roads in 1899, the next few years produced significant gains for the good roads movement at the state level. A crucial stimulus continued to come from the ORI, which was renamed the Office of Public Road Inquiry

(OPRI) in 1899, and where Martin Dodge became its director on October 23, 1899. In most respects, Dodge retained Stone's mix of technical and promotional efforts through cooperation, but altered the balance of these activities, not to mention the style of operation. Stone had used technical activities to further promotional ends, whereas Dodge reversed the order. It seems significant that Dodge was not an engineer but an Ohio lawyer with ties to President William McKinley. This difference may explain why Dodge's first priority at the OPRI was traditionally political—to raise the office's budget. He was not, however, ignorant about roads, having served on the Ohio state road commission and having become an advocate of brick roads. First replacing Stone temporarily when Stone was recalled to duty for the Spanish-American War, he significantly expanded the office's technical program as permanent director until 1905.[21] But Dodge's overtly political style eventually backfired, thus marking the contrast between the engineer and the bureaucrat.

The change was apparent in 1900 in Dodge's first annual report, in which he explained that "an attempt has been made to inform and stimulate the public mind by attending and addressing state legislatures, the legislative committees, road conventions, farmers' institutes, and other meetings."[22] This policy was not new, but Dodge's aggressive pursuit of state road legislation was. He hired three part-time special agents in part to supervise object-lesson roads but also to organize road conventions and meet with political and civic leaders. One drafted several bills after consulting with ten legislatures in 1899, while another agent distributed 200,000 copies of an OPRI circular on state aid to thirty-five Illinois farmers' institutes in 1901 before preparing a bill. In 1900, OPRI agents appeared in twenty-three states, for as Dodge's assistant later explained, "a great deal can be said in a speech that cannot be printed in a Department publication."[23] As a result, by 1904 the only state north and east of the Mississippi that did not provide assistance for roads was Indiana, although the states offered little money and even fewer controls to upgrade local work.

These successes mirrored growing public interest in roads after 1900. The OPRI also found other outside groups with which to cooperate, even as the abrupt end of the bicycle craze after 1897 removed the LAW as promotional partner. W. H. Moore, author of an OPRI circular in 1900 and ambitious head of a new organization, the National Good Roads Association (NGRA), quickly assumed the bicycle organization's role.[24] Moore proposed a new wrinkle for the demonstration road program—the use of railroads to carry machinery and lecturers to major towns, where a good roads convention could be held and an object-lesson sample could be built. Railroad interest in good roads was not new. *Railroad Gazette* had encouraged railroads to support good roads as early as 1891, and Stone had relied on them for data on railroad rates and other information. The logic that had been stated simply by Albert A. Pope as far back as 1892—better rural roads could increase railroad traffic—made railroads the largest corporate backers of good roads until 1916.[25]

The first Good Roads Train, organized by Moore with OPRI backing, ran on the Illinois Central Railroad from Chicago to New Orleans from April 20, 1901 to July 27, 1901. Another followed, traveling from Chicago to a good roads convention in Buffalo, but the most impressive covered 4,037 miles on the Southern Railway in five months during 1902, sponsoring eighteen object-lesson roads and conventions. Spurred by this success, the Great Northern, Kansas City Southern, St. Louis and San Francisco, Pere Marquette, and Burlington railways provided equipment for trains that operated without OPRI sanction. They attracted enormous attention everywhere they stopped, and Dodge increased this publicity by publishing the proceedings of the conventions held along the routes.[26]

In general, both the expanded state lobbying efforts and the support of the trains marked Dodge's growing preoccupation with the "education" of public opinion. Between 1900 and 1905, his office issued six bulletins, all convention proceedings. One circular listed good roads organizations, while others promoted state-aid laws. The topics were not new, but unlike Stone, Dodge ignored the pitfalls confronting a federal agency entering so obviously into state affairs. An Illinois citizen who asked, "Is it lawful for your department to employ men to work in the interests of any proposed state legislation?" never received a reply.[27] But the question was a good one, especially since Dodge, and to a greater extent Moore, seemed to see the movement as an opportunity for personal gain. Moore attempted to collect organization dues and speakers' fees from each town in which a Good Roads Train stopped, a move forbidden by the Agriculture Department. This led the secretary of agriculture to halt OPRI involvement in the trains after 1902, yet Dodge worked around this. Incredibly, in 1905 the OPRI chief aided Moore in a fraudulent scheme to collect fees for future object-lesson roads that both men knew would not be built. As described by one OPRI official,

> Moore always conveyed the impression that this was a government commission at large in the land to redeem the people from bondage to the mud. As the countenance of Dodge always beamed "Amen" to the most grotesque promises, pretenses, and philippics of these confidence artists the communities were prone to accept them at face value. . . . Only the mighty Dodge, representing the majesty of this supreme government and its beneficent Dept. of Agriculture, enabled this crafty gent to work his schemes upon honest people and avoid the jail.[28]

This was not the only indication of Dodge's limited understanding of professional conduct. He entertained several illegal proposals to provide official endorsements of patented materials and construction processes and even considered joining the companies sponsoring these products. He also courted political controversy while trying to increase the office's budget, offering to build object-lesson roads as *quid pro quos* for congressional supporters. Even OPRI staff engineer E. G. Harrison bragged that he was "gunning Congressmen when I get the shot." Dodge did, however, receive larger budgets: $14,000 in 1901 and $35,000 for 1904 and again in 1905.[29]

Even more clearly improper was Dodge's campaign for federal funds for state road construction and a national highway commission. This effort began in 1902 when Congressman Walter Brownlow asked Dodge's assistant M. O. Eldridge to draft a federal-aid bill for Congress. This was legal, but the OPRI's subsequent choreography of lobbying for Brownlow during 1903 and 1904 was less innocent. Special agents distributed copies of the bill across the country; one even drafted a memorial from the Colorado legislature to the state's congressional delegation. Dodge lobbied railroad officials and good roads groups, and reminded congressmen of the Republican Party's platform plank favoring good roads. But the worst indiscretion came when Eldridge, assigned to work with Brownlow, used $10,000 from the Automobile Club of America (later the American Automobile Association [AAA]) to hire a publicity agent, stenographer, and an office for the distribution of a million copies of the bill.[30]

Dodge, in cooperation with Moore's NGRA, added the finishing touch to this publicity campaign—a good roads convention at the Louisiana Purchase Exposition in St. Louis in 1903. Moore prepared the program, which included addresses by William Jennings Bryan and Theodore Roosevelt, while Dodge added a cover letter to Moore's invitation, provided 25,000 franked envelopes (Moore wanted 50,000 more), and hired a stenographer to transcribe the speeches into an OPRI bulletin. Afterward, Eldridge exulted that "the idea of National Aid is sweeping over the West like a tidal wave." But this wave crested well short of Washington, despite congressional hearings by Brownlow, a second national roads convention in 1904, and a successful lobbying campaign by Dodge and Moore that convinced both political parties to adopt good roads platform planks.[31]

The fallout from this venture was not long delayed. The secretary of agriculture exploded at Dodge's disregard of directives to limit ties to the NGRA because it advocated a particular legislative proposal. Eldridge almost lost his job, and all other promotional efforts, including a planned circular on state aid, were curtailed. Dodge regretfully told the automobile club that

> you perhaps know that many members of Congress thought last winter that we went a step too far in the circulation of literature. The skyrocket which you fired made a greater commotion in the halls of Congress than, perhaps, you are aware of yourself. The effect of this reached the secretary in many ways and I am sorry to say annoyed him greatly. As a consequence of this, he thinks it best to be extremely conservative.[32]

Clearly, Dodge had pushed the OPRI into a realm in which, as Secretary of Agriculture Morton had warned Stone in 1893, it was subject to hostile criticism. In addition, Congress tacked a provision to the office's appropriation in July 1905 that required that its director be a "scientist."

This move, of course, mirrored the dawning Progressive Era's penchant for achieving reform through apolitical experts. It also brought the OPRI into line with other offices in the Agriculture Department in which technical credentials were required. But even though this "reform" was imposed from outside, other activities within the OPRI itself showed that the enthusiasm for experts was already alive. Somewhat surprisingly, even while active in politics, Dodge had strengthened the technical capability of the office initially built by Roy Stone. Perhaps the most important improvement Dodge introduced was an expanded object-lesson program. This remained the OPRI's technical showpiece, and it certainly benefited from the larger budgets Dodge secured. By 1905, the OPRI had fielded four teams of engineers and had completed ninety-six demonstration projects totaling thirty-nine miles. This immensely popular work continued to demonstrate simple construction techniques such as earth roads and basic principles such as ditching and avoiding steep grades that local authorities could easily adopt. In the South, for example, the OPRI promoted techniques its experts had developed for improving clay soils by burning or mixing with sand. As described in OPRI bulletins, these ideas were widely adopted in regions that lacked road stone.[33]

The career of road expert W. L. Spoon demonstrated how the OPRI helped upgrade local roads after 1900. Spoon had served as a local road official, sold road-making equipment, and worked on the Southern Good Roads Train before joining the OPRI as a special agent in 1902. This background, but not his engineering degree from the University of North Carolina, was typical of other agents, who were usually trained by experience rather than by formal schooling. Spoon most often worked in conjunction with the North Carolina Geological Survey, as state geologist J. A. Holmes served as the OPRI special agent in the South. Spoon traveled widely in North Carolina, offering local road supervisors advice on convict labor, road locations, repair and construction problems, and general administration. Frequently, he drew simple plans and estimates for these untrained officials. Spoon became a regular speaker at roads conventions, and once spent ten days lecturing to engineering students at the North Carolina A & M College in Raleigh. The promotion and technical instruction he provided was less exciting but probably more effective than that offered by the Good Road Trains. In his wake, the state began to improve its roads, a result that also followed from the efforts of three other OPRI special agents, R. W. Richardson in the Midwest, Jay F. Brown in the central plains states, and James W. Abbott in the Rocky Mountain region. Along with three road experts working out of the Washington office, these men were the primary source of information available to local road builders. Engineering periodicals were devoting more attention to roads, but few local officials were trained engineers. Thus the visits by government agents brought practical advice directly to those who needed it most.[34]

No less important to the local good roads movement were the answers that OPRI engineers provided to thousands of questions received at the Washington headquarters. These ranged from queries on the use of materials to requests for guidance in forming local road agencies or good roads organizations. The OPRI also extended its technical investigations by adding experiments on brick paving to Stone's original study of steel-track roads. In addition, in 1904 a major project was undertaken—the first national road census of mileage, type of construction, state of repair, methods of administration, and level of expenditures. But the office's most important step in raising the technical level of American road construction was the addition of a laboratory in Washington, D.C., for physical tests on road materials in 1900. The secretary of agriculture supported this move, saying it was his "opinion that the states and territories should build their roads, but that the Federal Government should do scientific work for the whole country that all the states cannot do quite so well."[35] Again, this step moved the OPRI into line with other nationally recognized research bureaus in the department, and the new facility was operated in cooperation with the Bureau of Chemistry.

Dodge also had an institution outside the Agriculture Department in mind as a model for the new laboratory. In 1900, France boasted the best roads in the world, in part because responsibility for roads and bridges had long resided with a national engineering corps and in part because these engineers relied on an advanced laboratory for testing materials. Dodge accordingly understood the importance of the testing center, and in 1900 he hired as its first director Logan W. Page, an 1889 graduate of

Virginia Polytechnic Institute in Blacksburg and one of the first three graduates of Nathaniel Shaler's highway engineering program at Harvard. In 1891, Page had assumed responsibility for Harvard's Lawrence Scientific School testing facility, which inspected materials for the Massachusetts highway commission. The young engineer soon visited France and brought back French methods and equipment that he modified. He was the ideal choice to head the OPRI laboratory.[36]

The new facility began testing for suitability in road construction samples of stone, sand, gravel, oil, asphalt, and cement submitted from around the country. The procedures included physical tests of strength for rock and cement and chemical analyses of oil and asphalt. Railroads, contractors, government officials, and the public utilized this free service, frequently asking for construction directions as well. Nearly all results were returned with a short explanation of the analysis and, when available, an OPRI publication. In a short time, the laboratory received hundreds of requests each year. Few activities of the OPRI did more to improve American highways.[37]

Because of these expanded technical activities, Dodge's failure of judgment in the political arena was not translated into an indictment of the entire OPRI, which road builders already saw as embodying the ideal of impartial expertise. Thus an Army engineer who asked the office for advice about a particular material explained to Dodge that his superior attached "considerably more importance to your unbiased opinion on the subject than to the naturally somewhat prejudiced views of the manufacturers."[38] Thus an administrative style that would dominate reform during the Progressive period was well-established in the OPRI even before Congress placed an engineer in charge of the office. Ironically, these expanded technical programs rested on the larger budgets Dodge had cajoled from Congress.

In hindsight, Dodge's departure in 1905 marked a turning point for the good roads movement. After some fifteen years of sustained effort, few improvements were visible in the nation's highways. The OPRI's 1904 census, the first overall picture of American roads, would not be released until 1907, but everyone knew few states were engaged in serious road improvement programs. Most of the seventy good roads bills introduced in thirty states in 1903 failed to become law. Only Massachusetts had expended any appreciable funds on roads between 1894 and 1903, providing $6.75 million that improved 480 miles of state highways. The first attempts to drive a horseless carriage across the country in 1903 only proved how poor road conditions still were: While traveling by automobile for sixty-three days between San Francisco and New York, a Vermont doctor encountered apparently bottomless mud holes in the

East and nearly impassable trails without grading and bridges in the West. Even the annual Glidden Tours in the East between 1904 and 1910, which combined automobile durability tests with sightseeing, tested pioneering motorists' patience on truly awful roads.[39]

Yet in spite of the lack of visible progress, significant strides had been made in winning public support for roads. Promotional efforts and the growth of RFD had blunted rural opposition, and the gospel of "getting farmers out of the mud" was widely accepted. In addition, the basic legislative goal of the movement—state funds for road improvements by the counties—had been set early by the LAW and reinforced by the ORI and OPRI. Significantly, all these steps had been taken long before the automobile appeared as a factor in the good roads equation. A federal presence was similarly well established by the OPRI's ability to supply technical information and statistical data. This pattern was the hallmark of the increasingly scientific Department of Agriculture, but it was also clear that technical expertise extended the OPRI's influence beyond the strictly technical realm and into the development of state and local policies.

Overt political efforts to extend this influence to the national level, however, brought a sharp rebuff to Martin Dodge and his traditional political style. Dodge and Stone had differed more in degree than in detail over political activity, although the latter better understood the need for propriety. Still, Dodge's open indiscretions confirmed the growing belief that experts should manage reform. With the appointment of an engineer as director in 1905, the OPRI became a premier example of the Progressive philosophy of apolitical expertise. This did not mean, however, that the office ended its promotional efforts. Headed by one of the first American highway engineers, the OPRI assumed an even more prominent leadership role, for the country now seemed ready to begin building roads.

Chapter Two

The Progressives, Expertise, and Highways, 1905–16

There can be no such thing as a political engineer, because when an engineer becomes a politician he ceases *ipso facto* to be an engineer.

T. G. Dabney, 1916[1]

In 1905, the good roads movement entered a new era. To be sure, progress had been apparent earlier; a survey by the OPRI in 1902 had found eighteen state and forty local good roads associations, the National Good Roads Association, and thirty-two affiliates of the Automobile Club of America active in road promotion. But the OPRI's 1904 highway census found only $80 million available for construction, nearly all from local sources, while road administration at every level was poor. Fourteen states had highway commissions in 1905, but five were newly created. By 1916, however, American highway conditions had substantially improved. Stronger highway agencies in the eastern states had set minimum requirements for roads built with state funds, and in 1914 a second OPRI road census reported 257,000 miles of improved highway, an increase of 100,000 miles in a decade, with 32,000 miles of brick, concrete, or bituminous macadam paving. At the same time government spending for roads had jumped to more than $240 million and outstanding bond issues approached $350 million. Finally, Congress approved a formal highway policy and regular appropriation in 1916.[2]

These gains followed the application of the Progressive ideology of reform through apolitical expertise to the good roads problem, largely under the guidance of the renamed Office of Public Roads (OPR). Logan W. Page, the engineer who had headed the OPRI's testing laboratory, became director of the office in 1905 and transformed it into a model of Progressive reform. This chapter will examine Page's efforts, first tracing his attempts to strengthen the OPR's reputation for disinterested technical competence and then showing how that technical capability legitimized OPR political initiatives. Page the technical expert became the central influence in shaping the first federal highway policy between 1911 and 1916. The OPR was not the only source of highway construction or policy innovation, but the ideas widely adopted by road builders and legislators were generally encouraged by the OPR. This pattern appeared even before federal funding offered the office a financial lever, thanks to the Progressive respect for expertise.

Page's role in the good roads movement has been overlooked because the rapid increase in automobile registrations—from 78,000 in 1905 to 2.3 million in 1916—seems to explain the faster pace of road improvements and new policies after 1905.[3] Without doubt, the vocal lobbying of the auto industry and motorists did help convince Congress to spend money on highways. Indeed, the very number of cars probably made a federal policy inevitable. Nonetheless, to credit the changes of 1905–16 to the car alone reads the present into the past. The automobile was too new to have such an impact, and while Henry Ford may have anticipated massive numbers of automobiles, no one in 1916 envisioned our present automobile-based society. In fact, the first federal highway policy largely ignored the specific demands of the automobile industry. Instead, by adding the Progressive assumptions about expertise to the traditional gospel of good roads, Logan Page, more than the presence of the automobile, helped build a vision of rural highway improvement that informed the evolution of highway policy up to World War I.

Expertise and Highways: Technical Information

The label given to the first years of the twentieth century—the Progressive Era—reflects the then-prevailing optimism that governmental and social reforms could resolve the social, economic, and political ills created by the country's unsteady movements toward an urban, industrial culture. Although numerous contradictions threaten any generalizations, Progressive reforms were characterized by an attempt to replace political corruption with honest, efficient administration, a rational desire to base action on information, a reliance on voluntary cooperation, and a sense of social justice.[4] The good roads movement, which considered poor roads both wasteful and unjust to rural Americans, was a typical product of the Progressive equation of inefficiency with immorality. Equally typical, however, was the remedy increasingly favored after 1905 for road improvement and other reforms—the use of apolitical experts to build a better world.

Page's OPR became the leading promoter of highway reforms that sought both efficiency and social justice by embracing the ideology of expertise. No radical change in the OPR was required because the pattern of activity initiated by Roy Stone—control of technical information, promotion of the good roads gospel, and a cooperative approach—had almost exactly foreshadowed Progressive reform efforts. Nonetheless, as Page introduced a code of disinterested expertise in keeping with the times a significant change occurred in the rhetoric of the good roads movement and the activities of OPR. As the agriculture secretary explained in a 1905 statement, probably prepared by Page, "The necessity of demonstrating scientific and economical methods of road construction instead of mere agitation has been clearly established." Reacting to Dodge's loose ethical standards, blatant political lobbying, and lack of professionalism, Page placed limits on convention appearances by OPR representatives and frostily denied any interest in legislative promotion.[5] He also severed all ties to the NGRA, the good roads group with which Dodge had worked closely, and even started a mail fraud case against its head, W. H. Moore, that reflected the moral outrage of a typical Progressive. Page later boasted, "I fought Colonel Moore in the public press and he finally dropped his road work. . . . As you well know, there are a lot of human vultures feasting on the road movement."[6]

In short, Page claimed that he had replaced political intrigue at the OPR with disinterested, objective professionalism. The altruism evident in the outlook of many OPR engineers was another indicator of the new Progressive tone introduced by Page. While telling a new special agent that his daily salary was only $8 plus expenses, OPR engineer A. S. Cushman added, "It is safe to say that all of us in the office are underpaid, compared with what we should expect if we were working less on the *pro bono publico* idea." Page himself answered one complimentary letter by observing that praise for a job well done was the true reward for a public servant.[7]

This shift in outlook also explained the priorities Page set for the office. Central to the Progressive reform method was the desire to collect data as a guide to action. Robert La Follette's Wisconsin scheme of regulation, for example, used a public utility commission to develop a factual basis for making rate decisions, while Louis Bradeis, the reforming Boston attorney and Supreme Court justice, always stressed "the facts."[8] Both Stone and Dodge had built a solid technical capability at the office, but Page operated in an environment that identified information gathering as the hallmark of expertise. He therefore made the development of technical information the highest priority of the OPR, working steadily to add to its reputation for solving road problems. To be sure, other trade and professional groups concerned with road construction began to appear at this time, including the American Road Builders' Association (ARBA) in 1902, the American Society for Municipal Improvement (1894), and the Association for Standardizing Asphalt Paving Specifications.[9] None of these groups, however, developed the breadth of program or the technical stature of the OPR, for under Page's direction the office truly dominated the good roads movement.

Page's priorities were evident in his rearrangement of the OPR into three technical divisions.[10] The Highways Division was responsible for object-lesson and experimental road construction, while the Information Division collected and disseminated data. The last division, Tests and Inspections, conducted all laboratory and field investigations. Using the men formerly assigned to promotional lectures, Page expanded the object-lesson program, and construction skyrocketed from 79,203 square yards (about nine miles of fifteen-foot surface) in 1905 to a million square yards in 1910. By 1912, the OPR had built 616 miles of object-lesson roads; 500 miles were simple earth and sand-clay roads. Moreover, a review of twenty-eight projects in 1910 reported that visits by object-lesson teams had prodded fourteen communities to improve 730 miles of road. Between 1909 and 1917 similar results followed two major demonstrations of the patrol system of maintenance, under which local caretakers maintained each ten-mile section of road. OPR agents taught the patrolmen to use a split-log drag for smoothing earth roads and to perform basic maintenance chores. As one engineer explained, "Many sections have the necessary laws but they seemingly lack intelligent direction. When once *shown how* the greatest good can be done the greatest number."[11]

Additional efforts to improve local roads dovetailed with other Progressive reforms as the OPR continued to cooperate with other groups. Building roads in the National Forests, newly established by conservationists, was one notable example, and an even more extensive effort grew out of a 1906 agreement with the Post Office to help expand rural free mail delivery. After 1899, Post Office inspectors had rejected many local applications for RFD service because of poor road conditions. Under the agreement, OPR engineers could inspect problem routes and recommend improvements; hundreds of disappointed RFD petitioners requested this assistance.[12]

While the OPR spread technical information through these highly visible activities, the real sign of Progressive influence in the office after 1905 was Page's intensive effort to develop new information including the statistical data so dear to Progressives. Tabulation of the road census of 1904 was a major step, and the thirty-nine circulars that reported the census data during 1906 and 1907 provided the first comprehensive picture of national highway costs, benefits, and administration. The OPR institutionalized statistical collection in 1909 with an Economic Investigations unit that conducted new inventories in 1909 and 1914 and struggled to document the monetary benefit of better roads.[13]

Even more important to Page, however, was the expansion of the OPR's testing laboratory, which moved into a new four-story building in 1906. In 1912 *Scientific American* called the facility "the equal of any in the world," and the number of stone, sand, gravel, and asphalt samples analyzed at no charge climbed steadily. OPR investigators also developed new equipment and methods for testing road materials and soon moved beyond routine tests into studies of the chemical reactions in setting cement, the expansion and contraction of concrete, and the corrosion of iron and steel. Moreover, Page wanted more than simple laboratory analyses that identified suitable road materials, so he launched field studies to fine better construction methods. The usual procedure in such studies was to improve an existing road with several different materials or construction methods and then compare their durability and performance under actual conditions. Paving brick from several different clays, wood block, blast-furnace slag, oil-mixed concrete, and many styles of macadam construction were tested in this way. The opening of an experimental center in Arlington, Virginia, in 1912 enhanced these field investigations.[14]

From the start, Page assumed that highway engineers were responsible for providing a road that met the demands of vehicles. The longest-running OPR field study, an examination of dust control measures, marked the extension of this approach to the automobile. Dust, the first road problem specifically related to cars, was a seemingly minor nuisance until it became apparent that dust was a signal of damage to macadam surfaces, that is, packed stone bound together with earth. Page discovered that powered rear wheels loosened the earth binding, thus creating dust clouds and destroying the surface's integrity. Comparative field tests showed that a surface coat of oil protected the binding material, and other studies conducted with the American Society of Civil Engineers (ASCE) led to the complete replacement of mineral binders with bituminous materials.[15]

By 1912, the OPR's technical leadership of the highway field was best symbolized by its work in an area that epitomized Progressive efficiency goals—the introduction of standard specifications for road materials and construction. After 1880, large corporations in advanced manufacturing industries joined the railroads in discovering the value of uniform technical parts and equipment, usually achieved through voluntary consensus standards developed by users and producers. Engineering societies, notably the American Society of Mechanical Engineers and later the Society of Automotive Engineers, provided forums for reaching these agreements.[16] Because the highway materials industry was composed of small, decentralized firms largely immune to the technical and market pressures encouraging industrial uniformity, the OPR took the lead in preparing standards. It had provided instructions for using materials since 1895, but requests grew so numerous by 1909 that the office issued both typical material specifications and testing procedures in 1911, and construction guidelines and bridge specifications soon followed.[17]

Other trade and professional groups, including the American Road Builders' Association, the American Society for Municipal Improvement, and the Association for Standardizing Paving Specifications, also established highway standards. The Lincoln Highway Association, formed in 1913, developed an ideal pavement section, and California and a few other states introduced their own construction standards.[18] Nonetheless, the OPR remained the preeminent source for standards because no other agency conducted testing and practical field trials. The decade of experience of the OPR's laboratory proved crucial, because the large number of samples it tested provided both a performance scale and data for correlating measurements. These were central factors in setting standards acceptable to road builders. The timing and content of state standards that began to appear after 1913 show that they followed OPR precedents, a move facilitated by the close ties between Page and state highway officials. Indeed, two of the leading standard-setting states, Maine and California, employed former OPR staffers as state highway engineers. Thus as early as 1912, *Scientific American* could report that "routine tests have been standardized and the government standards have been largely adopted."[19]

Page, however, shared the Progressive belief that the role of government should be limited and thus disliked such national exposure. Earlier industrial standardization had shown that cooperation between government and business was essential to success; any government dictation guaranteed that some manufacturers would balk. Generally, leadership by big corporations produced acceptance of uniform specifications by smaller firms. But in decentralized industries without large national firms, such as highway materials, trade associations assumed more importance. Page relied heavily on trade groups to maintain the comfortable pattern of cooperation associated with consensus standards. Moreover, associations such as the National Paving Brick Manufacturers' Association and the American Association of Portland Cement Manufacturers often furnished free materials for cooperative research that led to OPR standards being widely accepted by industry.[20]

National engineering groups provided another important avenue for the widespread use of OPR standards since they brought together producers, users, and university engineers. Although few questioned or criticized the OPR's proposals because of the office's reputation, the societies' seal of approval smoothed their introduction and thus made the OPR's dominance in the preparation of highway specifications less obvious.[21] The most important cooperative linkage involved the American Society for Testing Materials (ASTM) in Philadelphia, then and now the nation's leading source of standards. The ASTM was organized in 1899 to develop interindustry consensus standards in all fields, but it had no research staff. It provided, however, a sanctioning body for standards developed by OPR engineers after Page became the first chairperson of the ASTM Road Materials Committee in 1904. He appointed his assistant and later the head of the OPR laboratory as secretary of the committee, a pattern that continued for years. Since the secretary prepared the ASTM agenda, every OPR standard for road materials and corrugated iron and steel culverts automatically received ASTM perusal and approval. Because the committee included manufacturers, city engineers, state highway commissioners, and university representatives, this approval guaranteed wide acceptance. Through such cooperative mechanisms, the OPR prepared the standards for the nation's road builders.[22]

With its array of technical activities, the OPR was a model of the Progressive goal of reform through the generation of information by experts. The selection of Page to head the American delegation to the International Road Congresses at Paris, Brussels, and London in 1908, 1910, and 1913 only confirmed that the office had no American rival as a source of technical information.[23] By following a program of demonstrations, testing, experimental construction, and publication based on an assumption that information was the basis for reform, the OPR slowly but steadily encouraged the construction of better highways for more demanding vehicles with standardized products, tests, and techniques. But reformers, especially Page, also knew that the technical information developed by the OPR was useless without better management by skilled supervisors.

Expertise and Highways: Management and Administration

The Progressive interest in efficient management was as central to their reform goals as the collection of information, with the two activities linked by the belief that only the providers of information were qualified to interpret and utilize it. In this way, so-called businesslike administration was to be introduced. Thus when highway reformers, led as usual by the OPR, targeted local and state road administrations for improvement, they again were working within Progressive patterns. From the beginning, weak local road-building agencies had been recognized as a major obstacle to improvement. The first state highway commissions had been created in the 1890s to raise the level of local competence, but most lacked the authority to force real changes. Even as late as

1908 OPR engineer Charles T. Harrison observed that "even in this age of progress, there is a lamentable lack of any well-defined system of management of county and township road work," and Page believed that "a very large percentage of road revenues is annually wasted through bad management." A deluge of requests for help from local officials confirmed his feeling that local highway administration had to be improved.[24]

Page began by forming a Special Advice and Inspections section in 1906. Like the object-lesson demonstrations, on which it was modeled, this program sent OPR advisers to counties and townships, but for longer stays better suited to teaching management and financial skills. These federal engineers provided advice that ranged over the entire field of highway improvement, including administrative procedures, financial planning, and standard methods of construction and maintenance. The visiting engineers always opposed the statute-labor tax system of road construction and encouraged the hiring of engineers instead. The OPR advisers visited 50 local road agencies in 1909, the first year of the program, and made 243 calls in 1910.[25] The obvious demand for visiting engineers inspired the creation of a second program in 1908, the Model Systems project, which extended the length of the visits of the office's engineers, thus giving them time to implement their frequently ambitious recommendations. By staying as long as two years, the federal advisers became de facto county highway executives who created, as advertised, model rural road systems. Plans for one Louisiana parish, for example, included a major bond issue, 225 miles of construction, patrol maintenance, and a 75 percent tax increase. By 1916, 144 counties in twenty-eight states had entertained Model Systems advisers.[26]

In 1913, the OPR assigned fifty-three field engineers to this work, but obviously they could not reach every American county and township. The real solution to poor local administration, Page believed, was to create stronger state highway departments run by engineers who could oversee local expenditures of state funds to insure efficiency. Although this idea had been implemented by Massachusetts in 1893, it was widely accepted only after 1905, when it was presented in the following Progressive terms. "The only way to build roads economically and efficiently is under the direction of honest and competent engineers. . . . As long as the men at the top of the system . . . are chosen for political reasons, so long the work of highway construction and maintenance will be tainted with corruption and will manifest inefficiency."[27]

The OPR played its usual central role in promoting this type of reform, acting in two ways. First, in an attempt to remedy a shortage of highway engineers, a major obstacle to placing skilled experts in state and local administrative posts, Page created a one-year postgraduate course in highway engineering at the OPR for newly graduated civil engineers. Between 1905 and 1915, the program trained seventy engineers, many of whom moved into state and local positions.[28] Second, the OPR worked directly to influence state highway legislation. Page initially was reluctant to engage in legislative promotion because of the problems caused by Dodge. In 1905 and 1906, his standard response to inquiries about state legislation was, "I have to say that it would be going beyond the province of this office to endeavor to influence road legislation by any direct argument or advice."[29] But after 1907, Page—the apolitical expert—could justify intervention in state politics through the ideology of expertise.

Renewed OPR promotion of legislation creating state highway departments was motivated in large part by the lack of direction in the states themselves. In most states, some group or a few individuals were pushing the cause of road improvement, but few made substantive progress. In North Carolina, for example, two state geologists nurtured the North Carolina Good Roads Association but were unable to form a professional highway agency. Yet even when such agencies were created, most lacked funds or authority, a fate that befell the efforts of the state legislator in South Dakota who struggled from 1907 until 1913 to create a highway commission only to have it receive no money. Even Iowa, generally considered an advanced state in highway matters, gave Thomas MacDonald, the young civil engineer who assumed control of the new highway commission in 1904, almost no money or power. He fought for engineering control over state funds for local road construction, while preparing and encouraging adoption of standard specifications by the counties. In 1913, the legislature finally granted him limited supervisory authority but offered no state aid. Many state programs, including an ambitious, four-tiered construction plan for New York and the creation of a road commission in Oregon after 1913, were totally hamstrung by political controversy.[30]

This confused, politically charged situation opened the door for Page to exercise his influence. He presented himself as a strictly nonpolitical, disinterested reformer seeking more efficient organization. The rationale for his action was explained by another engineer:

> It is not sufficient for an engineer, or any other educated man, to *know*. He must *act*. A part of the engineer's duty is to help in molding public opinion by educating and guiding the public in those matters in which engineers by virtue of their training and occupation have superior knowledge.[31]

This argument gave Page not only the right but also the *responsibility* to encourage state highway legislation, as the secretary of agriculture explained in perfect Progressive terms in 1910:

> Old systems of road administration, involving the principle of extreme localization, are fast breaking up, and new systems, involving the principle of centralization, are taking their place. Road administration is, therefore, in a transitional or formative stage, and it is of the utmost importance that the movement be directed along the right lines.[32]

The OPR launched its campaign for the "right lines" in state legislation at a Good Roads Conference at Denver in late 1906. Page argued that states should: (1) create road systems to insure orderly development, (2) furnish aid to counties in return for inspection privileges, (3) grant engineers authority over all technical questions, and (4) stop politically motivated appointments. Page steadfastly worked to implement this ideal of state highway administration, explaining in 1911 to Anson Marston, an engineer educator at Iowa State Agricultural and Mechanical College and a good roads supporter, "My whole object in creating a commission of this nature is to get the state highway department out of politics . . . and at the same time put it in charge of engineers."[33] Always, he attacked politics as the source of inefficiency and waste in road building, summarizing the problem in a 1916 presentation to the American Academy for the Advancement of Science. Politics had hampered improvements in highways, he argued, by prompting personnel turnovers and other changes in thirteen highway departments over the last decade. New York alone had had nine highway executives in fifteen years because of partisan political struggles. As a result, the BPR slowly grew bolder in its promotional efforts. By 1908, one staff engineer volunteered, "The opportunity frequently arises for us to be of great assistance to State legislatures and to Highway Commissions in furthering the object in view."[34] Such assistance included drafting bills and assigning OPR field agents to give lectures. Within a year, the office had sent a model state-aid bill to at least twenty-nine states, and its state legislation program soon surpassed Martin Dodge's.[35]

Page's efforts, in fact, closely resembled those of his predecessor. He rediscovered cooperation as a lobbying tool, for example. After 1909, he enlisted the Country Life Committee of the National Congress of Mothers in several schemes to promote good roads, and maintained ties with many other groups interested in better highways.[36] But his grandest cooperative venture exceeded any envisioned by Dodge. In 1910, state highway engineers agreed to support Page's plan for creating an umbrella agency to unite the good roads movement. He accordingly prepared a constitution and was elected president of the American Association for

Highway Improvement (AAHI) in November 1911, with the head of the OPR Information Division, J. E. Pennybacker, becoming the secretary. Page then transferred most OPR promotional efforts on behalf of the model state highway bill to the AAHI and forwarded all requests for information to Pennybacker. To all appearances the AAHI, which sponsored an annual American Road Congress, published *The Official Good Roads Yearbook of the United States* annually from 1912 to 1917 to highlight state road bills, and supplied its own lecturers, was the leading advocate of state highway legislation. But it remained Page's organization.[37]

The purpose, program, and ostensible independence of the AAHI recalled Stone's connections with the Wheelmen and the NLGR during the 1890s. But other Agriculture Department bureau chiefs also built links to national associations as a means of winning budget increases. And Gifford Pinchot's organization and leadership of the American Conservation Association, an active lobbying group, offered an even closer parallel.[38] The only real difference between Page and Pinchot was the former's greater success, for the model highway commission bill advocated by the AAHI provided the foundation for most state highway agencies while Pinchot failed to secure approval for his resource allocation policies.

To be sure, other agencies, notably the Lincoln Highway Association and the American Road Builders' Association, also encouraged state highway legislation that gave direction to professionals. By 1914, engineering periodicals contained numerous articles describing ideal state highway administrations on the same lines.[39] Moreover, state reforms continued to be based on grass-roots efforts, without which the OPR could not influence state legislation. But even in those states that were eager to develop stronger highway agencies, the OPR's model bill and presentations by special agents or even Page himself often provided the essential imprimatur of expertise that secured passage of a bill embodying the Progressive ideal of administration by professionals. Michigan, for example, had created a weak highway commission in 1905, and only required the commissioner to be an engineer in 1913, a change that matched OPR recommendations then circulating in the state.[40] In other states, the office's influence was similarly evident in both the timing and the content of the reforms that were passed. OPR assistance did not always guarantee success or a strong department. But even when the OPR's model bill was not passed intact, one cannot miss the OPR's guiding influence in the development and implementation of state highway organizations that reflected the basic Progressive goal of administration by apolitical experts.[41]

The Social Justification of Good Roads

The return to lobbying by Page was an important indicator that highway reform was not simply a technical matter, however much Page stressed this component of the OPR's work. Indeed, perhaps as important in defining the Progressive aspect of the good roads movement was the gospel of rural road improvement. The argument that mud-bound farmers deserved the same social, political, religious, and educational opportunities available to urban residents provided a social justification for the technical programs of the OPR. Poor roads thus became a moral wrong and their improvement a matter of democratic fairness. This idealistic sense of social justice, which was evident in other reforms of the day, gave highway enthusiasts the moral spark typical of the Progressives. This gospel of good roads had not originated in the OPR, but with the League of American Wheelmen in the 1880s. But from 1893, the office always supported it. The growth of RFD after 1899 and the efforts of the National Grange, the farm organization that by 1905 was a leading spokesmen for rural America, added other voices of support. An increasing number of popular magazine articles echoed these views.[42] The OPR, however, played a large role in convincing Americans to support better rural roads as a social necessity, and remained a leader in this work after 1905.

Given Page's insistence on professional conduct, it might seem surprising that the OPR played this promotional role. In fact, to distance the OPR from Martin Dodge, Page did ban promotional lectures and publicity efforts in 1905. But this ban lasted barely two years. Even Secretary of Agriculture James Wilson observed in 1905 that lectures "aroused people to the realization of the need for better roads."[43] By 1907, Page was cautiously renewing efforts to "educate the public," as OPR engineers prepared short, practical articles that were being carried in 2,500 local newspapers by 1908. Shortly thereafter, OPR engineers were again appearing on the lecture circuit, delivering 1,135 talks in 1912, up from only 150 in 1905. Page had concluded that "public sentiment is a powerful force and must in the end determine the course of public events. It is essential, therefore, the public sentiment be created and directed along the right lines."[44] Whatever the technical subject of the engineers' talks, the foundation was the good roads gospel, for that social message generated the excitement essential to this "educational" program. Thus Page the disinterested reformer built a promotional program identical in all but tone to that of his predecessor.[45]

One indication of this continuity was the OPR's rediscovery of fairs as a promotional tool in 1909. At the Alaska-Yukon-Pacific Exposition in

Seattle, the OPR first exhibited dioramas, some with working miniature machinery, to illustrate road construction with different materials. These models generated enormous attention, proving so popular that for a decade several sets toured state, regional, and county fairs. A 1909 bulletin describing the models was a "best seller" for the office.[46] Moreover, Page revived the Good Roads Trains in 1911. This came as the railroads, seeking to increase traffic, began using agriculture education trains to teach farmers new techniques and crops; the Agriculture Department alone sponsored seventy-one trains carrying lecturers and demonstrations in 1911. Page joined this program by assembling trains that transported the machinery models and stereo-opticon slide equipment in a lecture car. These arrangements paralleled those used by Dodge, for the AAHI served as the liaison between the OPR and the railroads. The first two Good Roads Trains on the Pennsylvania and Southern railways reached 65,000 people in 1911, while six railroads ran trains that were visited by 160,000 people in 1912. Always a source of great excitement, these ventures continued through 1916, as Page placed the OPR in the center of promotional efforts.[47]

With its extensive publicity program, the OPR was the unrivaled leader in the field of promotions, as it was in the areas of technical information and management. Indeed, in the eyes of newspaperman E. W. Scripps, Page was a model reformer. Writing to Page in 1909, Scripps said, "In all this great nation there are perhaps no other two men who have better opportunities to serve their country and who are making better use of them than are you and Mr. [Gifford] Pinchot. Despite the fact that neither of you have any high sounding titles or official positions which in themselves would give you great distinction—perhaps just because of that fact, you are epoch-makers."[48] The tone of Scripps's comments indicates how fully Page and highway reform had come to fit the Progressive pattern. Significantly, this combination of technical authority and moral justification provided the impetus for working toward the final goal of the road reform process—federal funds for highway construction. Again, the OPR's expertise gave it the preeminent voice in shaping the federal-aid policy that emerged in 1916.

Federal Aid for Highways, 1911–16

By 1911, highway reform had become a topic of national interest, with one area of new support coming from the small but growing influence of automobile manufacturers and owners. Good roads supporters had always recognized the automobile as a potential ally, and by 1910 the Detroit automobile manufacturers had formed a Central Good

Roads Finance Committee, which was pushing Page's state-aid bill in Michigan.[49] Given the general tendency toward centralized power evident in American society after 1880 and the Progressive belief that federal influence could encourage more efficient state administration, it was not a coincidence that this growing popular interest led to renewed efforts for federal highway funds. But for the first time, a campaign for federal highway reforms had a realistic chance of success, and this effort dominated the good roads movement from 1911 to 1916.

This legislative campaign points to a basic dichotomy of the Progressive era, however, for despite the emphasis on disinterested expertise, the apolitical stance of experts itself almost automatically opened for them the role of advocate. Gifford Pinchot, for example, waged an aggressive lobbying campaign to gain control of resource allocation policy, arguing that federal experts were not susceptible to state and local political pressure and other nonscientific factors. The development of federal highway policy was another instance of the advocacy of a particular policy by supposedly apolitical experts, but in this case the experts proved more influential. The legitimating cloak of expertise afforded Page an enormous advantage over both his predecessors and competing policy positions. Even while the OPR maintained that "in regard to the advisability of National aid in road building, this Office has no opinion to express either for or against," after 1911 Page promoted a Progressive vision of rural road improvement, rather than a highway system to serve the automobile, that in 1916 became the first federal highway policy.[50]

Interest in federal support for road building had never completely faded after Dodge's ill-fated plan collapsed in 1903–4, but no specific proposal emerged from the congressional committees.[51] By 1910, however, interest began to quicken. Page and his new lobbying group, the AAHI, played a central role in shifting the movement's goal from state-aid to federal funds, a move aided by the fact that there were 468,000 cars on the roads in that year. The opening session of the AAHI's first convention in November 1911 indicated the new direction Page favored. He had assembled the program, which began with welcoming remarks by the secretary of agriculture, who was standing in for President William Taft. Three congressmen then presented proposals for federal assistance for roads. Not surprisingly, given this quasi-official prompting, the convention approved a resolution endorsing such federal legislation. Even though he was president of the association, the OPR director attempted to divorce himself from this step, a move in keeping with his numerous earlier refusals to become involved with any group advocating federal aid. Yet since he had clearly orchestrated the meeting, Page's claim of impartiality rang hollow, despite the AAHI's assertion of nonpartisanship and freedom from special-interest ties. Moreover, the convention's timing—it came on the heels of the first revival of the Good Roads Trains—was perfect, for in its wake came an enormous outpouring of interest in roads.[52]

Signs of support for good roads were everywhere, but perhaps the best indication was from the movement for building memorial highways across the country. In 1912, Carl G. Fisher, builder of the Indianapolis Speedway, proposed "a coast-to-coast rock highway" that a year later he named the Lincoln Highway. Other national highway associations, all private ventures, quickly followed, including the Ocean-to-Ocean Highway (New York to San Francisco), the Dixie Highway (Michigan to Florida), the Meridian Highway (Galveston to Winnipeg), the Southern National Highway (San Diego to Washington, D.C.), and the Yellowstone Highway across the northern states. The National Highway Association presented plans for a 51,000-mile system with three east-west and three north-south trunk lines and a web of roads connecting every state capital.[53] The automobile's impact was readily visible, for when cross-country tours encountered bad roads, they generated fantastic headlines. But the real proof of change came with the introduction of sixty road bills in Congress between December 1911 and July 1912. *Engineering News* reported that by September 1912 a "widespread demand" existed for congressional action.[54]

In reality, Congress faced a flood of divergent proposals, including plans for national highway commissions, construction bonds, federal aid, and post roads for rural free delivery. Several bills called for rental payments of fifteen dollars to twenty-five dollars per mile to counties that maintained post roads. Recognizing the potential for confusion, an engineer in the American Road Builders' Association urged the ARBA to "do our utmost to influence the taking of the step in the proper direction."[55] Other groups also mobilized to "help" Congress decide, with the American Automobile Association launching the most ambitious lobbying campaign. Funded largely by the National Automobile Chamber of Commerce, an industry trade group, the AAA pushed for federal construction of a cross-country system of hard-surfaced roads.[56] But Logan Page and the OPR proved to be the most important outside influence, thanks to the office's technical stature. Page's contacts with Congress began with his role as the Agriculture Department's spokesperson on highways, a position that enhanced his image as the unbiased technical expert. Moreover, the OPR answered all congressional requests for opinions about legislation sent to the secretary of agriculture. Both roles helped guide congressional opinion while creating the proper image for the OPR, so that, in the words of one engineer, congressmen who lacked any special knowledge on roads were "relying largely on certain . . . officials for guidance."[57]

The policy Page presented to Congress paralleled his Progressive vision of state highway administration: a cautious position that provided limited government aid through voluntary cooperation. Page sketched a federal-aid plan, a cooperative venture in which the federal government provided technical and financial assistance for roads built by the states. He condemned the concepts of a federal highway commission and cross-country highways as economically and socially unnecessary and argued that the focus should instead be on farm-to-market roads, the roads for rural free delivery. Finally, Page opposed, on efficiency grounds, a rental scheme supported by Senator D. W. Shackleford and others, arguing that a $15 to $25 allowance per mile would produce few improvements.

Page's caution was advisable because even though RFD mail service had defused constitutional arguments that roads were not a federal matter, many opposed what *Engineering News* called a "raid on the Treasury . . . . It needs but the slightest knowledge of Congressional methods and precedents to perceive that the first grant of Federal aid to highway work will be the opening of a drain out of the Treasury which may easily menace national solvency."[58] In contrast to the plans of the National Highway Association for a 51,000-mile network, which carried a $30,000-per-mile price tag, Page's plan was simple and not extravagant. Of all the proposals, federal aid for rural post roads most clearly fit the Progressive mold. The basic ideas for these roads had been proposed by Roy Stone as early as 1893, but also drew on precedents in several other Agriculture Department programs. Agriculture experiment stations, created in 1887 and expanded in 1906, relied on a similar sharing of funding and responsibility with the states, as would the Agriculture Extension Service and the Forest Fire Protection Service, both created in the 1910s.[59] Page's real contribution was shaping this general cooperative policy to fit highways.

In doing so, Page did not limit his advocacy to official contacts. He involved the board of directors of the AAHI in congressional hearings so this group could lobby congressmen. Moreover, Page had a special relationship with Senator J. H. Bankhead, who served on the Agriculture and Forestry Committee considering the road bills. As early as 1909, he assisted Bankhead in preparing an amendment to the Agriculture Department's appropriation that provided funds for the OPR to improve post roads experimentally. A year later, Page wrote a committee report on this plan for the senator, who argued, "You can do it much better than I could, and since it will be made a Senate Document, I am anxious that it be made as thorough and convincing as possible."[60]

Page's influence became apparent when the only action taken by Congress in 1912 matched his positions in every respect. It added $500,000 to the Post Office Appropriation Bill for fiscal 1913 specifically for post road improvement, rejecting both a national commission and rental payments in favor of a pilot program to demonstrate cooperative post road construction. It also created a joint congressional committee to study roads further. Page had proposed the post-road experiment to Senator Bankhead as early as 1909, but the AAA claimed credit for the bill, and its role in fact cannot be ignored. The AAA had sponsored its own good roads convention in January 1912, fought the Shackleford rental plan, and probably proposed the joint congressional committee as a means of sorting out the federal government's place in road building. But the automobile industry's national construction plan failed to win approval.[61]

This episode set the tone for the development of federal highway policy over the next four years, as Page guided the formation of the first such policy along lines proposed by the OPR. The pattern was evident in the activities of the joint committee, which deliberated through 1913 and 1914. It immediately turned to the OPR for information as it worked to delineate a workable highway policy. The move was logical, as only the OPR possessed a statistical picture of American roads, comprehensive descriptions of state and local highway organizations, and analyses of expenditures. Further, an OPR topographer had prepared maps of state post roads, and the committee used J. E. Pennybacker, head of the OPR's Information Division and secretary of the AAHI, as its statistician. Moreover, Page and at least three directors of the AAHI testified before the committee, which later singled out the OPR for praise.[62]

Not surprisingly, the final document issued by the panel in early 1915 conformed to principles sketched by the OPR. The committee avoided endorsing specific proposals because three committee members had introduced highway legislation. But the document did propose two main guidelines for federal expenditures, which reflected Page's ideas. First, the committee stated that the federal role in highway matters should be technical and advisory, focusing on raising state construction standards and encouraging efficiency. Second, cooperation, rather than coercion, was desired between the states and the federal government, as the committee made clear its opposition to a strong federal agency for building roads.[63]

The actual debate in Congress over a federal highway policy began before the joint committee's report was released, however. Ten bills appeared in the Senate and thirty-nine more in the House in 1914, with discussion centering on the Shackleford bill, which authorized $25 million for rental payments to the counties, and on the AAA's plan for a national system of main roads. Shackleford's proposal actually passed the House in 1914 by a vote of 284 to 42, but both measures encountered strong criticism. *Engineering News* scoffed that "very little benefit" would follow the Shackleford plan, claiming that his bill and a bond-financing bill by Senator Jonathan Bourne, chair of the joint committee, "were intended rather for political effect than as measures intended for enactment into law."[64] The Senate apparently agreed and rejected the plan.

The AAA scheme fared little better because motorists had inherited the "idle rich" label that had been applied to bicyclists in the 1890s. Again, *Engineering News* captured the prevailing popular feeling. "We may say frankly that to propose such an expenditure of the public money for the benefit of a few wealthy pleasure seekers at a time when the masses of the people are crying out in a protest against the increased cost of living seems like trifling with a grave situation."[65] The automobile industry had recognized this problem, which led one auto maker to say "the only sure way to success" with state legislation was to take no action until the Grange had endorsed the bill, for "the initiative should be taken by the farmers." But even this strategy proved ineffective because of the immensity of the task of promoting national legislation. A later *Engineering News* editorial argued that "one is staggered by the problem of creating an organization to carry on, from a central office in Washington, the economical maintenance of a million and a quarter odd miles spread from the Atlantic to the Pacific."[66]

In this situation, Page's reputation as the ranking highway expert and the congruence of his package of reforms with basic Progressive attitudes gave the director substantial political influence. When Congress passed the Federal-Aid Highway Act, signed by Woodrow Wilson on June 11, 1916, it was clear that Page had been the arbiter of the first federal highway policy. In line with the accepted apolitical style, his efforts were circumspect. Moreover, a highly visible promotional campaign by the AAA and its newly created ally, the American Association of State Highway Officials (AASHO), dominated the proceedings. With membership limited to senior state and federal highway officials, the AASHO was the product of several eastern highway officials' unhappiness with the rural roads focus of Page's American Highway Association (the AAHI after 1912). Since their states had well-developed state highway systems, highway engineers H. G. Shirley of Maryland, G. P. Coleman of Virginia, and Joseph H. Pratt of North Carolina saw a federal road network as the next logical step. With the encouragement of the AAA's president, A. G. Batchelder (who also understood the importance of marshalling expert opinion), these men organized the AASHO in Atlanta in November 1914, assigning first priority to the preparation of a federal roads bill. Batchelder helped the engineers draft a plan for a national system that was sent to Congress, the AAA, and the National Association of Manufacturers before most other highway engineers saw it. As the AAA president explained, "This hiding behind the wood pile is no longer necessary, though I do think at the present moment it is just as well that we brought about the organization of [AASHO] and got it to introduce a bill behind which we have put our strength. . . . Even some of its own members are not entirely aware of our activities."[67]

Unfortunately for the AAA, midwestern highway engineers protested both the plan for a national system and its hasty preparation. They also resented the deliberate exclusion of AAHI officers from the AASHO's executive committee, a move apparently intended to insult Page. But the OPR director steadily supported the new group, and a meeting to settle the dispute was arranged for September 1915 in Oakland, California. A group led by Thomas MacDonald of Iowa used this occasion to transform the bill into a federal-aid plan congruent with Page's ideas. MacDonald and the OPR director, who attended the meeting, were friends, but Page's involvement in the issue cannot be documented. In any event, the AASHO's convention in December 1915 ratified the revised bill, which Senator Bankhead introduced.

During the next six months of congressional maneuvering, Page's efforts were more open to observation.[68] He worked with Senator Bankhead in an official capacity, furnishing both information and opinions on other legislation. The OPR staff also prepared detailed memoranda on the general principles behind federal involvement in road work and suggested many changes in the legislation. Page answered questions from the floor about the committee's amendments for Bankhead and prepared a line-by-line comparison of the Shackleford and Bankhead bills. He served, in effect, as an adviser to Bankhead's Senate committee and his ideas were generally accepted as *the* federal highway policy. In fact, federal-state cooperation through federal aid was the only principle seriously considered by Congress.

The specific provisions of the bill demonstrated the extent of congressional acceptance of Page's suggestions.[69] First, the federal government agreed to match state expenses up to $10,000 per mile. But because the OPR had experienced difficulties administering county road construction using the $500,000 post road fund of 1912, Congress accepted his argument that the federal government should deal only with states, not with hundreds of counties.[70] Page also convinced the committee to give the Agriculture Department strong, although not domineering, supervision of state programs, requiring that the establishment of a state highway department that met with OPR approval be a condition for obtaining federal funds. Thus states that had not adopted the model state highway bill were now forced to follow its guidelines. These stronger state highway commissions were supposed to prepare all plans and specifications and control all construction and maintenance, with the federal government reserving inspection rights.

This focus on efficient administration by experts was, of course, typically Progressive and vintage Page. The nature of the actual appropriations showed the same influence. Only $5 million was authorized the first year, escalating in annual steps of $5 million for five years to a total of $75 million. The distribution of the funds among the states was removed from the political process by an allocation formula that determined each state's share based on its population, area, and road mileage. Finally, the bill embraced the gospel of rural roads by limiting federal aid to rural post roads.

This last requirement was as important as the federal-aid principle in defining the government's road program. Moreover, acceptance of the Progressive argument that roads should first serve a social function in rural America marked the victory of Page's influence over that of the automobile industry's. His plan did not create a national system of main roads but rather reflected his idealistic commitment to improving the quality of rural life. This concern is evident in Page's writings and was recognized by the secretary of agriculture, who transferred responsibility for rural engineering projects—irrigation, drainage, power machinery, building construction, and domestic water supply—to a renamed Office of Public Roads and Rural Engineering in 1915.[71] Importantly, Page never opposed the automobile; on the contrary, his investigations led to the more durable roads that the new vehicles required. He was convinced cars would replace horses, and he almost certainly agreed with the sentiment expressed in a 1917 *Engineering News-Record* editorial that "to Henry Ford belongs the honor and glory of having furnished the only good roads argument which thousands of people in the backwoods sections of the country are capable of absorbing."[72] But like most people in 1916, Page failed to anticipate a transportation system based on cars and roads. Implicitly, he and many others saw railroads as the means of long-haul transportation, with roads serving local traffic and feeding freight to railroad depots.[73] By incorporating this vision, the nation's first highway policy combined the Progressive goals of economic efficiency and social betterment. In comparison, the automobile industry's goals appeared to be those of a narrow self-interest, especially since the democratizing tendencies of the automobile were not fully evident before World War I.

Nonetheless, Page's accomplishments were impressive by any measure. He not only turned the OPR into a leading technical facility, but also put its information into practical use, with noticeable effect on the nation's highways. In the Progressive tradition, he instilled a rigorous sense of professional pride and public service in the office's activities while expanding its budget from $50,000 in 1905 to $279,000 in 1914, when it employed 218 people. His efforts, more than those of any other individual, extended the pattern of administration by experts, leading to the creation of professional highway departments under the control of engineers. Moreover, Page's influence may be seen in other Progressive elements of highway policy that still exist in the 1980s, including the basic procedures by which states build roads under federal inspection, using funds allocated by a formula to limit political scheming. He also ensured that the OPR, a prototypal Progressive agency, remained the nation's central highway authority. He did so by working cooperatively instead of dictating. This was another long-lasting pattern, one developed before the OPR had the funds or the power to demand compliance with its ideas. Certainly he benefited from increased public support, especially as the automobile became a fact of life in American society. But the decisive factor in shaping the initial highway policy for this country was Page's position as highway expert in a country that accorded special importance to such expertise.

It is possible, therefore, to see highways as the perfect reform and Page as the perfect Progressive, for he combined the age's emphasis on efficiency through apparently disinterested professionalism with its idealistic and crusading efforts. The editor of *Engineering News* described him as "an idealist whose imagination clearly pictured better conditions for a whole people, an engineer who knew how to reduce his dreams to practicable plans, a man of such forceful personality that he wrested from an uninterested public the necessary initial support for those plans, an executive who finally carried them forward by administrative skill so successfully that the entire nation calls urgently for more of this service."[74]

Yet Page also personified the contradictions that frustrate the attempts of historians to understand fully the Progressive period.[75] It was typical of Page and his time that he possessed a reputation as an expert who stood above the political fray, despite intimate involvement in state and national politics. In March 1916, when the secretary of the Tennessee Department of Highways offered to lobby for the road bill, Page appeared as the model of disinterested rectitude as he stiffly replied, "I . . . will say that this Office never participates in any way to create sentiment in favor of Federal legislation of any sort. I am glad that you approve of the Federal aid bill now before Congress."[76] Yet this same individual founded and ran his own lobbying group solely to "create sentiment" favoring his vision of highway reform. This contradiction points up a fundamental dichotomy in Progressive attitudes, for implicit in their reliance on experts was a distrust for "the masses." Obviously, this viewpoint stood in direct opposition to the democratic values at the core of their moral system, so that the campaign to "educate the public" about roads was different only in style not intent, from the later Progressive propaganda associated with the war in Europe. Both informed the public of the correct way to think.

It is not possible to determine from a single case study which approach—that of the moral democrat or the strident advocate—the Progressives favored. Page's behavior was not corrupt, and he steadfastly believed that his policy best served the nation. Yet the ease with which experts carried their arguments, thanks to the legitimacy granted by their expertise, undercut the democratic goals they claimed to be advancing. The greater significance, however, may lie in the failure of most Americans to recognize that disinterested expertise involved far more complex and ambiguous issues than initially met the eye. Highway reform was one result of the Progressive combination of efficiency, altruism, and idealism; however, it also involved advocacy and more sophisticated political manipulation than the political machines that the experts were intended to replace. Although rarely used cynically, the power of experts was greater, and thus more disturbing, because of the certainty and moral superiority that they projected. By mixing these elements, experts in the Progressive era played the central role in shaping the first national highway policy in 1916. As a result, after twenty-five years, these same experts were ready to build roads.

Chapter Three

The Creation of a National Highway System, 1916–21

Sentiment has been growing all over the country in favor of the construction and maintenance by the Federal Government of a highway system that shall transcend state bounds and that shall do nationally what state control has done within state borders—place the through routes under a single competent authority.

E. J. Mehren, 1918[1]

[The Federal Highway Act of 1921] permanently laid to rest the idea of a national highway system under Federal control.

Federal Highway Administration, 1976[2]

The Federal-Aid Road Act of 1916 was a milestone in American highway development, for it directly involved the federal government in highway building. In the best Progressive fashion, the law provided a $75 million federal carrot to encourage the slow but uneven movement by states toward centralized highway administration by experts, promising improved quality and greater efficiency. The Office of Public Roads, assigned to administer a program it had helped to create, believed that most highway problems could be corrected by securing sound local administration. Remaining obstacles would then be overcome in the same way OPR investigations and tests had alleviated technical difficulties.[3]

Unfortunately for the OPR, between 1916 and 1921 the federal-aid program was plagued by wartime shortages, construction delays, and tangled administration. This situation provoked a rancorous federal policy debate that challenged the leadership of the government's road office, renamed the Bureau of Public Roads (BPR) in 1918. The war significantly altered the general perception of the role of highways in the United States, inspiring attempts to jettison Logan Page's policy. Stability only slowly reemerged after the war, and not until November 1921 was an amended policy set in place. Despite this confusion, federal experts ultimately remained the arbiters of American highway policy, and the basic pattern of disinterested expertise as well as other Progressive tendencies continued to undergird their actions. Working with important elements from Logan Page's vision, a new chief of public roads established the basic rules of highway construction that survive even in the 1980s. Yet as the Progressive Era ended in disillusionment after the war, new attitudes that would mark the decade of the 1920s were also influencing the shaping of a revised highway policy. Thomas H. MacDonald melded elements from both the new and old to form a consensus that united road builders for almost twenty years.

False Start for Federal Aid

In 1916, road building seemed ready to move ahead quickly, for in addition to the first $5 million in federal aid, the states provided more money to match the federal grants, and state highway department budgets increased from $41 million to about $60 million. By 1917, total funding for state roads had risen to $116.5 million, compared to $87 million a year before.[4] But for several reasons, the new federally supported road-building effort made only fitful progress over the next several years. Some problems were beyond the control of federal administrators, but others might have been prevented. One pattern that emerged quickly, however, was the continuing dominance of federal highway engineers within a program that had been deliberately designed as a cooperative venture between the state and federal governments.

The most immediate obstacle to inaugurating the road building program quickly was imposed by the federal law itself, which required that a sound administrative structure be enacted before actual construction began. First, the states had to form strong highway departments that met with OPR approval. These commissions thus had to have the financial resources, legal authority, and engineering skill to initiate federal-aid construction and maintain the completed highways. The OPR intended to end the pattern of counties constructing all roads under loose or nonexistent state supervision, a goal in line with the steady centralization of power in the United States after the 1870s. This move, however, forced wholesale alterations in state highway laws, since only those of California conformed to all OPR guidelines in 1916. Eight states had to form highway departments, with three needing constitutional amendments to do so, while nine existing departments benefited from major legal overhauls. Another eighteen states strengthened or reorganized their highway commissions. The office furnished model laws and expert testimony in twenty-three states, and OPR engineers helped organize highway departments in Texas, Missouri, Indiana, and South Carolina.

The second step was the implementation of rules and regulations for using federal aid. Given the bill's federalist structure, which delegated control of construction and maintenance to the states, subject only to federal inspections, Page claimed that the rules would not "disturb in the slightest degree the present powers and duties of the states." The OPR, he said, would only encourage states to develop orderly five-year plans for road work.[5] But its superior expertise again extended the OPR's influence, setting a long-lasting precedent for federal-state interaction on highway construction. Senior OPR engineers prepared a multistep procedure for inspections of state routes, plans, and estimates even before final congressional action, presenting the plan to state officials within a month of the bill's passage. *Engineering News* correctly commented that "the rules and regulations governing surveys, plans, specifications, estimates, contracts, construction work, costkeeping and payments, are so comprehensive that the [OPR] will have a dominating influence in all work for which application for federal aid is made."[6] The office exercised similar influence in developing the actual construction guidelines for federal-aid roads, although cooperative mechanisms masked its role. Technically, the AASHO prepared standard specifications for construction by the states, and uniform testing procedures and requirements for drawings and plans emerged from two conferences of state engineers. But in reality, state engineers accepted the established procedures of the OPR; the conferences merely rubber-stamped OPR proposals. In other words, the OPR developed both the institutional structure and the basic technical policies of federal aid, decisions the 1916 legislation had supposedly left to the states.[7]

Despite the OPR's dominance, harmony prevailed among state and federal highway officials through 1917, even when construction lagged. But unity faded as several factors conspired to limit serious highway construction in 1918; they also destroyed Page's vision of a federal-aid program. One difficulty concerned the OPR's inspection of state plans. OPR engineers were convinced "the present system is generally a sort of rat hole through which to squander the road fund,"[8] especially after one state submitted plans for a road that would have frequently been under water. As a result the OPR conducted very thorough inspections of state plans in both the district and national offices. Such minute attention to detail and requests for occasionally expensive and needless changes irritated the states, but even more troubling was the federal attitude. In one instance, federal engineers demanded twenty-four pages of "essential" modifications in Michigan's tentative material specifications, and they refused requests for compromise.[9] This rigidity stymied the federal examination of state proposals, and in 1917 federal funds were released for only six projects. Some of the blame rested on the states, which provided complete drawings for only half of the proposed projects. But equal responsibility belonged to overzealous and inflexible federal engineers. Only by late 1918 was genuine cooperation between state and federal officials appearing after this rocky beginning.[10]

Other problems remained, however, and perhaps more vexing to state engineers was the Progressive core of Page's policy that federal funds should be used primarily to improve rural RFD roads. Confusion stemmed from OPR requests that states avoid constructing "wiggle worm" patterns of disconnected highways and consider creating state road systems. For western states such as Nevada, which had only four post roads that failed to provide the basis of a system, this suggestion posed special difficulties. But eastern states also complained, since post roads almost by definition followed circuitous routes.[11] The ambiguity of the law did not lessen the confusion. One later observer claimed that "the Federal-Aid Road Act of 1916 was a distinct triumph for those favoring a national system of highways under national supervision," but a ruling by the attorney general had limited funds to RFD routes. Even after the agriculture secretary told district engineers, "I don't see that we are primarily concerned about the main roads," the department's Agricultural Advisory Committee complained that funds were being used improperly to create a national system.[12] In other words, the national system versus rural roads argument that had surrounded the 1912–16 congressional debate had arisen again, and would remain a matter of debate for four more years.

Compounding this federal-state conflict was the $10,000-per-mile limit on federal contributions for state highway construction. This requirement further reflected the rural orientation of the federal policy, because only simple country roads cost so little. To states building more expensive paved highways, this limitation seemed unfair. As his office was criticized from both sides, Page amended the rules but could not eliminate these two sticking points. He eventually proposed amendments to the original bill that expanded the number of roads qualifying for federal aid by defining RFD roads as any road that might be used, even at some point in the future, for mail delivery and that lifted the limit on the federal contribution. But the damage had been done, as discord emerged in the road-building community over the technical features of the federal policy.[13]

The greatest obstacle to the start of American highway construction, however, was the conflict in Europe. The war not only jarred many industries out of their usual patterns but also, and more significantly, destroyed both prewar attitudes about highways and Page's policy. Problems began with mobilization efforts that gave road construction a low priority and then banned rail shipments of road materials by November

1917. Intended to ease rail congestion and threats of coal shortages, these actions and later restrictions on the use of bituminous materials effectively halted road construction. As a member of several priorities committees established by the President, Page eventually found a way to end the total ban on rail shipments and bituminous materials, but the loss of state and federal engineers to the Army negated even these moves.[14]

These disruptions crippled road construction even as American highways were facing their first real test. The war completely overtaxed the ability of eastern railroads to move freight, and the resulting congestion that produced freight embargoes in some cities eventually led to government operation of the railroads. Shippers accordingly turned to motor trucks, previously used only for local deliveries, and gradually extended their range as freight cars backed up one hundred miles from New York City. Manufacturers experimented with truck service on routes between Boston and Philadelphia, while farmers introduced long-haul milk and produce deliveries. Roy Chapin, head of the government's Highway Transport Committee, proposed the ultimate test in late 1917—to drive trucks from midwestern factories to Philadelphia and Baltimore for shipment to France. Despite heavy snowfalls, 30,000 of these trucks with three-ton loads reached the east coast after December 1917.[15]

The performance of trucks, both at home and in France, surpassed all expectations, and this had two immediate effects on the country's highway development. First, it radically altered American attitudes about the place of roads in the nation's transportation picture. An *Engineering News-Record* editorial of 1918 described what it termed "A New Era in Highway Transportation":

> Before the war who could have suggested without incurring ridicule the transportation of an army division of 38,000 men, with all its equipment, over a distance of 160 miles in 16 hours? . . . Who could have said without being dubbed a dreamer that the inadequacy of our rail transportation system would force the use of motor trucks for long-distance express service between points as far apart as Washington and New York? Such developments were beyond the expectations of even the most enthusiastic highway advocates five years ago; today even the man in the street regards them as commonplace.[16]

The wartime experience with trucks thus provided the first potentially viable economic rationale for intercity roads.

Second, the other product of wartime trucking—the destruction of many roads used by trucks—could not be ignored. Such destruction was stunning in its magnitude; as one OPR engineer explained, "Hundreds of miles of roads failed under the heavy motor truck traffic within a comparatively few weeks or months. . . . These failures were not only sudden but also complete, and almost overnight an excellent surface might become impassable."[17] This situation earned additional criticism, much of it undeserved, for the renamed Bureau of Public Roads, since priorities, not red tape, now held up federal-aid work. Despite their reduced numbers, BPR engineers had managed to approve 557 state projects and release $5.7 million for 218 others in 1918. Yet by March 1919, only thirteen miles of federal-aid highways had been completed.[18] Inevitably, the BPR was blamed for this poor performance.

Thus as 1918 drew to a close, dissatisfaction with the federal-aid program was rising. Even the *Engineering News-Record*, always a Page supporter, unlimbered a stinging editorial in November 1918: "It is too much of course to expect from Washington . . . that strong type of leadership which befits one of the greatest constructive programs the country has ever known. The Federal road administration has too long been in a subordinate capacity, lost in a department—that of Agriculture—whose main interests are foreign to road work."[19] The credibility of the BPR and its experts had tumbled.

Federal Aid versus a National Highway Commission

The challenge to federal leadership in the highway field may have sprung in part from the decay in support for government agencies and general disillusionment with the "make-the-world-better" goals of the Progressives at the end of the war.[20] Thus even though the armistice promised to ease some of the problems facing road builders, frustration over the BPR's administrative thoroughness remained. The contradictory combination of wartime changes in public perceptions that created more support for highways and postwar sentiments against an intrusive government agency (the BPR) led to a vigorous attempt to alter the nation's highway policy. The same groups who had supported a national highway commission from 1912 through 1916 again put forward such a plan, and this time it seemed they would win. They failed, however, to see the irony in their attempts to replace a federal bureau that acted cooperatively with a new national bureaucracy that would exclude the state highway departments.

As late as June 1917, most highway supporters agreed with the *Engineering News-Record* when it argued that road supporters should "bend their attention to promoting the construction not of main-line thoroughfares but radial highways connecting the all-necessary farmer with the market."[21] But in October 1918, this same journal laid the foundation for renewed efforts to install a national highway commission when its editor

E. J. Mehren asked two questions: "The states have highway departments. Why not a United States highway department? The states have highway systems. Why not a Federal highway system?"[22]

Others had already reached this conclusion, and by late 1918 Page's efforts to amend the federal-aid bill, broadening the definition of RFD routes and streamlining procedures, were being swamped by calls for a national highway commission. The leadership of this movement came from the auto and truck manufacturers, operating through the American Automobile Association and a trade group organized during the war, the Highway Industries Association (HIA). The momentum for a national road agency was strong enough that when the state highway officials held their annual meeting in December 1918 in conjunction with the Highway Industries Association, Page's plan failed to win endorsement, especially after Page died of a heart attack during the proceedings. To salvage the situation, Iowa's Thomas MacDonald and other midwestern highway officials offered a compromise resolution that proposed using federal aid to build a limited system of primary roads. The HIA, however, unanimously accepted E. J. Mehren's proposal for a 50,000-mile, federally constructed highway system.[23]

This open split in the highway community over the structure of federal involvement in road building clearly seemed to favor the supporters of a national commission, who seemed far more powerful than the backers of federal aid. In response to the road failures of 1917–18, Congress, for example, provided an additional $50 million for federal-aid construction in 1919 and $75 million each in 1920 and 1921, but pointedly chose not to extend the federal-aid plan beyond 1921. It also rejected the new definitions of post roads proposed by Page to remedy complaints about the system.[24] Clearer evidence of the strength of the national highway commission position was the support that the HIA attracted for its full-scale campaign for such a body. The opening salvo was a national highway system bill introduced by Senator Charles Townsend of Michigan in February 1919. Few believed this measure could fail, and the resignation of the two most respected BPR engineers with long experience in the development of federal legislation, J. E. Pennybacker and M. O. Eldridge, to join the AAA to work for this legislation in March 1919 was a clear signal of confidence.[25] By April 1919, the campaign had been endorsed by some 775 civic organizations. At least thirty-eight state highway departments supported this effort, as did tire manufacturers, the auto makers, the National Grange, the Portland Cement Association, and Rotary International. High-powered publicity efforts began that summer with magazine articles, 200,000 posters distributed by auto dealers, and a booklet published by the AAA. The highlight however, was an

Army truck convoy from Washington, D.C., to San Francisco on the Lincoln Highway.[26]

Yet in the end, the national highway commission forces failed. Quarrels among the various backers of the Townsend bill over who should lead the campaign hampered genuine coordination of the public relations effort. A more serious problem was the continuance of federal aid, for even highway engineers who supported a national commission hesitated to jeopardize the money already appropriated. Congress in 1919 certainly had no desire to consider a plan that would not take effect for two years. So in spite of warnings that waiting until federal aid expired in 1921 would jeopardize their chances, most supporters of the commission favored such a delay. As a result, the sense of urgency about a national highway commission palpable in early 1919 was frittered away. Townsend reintroduced his bill in January 1920, only to have debates over the peace treaty and League of Nations delay committee hearings until May. Publicity efforts resumed with the Firestone Ship-by-Truck transcontinental caravan and an accompanying program of speakers, films, tours, advertising, and handbook. But with a presidential election upcoming, Townsend saw little prospect of pushing the bill to the floor in 1920. Then, in June 1920, the Townsend forces were shaken when the senator's Republican Party adopted a federal-aid plank.[27]

This was an ominous sign for the supporters of a national highway commission, for it indicated that the federal-aid concept had found new life. Its rescuer was Thomas H. MacDonald, Page's successor at the BPR. He had served as state highway engineer in Iowa for fifteen years after graduating from the Iowa Agricultural and Mechanical College in Ames. His senior thesis on highway needs proved influential in the establishment of a state highway commission, and he became its first and only employee in 1904. From the start he emulated the OPR's advisory pattern, helping local officials develop standard specifications and improved construction methods. Convinced that cooperation was the best way to build roads, MacDonald's unstinting support of the federal-aid idea within the AASHO had saved Page's bill in 1916. Furthermore, MacDonald commanded the respect of other state highway engineers. Before accepting the BPR job, MacDonald discussed it with George Coleman, Virginia's consistent advocate of a national commission. Coleman replied,

> I know of no highway commissioner whom I would rather see in the office at Washington than yourself. . . . I think that I can say without any hesitation that you, as Director, would be able to bring about a better feeling and establish a spirit of cooperation between the Federal Department and the highway departments.[28]

If Page, with his laboratory background and moral crusader's outlook, had been the perfect choice to head the OPR in 1905, MacDonald brought traits that were equally well suited for running the BPR in 1919. His years in a state highway department gave him the insights essential for managing a cooperative federal-aid program. But perhaps as important was his narrow view of road work. As the moral emphasis of the Progressive Era died out, killed largely by the war, MacDonald brought to the office a justification of road work that stressed more narrow economic and technical terms, a shift that seemed in tune with the times. Indicative of this businesslike approach was his refusal to accept the position until the secretary of agriculture had convinced Congress to offer salaries commensurate with the responsibilities given the chief of the BPR and his district engineers.[29] The shift in tone was fully evident in MacDonald's first annual report as head of the BPR, in which he wrote, "It is very apparent that the whole highway program has passed beyond the propaganda stage, and that there is now the greatest possible need for highway organizations to approach the task of actual road building in a responsible, sane spirit that will result in the production of roads rapidly but without extravagance and without loss of faith on the part of the taxpayers."[30]

Missing from this explanation was the "get-the-farmers-out-of-the-mud" rhetoric that had characterized the policy of Logan Page. MacDonald, in fact, had attempted to modify Page's overtly social justification of roads even before he arrived in Washington in May 1919. At the 1918 AASHO convention, it was MacDonald who suggested that federal funds be limited to a definite system that reached beyond post roads, and, with the support of other midwestern highway engineers, he used this idea to revamp Page's federal-aid mechanism to produce a national network of highways. He had not, however, abandoned the basic idea that rural roads needed to be improved. He remained convinced for reasons of efficiency that the primary use of roads would continue to be local and that a special transcontinental highway system would only be an extravagance. MacDonald believed that commercial traffic would find a natural radius of about one hundred miles from populated areas. "There would be no advantage gained by considering the local uses of roads as secondary," he wrote, and he rated agricultural uses at the top. In any event, technical considerations should decide the issue, with the first step toward any national system being the classification of roads by use. By early 1920, MacDonald had started this classification process, drawing together members of the AASHO and representatives of the War Plans Division of the Army to plot all national, strategic, and military routes. The resulting map, often called the Pershing Map, confirmed Mac-

Donald's assertion that not even a military necessity existed for cross-country highways. Clearly, the new tone at the BPR replaced idealism and calls for social justice as the justifications for highway building with pragmatism and economic efficiency.[31]

MacDonald realized that he had to rebuild state confidence in the BPR if his ideas were to triumph over efforts to create a national highway commission. Page had shown in 1916 that technical expertise granted legitimacy to the OPR in policy debates, but when state engineers supported a competing position as experts in their own right, his influence with Congress had declined precipitously. Only unified support by State highway officials based on confidence in the BPR's technical leadership could insure MacDonald's success. The task of restoring cordial state-federal relations was formidable, but Page had already moved to ease the strains within the federal-aid program by late 1918. Further improvements came with revised rules for receiving federal aid that the BPR presented to the states in March 1919. At MacDonald's insistence, the secretary of agriculture transferred greater authority to district engineers, thereby permitting states to start construction after plans had received district approval. This decentralization did not end inspections and requests for changes by the BPR's headquarters staff, but the new procedure eliminated a major cause of delay. The new rules also allowed states to divide projects into sections and submit engineering plans for each part separately. Finally, three western district offices that also oversaw federal road building in National Forests were split into six to prevent delays in project approval.[32]

These changes combined with the end of wartime material restrictions to improve significantly state-federal relations. A March 1919 survey in the *Engineering News-Record* reported no delays in BPR inspections, and the record of approved projects for 1919 and 1920 continued this pattern. During fiscal 1919, the BPR approved plans for 736 projects and released $18 million to the states for 454 others. In 1920, 1,670 plans were accepted, 90 percent within four days of arriving in the Washington office, while federal engineers obligated almost $86 million for 1,286 projects. Federal engineers accepted more work in 1920 than in the previous three years combined. By the end of the year, construction had started in 1,835 projects covering 14,940 miles of road; 1,677 miles were completed. MacDonald was proud of the BPR's record through 1920—2,986 projects approved totaling 29,319 miles—and boasted that this even surpassed the achievement of the Panama Canal.[33]

But changes in the justification and rules for obtaining federal highway funds were not the only factors that produced these results. Equally important was the BPR's perpetuation of the Progressive belief in the importance of cooperation that the OPR had exemplified. MacDonald made this clear in his initial communication to all BPR engineers:

> It requires unusual tact and ability on the part of our engineers and our organization to act in harmony with so large a number of officials and under such a variety of conditions. Our success will depend largely upon the attitude of mind and confidence we establish on the part of the State officials.[34]

He expected district engineers to build such confidence through personal contacts, and followed this dictum himself. One of MacDonald's conditions for accepting the job was creation of a Federal-Aid Advisory Committee of six state highway engineers, which was expanded in 1920 to include the AASHO executive committee, in order to develop "the most cordial and mutually helpful relations possible between this department and the several state highway departments."[35]

MacDonald's goal was a partnership between the states and the BPR, and his success in building it was evident in the remarks of later commentators who presented the federal-aid highway program as a model of federalism.[36] His handling of the potentially troublesome requirement that state work be of "substantial construction" demonstrated his desire to avoid confrontations. He defined "substantial" to mean that roads should meet local traffic requirements and introduced also the concept of state construction. States could thus build a road with a simple surface such as gravel for little-traveled federal-aid highways by calling it the first stage of construction and promising to add a second-stage hard surface when needed, using additional federal aid. By recognizing that the different conditions in Nevada and New York required BPR flexibility, MacDonald prevented serious disputes.

This is not to say, however, that the bureau permitted sloppy work. It refused to build wooden bridges and assigned federal engineers to states with weak departments. BPR inspectors turned down contracts granted under suspicious circumstances, and federal funds to Arkansas were cut in 1921 because of political interference in the state's very weak highway department.[37] Obviously, MacDonald's BPR took its inspection and supervision role no less seriously than had Page's OPR, but these actions now won applause rather than complaints thanks to the cooperative attitude. Whereas OPR inspectors had *demanded* changes in Michigan's specifications, BPR engineers *requested* alterations. By the end of 1919, George P. Coleman observed that "cooperation between the various State departments and the Bureau of Public Roads has improved tremendously in the past 12 months. Mr. MacDonald and his bureau would, I am sure, feel very much flattered and encouraged."[38]

Not coincidentally, this cooperative federal leadership that eased tensions with the states also restored the BPR to its position as technical adviser to the entire highway community. Technical leadership after the war required a very different touch as the stronger, more competent state highway departments envisioned by the Federal-Aid Road Act were realized. Not all were run by engineering staffs without political interference, but a growing number deserved to be treated by the BPR as partners rather than as children in need of handholding. Several developed competent testing and research capabilities, and all began to use mechanized construction equipment. The leading states, such as Iowa, Michigan, Pennsylvania, and California, laid out highway systems of 6,000 to 8,000 miles. Moreover, state road construction budgets jumped significantly, partly through the provision for regular revenues and partly through highway bond issues that totaled $315 million in 1919 and $404 million in 1920. To this, counties added another $379 million, so that by the end of the 1920, outstanding bond issues for road work totaled $999 million. Actual state construction budgets of about $400 million in 1919 jumped to $633 million for 1920.[39]

The federal-aid requirements materially assisted such developments, but some states still required BPR help. This situation became evident as the resources available to state highway departments mushroomed after 1918. The best departments, such as that in Illinois, could handle this sudden increase in funds, arranging for 400 miles of concrete road construction in 1919. But newer, less experienced departments encountered serious administrative problems in making the transition from an annual budget of $2 million or $3 million to one of $50 million. Because the federal-aid program offered a model for creating state-aid systems built by counties, the BPR became a stabilizing influence, helping states resist the pressure to build roads quickly and haphazardly in the face of a rapid rise in automobile registrations from 3.3 million in 1916 to 8.1 million in 1920. Moreover, the BPR provided direct construction assistance with its expeditious distribution of surplus war equipment to the state highway departments, without charge if it was used in actual road building. Using the federal-aid formula, between June 1919 and October 1920 the BRP allocated 20,519 trucks along with 20,000 tons of TNT, tools of all kinds, and other equipment.[40]

The BPR's return to leadership showed clearly in the crisis that afflicted the highway construction industry in 1919 and 1920. In simple terms, in these years, the states planned much larger road-building programs than they had before the war, and were then saddled with the problems of securing and transporting road materials and rapidly rising inflation. Most states made slow starts due to inexperience, a shortage of contractors, equipment, and engineers, and late preparation of plans and drawings. Midwestern states finished only 53 percent of the work planned for 1919, and other regions had a worse record. But with the inflationary increase in wages and materials accelerated by railroad car and Portland cement shortages, 1920 proved to be an even worse year as only 25 percent of planned construction was completed nationwide.[41]

Thomas MacDonald quickly emerged as the industry's spokesperson during this troubled period, first ordering district engineers to expedite project approval and then interceding with the Interstate Commerce Commission to obtain cars for shipping Portland cement. In tandem with direct requests to some western railroads, this approach facilitated many shipments of stone and cement. Equally important, the BPR's inspection system prevented states from either building lower quality roads or letting exorbitant contracts. Both policies acted as a brake on federal-aid construction while preventing fraud and waste and pushed states to prepare engineering plans for future work. Reports on these problems in engineering periodicals portrayed the BPR as both the voice of the states and an anchor to sound construction practice. Declining wages and the end of railroad shipping problems finally made 1921 a better year, but equally important was the improvement in the engineering plans prepared by the states and the ideas recommended by the BPR, such as the use of local materials and phased construction, which halted the mad dash for concrete roads. By such means the BPR reestablished its influence in road building beyond federal-aid work.[42]

The final step in improving relations between the BPR and the states involved a variety of federal efforts to help the states with problems other than construction. MacDonald worked with western states, for example, on an amendment to the 1916 bill permitting the federal contribution for highways to exceed 50 percent in states where the extent of federal landholdings significantly reduced property tax revenues. While this proposal could not be immediately introduced in Congress, MacDonald's efforts, which ran counter to the opinion of the department's solicitor, demonstrated his desire to work for the states.[43] Other assistance was more overly political, but conveyed the same message. In 1920 the Southern California Automobile Club accused the California Highway Department of waste and inefficiency because of concrete pavement failures. The state's highway engineer turned for help to MacDonald, who launched an investigation that quieted demands for a legislative inquiry by presenting the California department as a victim of unexpected changes in traffic. The BPR study suggested that when the state had adopted concrete pavements, thin surfaces had adequately carried motor vehicles. These began failing after 1918 not from administrative inefficiency but from unforeseen increases in traffic volume and truck weight.[44]

By providing such assistance, MacDonald had repaired most of the damage to state-BPR relations by 1920. He received the credit for removing bottlenecks in the inspection and approval process, and largely ended the adversarial atmosphere that had appeared in federal-aid administration without reducing the bureau's ability to enforce standards. He also restored the feeling, which had existed before 1916, that the BPR was the place to turn for assistance with any type of highway problem. Thus the basis of success in the emerging federal highway program continued to be the technical expertise of the BPR and the cooperative attitude of these federal experts.

The Federal Highway Act of 1921

This restoration of harmony between the states and the BPR in the federal-aid program dimmed the prospects for the national highway commission advocated by Senator Townsend, because a working federal-aid system undermined the major changes called for in his bill. Supported by the states once again, the BPR argued that only a few minor alterations were required to improve the 1916 legislation. Significantly, Congress accepted this argument as well as MacDonald's ideas about these alterations in 1921. In the process, the importance of the BPR's ties to the states became especially clear. Even more than it had for Page, the AASHO provided a mechanism by which MacDonald could unobtrusively influence policy by presenting itself not as a lobbying group but as an impartial agency furnishing Congress with expert information. In other words, the Progressive ideology of expertise remained an effective means of shaping policy. MacDonald understood this process, having been part of it in 1916, and he continued to follow it as chief of the BPR, thereby remaining the arbiter of American highway policy.

By 1920, little effort was required for MacDonald to secure the highway officials' support for the federal-aid concept, thanks to the midwestern highway officials' organization, which he had headed until 1919. State engineers slowly realized that a national commission threatened both their autonomy and the prospect of construction work by state departments. In August 1920, the Mississippi Valley Conference of State Highway Officials drafted a memorial against the Townsend bill, proposing instead a four-year extension of federal aid. MacDonald carefully observed the conference through the reports of Iowa's new highway engineer, Fred White, who encouraged other state officials to notify the BPR of their opinion on federal aid. By November, only one state had reported opposition to the present policy, thus confirming an earlier poll by the Portland Cement Association.[45] Townsend's supporters tried frantically to stem the erosion of support, but were forced to cancel hearings and then watched helplessly as half of the states signed the anti-Townsend bill memorial before the AASHO convention in December 1920 and twelve others signed during the meeting. The convention also passed a resolution favoring federal aid, and in March 1920 state engineers on the BPR's Federal-Aid Advisory Committee endorsed federal aid.[46]

This movement toward uniformity of opinion among the states of course paralleled MacDonald's efforts to streamline federal-aid administration and remove obstacles to road construction. A letter from Maryland's highway commissioner, J. N. Mackall, to MacDonald in June 1920 asking for an opinion on a bill raising the federal contribution to $30,000 per mile, reflected the improving atmosphere as Mackall concluded by telling the BPR director to "be assured that whatever is your wishes in the matter will have our heartiest support."[47] Not surprisingly, the 1920 AASHO meeting was a harmonious affair that endorsed a legislative proposal prepared by the executive committee that embodied MacDonald's general ideas. This bill, introduced in Congress shortly thereafter, called for a federal-aid system and a larger federal share for western states in which the federal government owned a large percentage of the land. MacDonald presented the bill as strictly an AASHO initiative, but he had been actively involved in its drafting as a member of the organization's executive committee. Clearly, the BPR's chief engineer, P. St. J. Wilson, spoke correctly when he noted that the state engineers formerly favoring a national commission had changed their minds.[48]

Moreover, MacDonald also worked to establish BPR ties with the leaders of the national commission bill. He began this as soon as he arrived in Washington, meeting with Pyke Johnson, the lobbyist for the National Automobile Chamber of Commerce, and former OPR engineer J. E. Pennybacker, then with the AAA, to present his federal-aid system as a compromise alternative to their national commission. Thereafter he maintained regular contact with Johnson, who became one of the MacDonald's closest friends for the next forty years. He also met with the good roads committee of the National Automobile Chamber of Commerce (NACC) in July 1920, using the meeting to form closer ties with Roy Chapin, head of that committee and leading promoter of the Townsend bill. In short order, Chapin, who was president of the Hudson Motor Car Company, and MacDonald were working together to improve highway engineering and highway economics education and to win congressional support for American membership in the Permanent International

Association of Road Congresses. After their meeting, Chapin wrote to MacDonald, "We were able to report to the directors of the chamber . . . the very cooperative attitude which you showed, and I feel confident that the industry will always be willing to meet you more than half way in any problems which you may have."[49]

Opening such avenues of communication not only kept the fight over the shape of federal involvement in highways from degenerating into name-calling, but also increased the entire road-building community's respect for MacDonald. By establishing cordial relations with such leading opposition figures as Chapin, MacDonald was demonstrating the cooperative style of leadership that had convinced the state highway departments to change their position and support the federal-aid system. MacDonald also seemed to be saying that he was not concerned with politics and that he was a technical expert concerned only with building better roads. He would and could work with anyone sharing this general goal. By early 1921, this attitude had made MacDonald, along with the BPR, the clearly recognized leader of the road-building community. He had largely eliminated the red tape and other delays in the existing system and had worked to resolve the problems underlying the construction delays of 1919 and 1920. Moreover, he was constantly consulted by Congress, another sign of the BPR's return to a position of respect. It was not surprising, then, that in January 1921 the *Engineering News-Record* recanted its position on a national highway commission and endorsed federal aid instead, noting that "operation has disclosed no major direction in which change is needed."[50]

After this point, the battle between supporters of a national highway commission and advocates of federal aid was largely anticlimactic. Both sides continued to lobby hard for their positions, with the AASHO leading the charge for federal aid, while the National Automobile Chamber of Commerce and the AAA pushed the Townsend bill. But the House passed the AASHO's federal-aid bill in May 1921, as the senate passed a version of Townsend's bill amended to look like AASHO's plan in September. A difficult conference finally produced the Federal Highway Act of 1921.[51]

Several factors contributed to this victory for federal-aid supporters. First, President Warren G. Harding had announced he would accept only a modest bill, hopefully in the $50 million range. Only the federal-aid proposal matched this prescription; the national commission plan required substantially more money. Second, the fear of big government worked against Townsend. Representative Sam Rayburn summarized this position in the House debate on the bill, announcing he was "sick and tired of the federal government's everlasting sticking its hand into the affairs of my state. I am against any building up of more bureaucracies in Washington to reach out into the states and tell the people what they shall and what they shall not do."[52] Finally, hard, behind-the-scenes work of MacDonald was essential in preserving the federal-aid approach to highway building.[53] Beyond his constant contact with Congress regarding the various legislative proposals as chief of the bureau and administration spokesperson on highways, he also guided the lobbying tactics of the AASHO's Washington representative, W. A. Markham. MacDonald not only kept Markham and all AASHO members informed of events but also helped the AASHO's executive committee redraft the bill in mid-1921 to add two new points that insured its acceptability to Congress. One called for each state to designate 7 percent of its roads, 3 percent primary and 4 percent secondary, to be part of the federal-aid system. This clause satisfied MacDonald's arguments for developing a road classification system; it also created a genuine national highway system, the central goal of the supporters of the Townsend bill. This system, however, also reflected MacDonald's continuing belief that rural roads merited the highest national priority. The second point required each state, not the counties, to maintain these roads, a move to further strengthen state highway departments.[54] These two changes demonstrated nicely MacDonald's role in shaping this bill, for without disturbing the general administrative framework created by Page, he had created a way to build a national system of roads.

MacDonald's Policy and Consensus

This victory in Congress represented a stunning turnabout for MacDonald, especially if one remembers the obituaries prepared for federal aid in early 1919. As the *Engineering News-Record* commented, "The roll call was an overwhelming vote of confidence for the U.S. Bureau of Public Roads."[55] MacDonald's success in shepherding the federal-aid revisions through Congress rested on his cooperative philosophy more than any other factor. That he had restored the unity of the states and the BPR was apparent in an October 1921 poll that found that thirty-six state highway departments favored federal aid, four supported the Townsend measure, four were neutral, and eight had no opinion. Equally important, he had almost completely reunited the road-building community, regaining the support of the Grange, the American Road Builders' Association, and the *Engineering News-Record* for his approach to highways.

Just as Page the reformer had been in tune with the Progressive Era, so MacDonald was in line with the emerging attitudes of the 1920s. Cooperation, an idea with clear Progressive roots, would become the central attitude of both intergovernmental and business-government relations as rigid governmental regulation fell out of favor. The contrast was evident in the differences between the OPR, which had wielded its power of the purse very prominently from 1916 through 1918, and MacDonald's BPR, which avoided dictating to the states. Whenever possible, he preferred to let the AASHO assume the prominent role in policy questions, secure that the BPR's superior technical expertise had shaped that position. MacDonald, the former state highway engineer, abhorred heavy-handed federal action, and the 1921 legislation reiterated the federal-aid principles adopted in 1916 that had formally limited the power of the BPR. Yet the 1921 bill assumed that the bureau's expertise would produce improvements in technique and state administration as well as promote efficiency. The newer influences of the 1920s were evident in the prominent place of cooperation in both the bill and the tactics that had won its approval.

MacDonald's goal throughout was another indication of the changing nature of American society after the Progressive period. He clearly sought agreement among all the major players about the purposes and approaches to American highway construction. Although federal policy applied only to the federal-aid system, he understood that practices used on these roads would influence all other state work. Thus a consensus about priorities and techniques would guide nearly all highway construction in the country. To achieve a consensus, of course, required compromise; MacDonald never expected to have his ideas adopted in every detail. From the start, he showed his concern for creating a policy all could support by proposing to use federal aid to create a national system of some sort. By 1921, this system assumed the form of the 7 percent system instead of a post road network.

Clearly, this concept was intended to meet the changing use of roads after the war; it also fulfilled some of the objectives of the automobile industry. Yet the new federal policy did not abandon the emphasis on rural roads installed by Page. More than half of the federal-aid mileage was to come from the most important farm-to-market roads, and these had to receive 40 percent of the federal funds. This division was based on BPR studies such as one in Iowa that showed that 48 percent of all traffic was farm-to-market while only 7 percent was interstate. MacDonald believed the new system would benefit both types of traffic; it would also help the entire country by slowly producing a genuine national road network. Moreover, technical considerations of road usage, not political "pork-barrel" maneuvers or grandiose schemes, would determine the pace and size of the system. Nonetheless, MacDonald expected the even stronger state highway departments required by the 1921 legislation eventually would produce marked improvements on the local farm-to-market highways that comprised 80 percent of the nation's roads. It was clear, however, that MacDonald's efficiency goals were narrower than those of Page, who had meshed efficiency with a set of social reforms. MacDonald desired most of all an orderly development of the country's main roads.[56]

His success in building a consensus around these ideas showed in the response of the Townsend forces to his victory. Initially, Pyke Johnson was disappointed at their defeat, but he could soon accept Chapin's assessment:

> I am tickled to death with the passage of the Federal-Aid Bill. While it does not go far enough, it is a great step in the right direction, and I think everyone has a reason to feel that highway funds are going to be better administered in the future than they have been in the past.[57]

After 1921, MacDonald enjoyed significant support from all groups that had initially opposed the AASHO's bill. Through M. O. Eldridge, the bureau built strong ties to the AAA, so that, in Eldridge's words, "the Association will avoid taking any stand at variance from that of the highway officials and the bureau in important matters."[58] The National Automobile Chamber of Commerce also endorsed BPR ideas routinely over the following years. MacDonald had thus convinced his opponents that they had not lost after all. A. R. Hirst, Wisconsin's highway engineer and one of MacDonald's staunchest supporters, may have gloated, "I am very glad that our friend Townsend got it 'where the chicken got the axe' relative to his proposed federal highway commission. . . . I think the outcome will put a quietus upon the commission business for several years to come."[59] But MacDonald never responded in this way, although Hirst's prediction that a national highway commission was dead proved correct.

After the discord of the 1910s, the ensuing two decades were notable for their quiet, as state engineers settled down to construct a road system that could handle the influx of drivers. In a number of respects, the golden age of highway engineering was beginning, with Thomas MacDonald and his Bureau of Public Roads once more the recognized leaders of the field. The basic strategy he had devised between 1919 and 1921 would guide this technical community for the next fifteen years. It might be argued that MacDonald was only a skillful politician. Certainly he recognized the importance of cooperation, and he was an able administrator. Certain elements of his consensus, especially the reliance on federal forms of administration, were determined more out of political concerns than technical needs. But it was also clear that other road builders followed MacDonald's lead in these matters because of their respect for his and the BPR's technical reputations. This might not yet be apparent, given the focus to this point on legislative policy. But an examination of the implementation of the federal-aid program after 1921, including the BPR's role in standard setting and highway research, indicates the extent to which MacDonald's cooperative style relied on the BPR's image of technical superiority. By the mid-1920s, the manner in which technical expertise influenced highway policy was abundantly clear to the participants in the road-building boom that followed the war.

PART II

The Golden Age of Highway Building, 1921–36

Until 1921, the condition of American Highways slowly improved as experts developed better statistical information and construction practices. During the two decades after 1921, however, a genuine boom in highway construction occurred; the states alone built more than 420,000 miles of roads between 1921 and 1936. Road building between the two world wars took place within the federal-aid structure of shared power, responsibilities, and finances, yet boundaries that were so clearly defined in the legislation tended to vanish in everyday practice. If the law provided state officials with a large voice in highway policy, federal experts quietly came to dominate the field by their technical expertise. Senator William E. Borah had predicted in 1914 that such centralized authority "would make the head of the bureau the practical dictator of road matters throughout the United States."[1] But Thomas MacDonald never became a dictator; in fact, the BPR went out of its way to develop cooperative mechanisms that gave all members of the highway community a say in policy decisions. Nonetheless, after 1921 the Bureau of Public Roads emerged as the stronger partner in the state-federal road-building venture as its cooperation and expertise gave it almost unlimited influence over highway building despite its limited statute authority.

The next chapters examine the complex, even paradoxical position of the BPR in the federal-aid framework, and the perspective of the narrative shifts from the making of policy to the question of administration. This topic is of some importance, according to historian Samuel P. Hays, who has recently suggested that it is "the emerging central focus of the entire political system. . . . The arena of continuous choice which affects the larger society on a day-to-day basis is administration. And the actors well know it."[2] In the case of highways, it is clear that the BPR's cooperative administration of the federal-aid program is the most important facet of the history of the country's highway program during the fifteen years after 1921. Three different components of the program must be considered in reaching an understanding of the BPR's administrative style. The fourth chapter examines the BPR's oversight of state construction of federal-aid highways from 1921 through 1936, in which the crucial cooperative ties were to the AASHO. This task was the largest faced by the

BPR and directly related to its mandate from Congress to distribute funds to the states while insuring the states used them efficiently. But the administration of the construction funds can only be understood by considering the BPR's handling of two other aspects of highway construction during this period. Thus the fifth chapter turns to the highway research program, in which the cooperative ties to the Highway Research Board of the National Research Council masked the BPR's dominance. The sixth chapter considers the development of construction and material standards, a task that the BPR guided by operating through a number of professional and technical associations, including the American Society for Testing Materials (ASTM). In all three areas, states were supposed to make the major choices subject only to federal approval. Yet by combining cooperation and expertise, the bureau guided these decisions by the highway community while maintaining harmony. Thus for the BPR, the years between the wars were the "golden age" of American road building.[3] They also form a case-study in governmental administration.

In exploring the significance of the BPR's position, its administrative style, and its accomplishments, two themes related to American political life provide important frames of reference. The first is federalism, since the most basic feature of the national highway program was the sharing of power between the state and federal governments. The federal-aid highway policy has been widely considered a model of this political philosophy. But in fact a much more complicated interaction between governmental units took place. Although the basic parameters of road building were delinated by the precepts of federalism, the BPR's expertise blurred the sharp lines between state and federal responsibilities drawn by the legislation while at the same time preventing fears of federal interference from disrupting the harmony among road-builders. An essential element in this outcome was the attitude of BPR engineers, who saw themselves not as ordinary bureaucrats but as apolitical experts sharing a mission with state highway engineers. Rather than influencing the states through the power of the federal-aid purse, MacDonald relied on the BPR's continued technical leadership, for the bureau remained the clearinghouse for all types of data as well as the preeminent research center, thereby developing a means of breaking the political boundaries of federalism. Thus one of the themes of this section will be concern the interaction of expertise and federalism.

The other theme of this section is the way that the BPR's domination of highway construction after 1921 fits into the political philosophy of the "associative state," which emerged during the 1920s. First as commerce secretary and later as president, Herbert Hoover was the leading spokesperson for this idea, which held that government should act as a stabilizer and moral guardian for industry without directly interfering in its operation. Cooperative contacts were maintained through trade and professional associations. Scholars have traced the origins of the associative state to Progressive outlooks, with the differences evident in the way Hoover tempered the reformers' zeal by relying on the voluntary avoidance of predatory behavior by business rather than regulation. Similarly, new studies have stressed that Hoover's associative efforts foreshadowed important elements of the New Deal.[4] Crucial differences existed between the New Era of Herbert Hoover and the Depression policies of Franklin Roosevelt, most notably the magnitude of the federal presence and the deliberate attempts at social engineering in the New Deal. But rather than presenting totally opposite conceptions of government, the 1920s and 1930s can be seen as connected by a Progressive thread.[5]

Road building after 1921 offers an opportunity for a detailed examination of the workings of the associative ideals of the 1920s. The groups that the BPR used to involve roadbuilders in the decision-making process exemplified the cooperative style of business-government relations, for they were the crucial links that bound the highway community together. They provided the forums for spreading ideas and information from the BPR in an unthreatening manner, thereby preserving the basic agreement about the purposes and approaches to highway construction even as the BPR largely determined these issues. Moreover, the history of highways in this period demonstrates the basic political continuity of the first half of this century. In line with the scholarship emphasizing the ties between decades rather than the differences, the 1920s and 1930s may be seen as a single period in road building, even though the Depression altered particular aspects of highway activity. In short, this section will examine a period of road building marked by a surprising degree of harmony over the purposes of and approaches to highway construction, and while explaining this phenomenon, comment on the nature of both American federalism and the associative philosophy of the 1920s.

Chapter Four
Building the Federal-Aid Highway System, 1921–36

Each level of government had its own responsibilities and authority and the agreement of each was necessary to carry out a project. This provided a balanced partnership between the States and the Federal government with neither being dominant.

Sherwood K. Booth, 1965[1]

The Federal Highway Act of 1921 ended five frustrating years for American road builders and launched a basically unending engineering project. To the states fell responsibility for building both state and federal-aid roads, as the Bureau of Public Roads could only inspect state plans, construction, and maintenance on the federal-aid system. Little friction marred this program, which has been considered a model of both federalism and the associative philosophy of the 1920s. Yet the federalist delineation of state-federal responsibilities failed to describe the actual process of highway construction between the wars, as the BPR's expertise enabled the agency to transcend these boundaries and reach beyond the federal-aid program. In short, federalism was visible in the BPR's cooperative administrative style, which limited the activities and influence of federal engineers. But the way in which the BPR actually cooperated with the American Association of State Highway Officials suggested that it would be misleading to call the road-building program a model of federalism. Using cooperative mechanisms that were in theory intended to limit the influence of the federal government, federal engineers guided state highway building and led the highway community in the 1920s and 1930s.

Technical expertise and the political considerations of cooperation were intertwined in the BPR's domination of the federal-aid program. This point emphasizes again that the claims of experts to be apolitical were based on unrealistic expectations. The BPR never operated on strictly technical grounds in administering the federal-aid program; compromises were frequently made to maintain harmony. Nonetheless, the gap between the rhetoric of expertise and the activity of the BPR was surprisingly small through the mid-1930s, for the BPR relied as much as possible on its technical superiority to shape national highway policy. This point becomes clear through an examination of three different aspects of highway construction during 1920s and 1930s: the development of the federal-aid system and its administration by the BPR during the 1920s, the development of state highway programs, and the survival of the federal-aid structure during the early years of the New Deal. That expertise was more important than politics in all of these areas was another reason for considering this a golden age of road building.

Cooperation and Federal Aid

By late 1921, both the postwar economic problems and the political confusion surrounding roads gave way to a burst of highway construction that finally enjoyed broad public support as America's love affair with the automobile grew passionate. Thomas MacDonald argued in 1920 that "not until the present day traffic descended upon the highways like an avalanche has there been any general realization on the part of the public that here was a truly national burden which must be taken up."[2] This avalanche continued unchecked until the Depression, with vehicle registrations swelling from 10.5 million in 1920 to 26.7 million in 1929. Even during the 1930s, new car sales fell below 2 million only in 1932 and 1933, although registrations did temporarily decline.[3] Almost universal support for any and all highway construction followed from this explosion in auto ownership, and highway engineers translated this amorphous popular demand into highways about which they had long dreamed.

As a direct result of federal-aid requirements, stronger state highway departments took control of construction and maintenance and displaced the counties as the leading road builders. This had been a goal of the good roads movement from the start, because while counties built roads serving local needs, these were often cheaply constructed by nonprofessionals. By 1925 this goal was being realized, as MacDonald argued that "the bulwarks of the system of highway administration in the United States are the state highway departments. . . . The county road administrations are, with conspicuous exceptions, the weak links in the chain; in form and efficiency but little improved since the days of the horse and wagon."[4] The key to the decreased importance of county road-building agencies was the transfer of the funds and responsibility for constructing the primary highway system to the state highway departments. Hence between 1921 and 1930 annual construction by the states almost doubled from about 20,000 miles to 35,000 miles. State systems increased in size from 203,000 miles to 324,000 miles, while the mileage of surfaced roads rose from 387,000 miles to 694,000 miles. The state highway budgets that paid for this activity jumped from $430 million in 1921 to $1.3 billion in 1930, as the universal adoption of the gasoline tax generated $495 million by the end of the 1920s.[5]

Some state highway departments grew into their new responsibilities better than others. Departments organized before 1910 in the industrial states of the Northeast and Midwest showed their vitality by launching impressive construction programs. No state built as many roads as motorists might have liked, but the best departments completed a framework of adequate state roads during the 1920s. Moreover, highway organizations in rural states without a history of road improvement for the first time replaced piecemeal construction with state road systems.[6] In every state, the introduction of mechanized construction techniques speeded road work. Army surplus motor trucks, Caterpillar tractors, the earthmoving innovations of Robert LeTourneau (including the self-propelled scraper or earthmover), and the mechanization of paving operations completely redefined the technological capabilities of road builders. These machines eventually transformed road building into an assembly line process, although the earthmoving revolution was fully implemented only during and after World War II.[7]

Given this combination of technical advances and rapidly growing funds, one might expect to find less state dependence on the BPR, especially since federal funds did not increase while state revenues exploded. In fact, Congress reduced federal aid from $75 million to $50 million in 1923 and $65 million in 1924, although the usual $75 million was available from 1925 through 1930. Between 1921 and 1930, federal aid provided only $839 million of the $7.9 billion spent by the states on roads, and the percentage of federal funds in state road budgets declined from 20 percent in 1920 to less than 8 percent in 1930. These statistics, however, understate the importance of federal aid to state highway programs and therefore mask the BPR's continued role as arbiter of highway developments. Federal-aid projects were the only element common to all states, which faced enormously varied traffic and financial conditions. States in the West, for example, used minimally acceptable gravel surfaces for primary roads that carried less traffic than country roads in eastern states. Pennsylvania and North Carolina, on the other hand, expanded their state systems to build rural secondary roads to standards higher than federal minimums.[8]

Amidst this diversity, every state shared in building the federal-aid highway system, which slowly grew from 169,000 miles in 1921 to 194,000 miles in 1930. The rate of construction varied, although federal-aid roads formed the core of every state system. Many states even improved sections of the 7 percent system without waiting for matching federal funds. Herein lay the importance of MacDonald's consensus. By 1925, 49 percent of all state highway construction funds were assigned to these roads, even though federal aid formed only 8.3 percent of state revenues. In the West, from two-thirds to three-fourths of all state activity was on the federal-aid system.[9] Clearly, most state road building was affected by the BPR, but no outcry against federal usurpation of states' rights followed. The efforts of Thomas MacDonald were central to this situation, for his cooperative style eliminated state-federal friction without lessening the BPR's influence.

An examination of the BPR's efforts to implement the changes called for in the 1921 legislation demonstrated MacDonald's philosophy of stressing cooperation within a framework of federalism that limited the BPR's formal authority. The bureau was the watchdog of federal funds, ordered to prevent waste but not to tell the states how to build roads. MacDonald asserted that this mission safeguarded states' rights, because "the Federal Aid Act is founded upon the principle of local initiative, and the only infringement upon the exercise of such initiative is [the secretary of agriculture's approval]."[10] A month after passage of the 1921 law, his annual speech to the AASHO identified only two major changes affecting the states. First, if states failed to maintain federal-aid roads, the secretary of agriculture could repair the roads with the states' share of federal aid. MacDonald quickly added that "the bureau earnestly hopes that it will not be required to take over a single mile of highway for maintenance." In this conciliatory tone he continued:

> The bureau does not seek initiative. It does not seek to direct the states but to co-operate with them. There is now a plan of action for the guidance of both organizations [the BPR and the state highway departments] that is so clear and so explicit that neither can escape the responsibilities . . . and if difficulties develop in the administration of this law, they will come from a wrong attitude of mind on the part of one or the other organization. We can and will eliminate any possibility of such only through a mutual consideration and respect each for the viewpoints and obligations of the other.

Every BPR annual report in the 1920s repeated this message.[11]

The 1921 federal-aid bill also required the states to designate 3 percent of their primary roads and 4 percent of their secondary roads as the federal-aid highway system within two years. Again, MacDonald stressed that states had the initiative in this process even though the BPR had to approve the state selections. In a gesture that reflected his cooperative style, MacDonald outlined his general ideas about choosing such systems at the AASHO's 1921 meeting before circulating the BPR's more formal rules. First, the states had to submit tentative maps showing their systems for BPR comment, after which conferences were held with adjoining states to insure connections at state boundaries. Any requests for changes made by federal engineers were also made in these conferences. Next, district meetings of states confirmed regional connections, with final approval of state maps coming from the secretary of agriculture, which in fact meant from Thomas MacDonald.[12]

The formal elements of this procedure, however, failed to indicate the depth of the BPR's involvement in the federal-aid highway system. Even before the 1921 bill emerged from Congress, the BPR had begun to identify such a system. District engineers met informally with state road officials to study potential routes and then submitted confidential maps to Washington that presented their ideas for federal-aid routes. MacDonald also assigned E. W. James, the author of the first rules in 1916, to devise indices of population and economic development as markers for traffic demand in each county. James plotted routes that provided a standard for comparison with state maps. MacDonald told his district engineers "that the approval of the Federal Aid highway system is one of the most important duties ever entrusted to [the BPR]." He warned the states that selections were not final; more importantly, the BPR did not just wait for state maps, for federal engineers visited most states and offered advice in early 1922.[13] Significantly, the strategy that would guide highway construction for the next half century—a narrowly technical effort to accommodate an increasing volume of vehicle traffic rather than real planning—was already visible.

The importance of the formal approval process becomes evident here, for it prevented any appearance of federal dictation yet still produced a system in line with BPR goals. At the district conferences, federal reviewers not only conveyed suggestions but also appeared as voices of compromise in resolving disputes over connections at state lines. MacDonald attended several of these meetings, and his remarks to state officials gathered in San Francisco in July 1922 demonstrated the BPR's kid-gloves approach. The bureau had certain legal responsibilities, he noted, but desired to work with the states rather than coerce them. He estimated that 75 percent of the state-proposed routes were acceptable, and said that work could begin on them even before the whole system had won approval. In the end, the final maps differed little from those prepared by E. W. James.[14]

To observers such as Senator Townsend and Roy Chapin, this process indicated significant weakness on the part of the BPR. But Pyke Johnson, the National Automobile Chamber of Commerce lobbyist in Washington, knew better:

> Mr. MacDonald's plan is to secure proper location wherever and whenever possible by consultation between the district engineers and state highway departments. That failing, he can as a last resort veto. His whole strategy, as I gathered from discussion with him, is to handle all of these matters through cooperation rather than the laying down of rigid rules which would only afford a starting point for opposition.[15]

The success of this approach was evident in the tone of the AASHO's 1922 convention. When a consulting engineer from Missouri criticized detailed federal inspections, Illinois Superintendent of Highways Frank T. Sheets defended the BPR's review procedures: "If I should try to voice a criticism of the way our work has been handled for us by the Federal office, I would have to hunt quite a while to find just grounds."[16]

MacDonald also applied this cooperative style to overseeing state construction of federal-aid highways. He relentlessly demanded that his engineers maintain cordial relations with state engineers, criticizing not only delays in the approval of state projects but also, "what is more serious, lack of the proper attitude of helpfulness toward the States." He wanted men who understood construction to handle these inspections:

> There is too much unnecessary antagonism toward the Bureau. It is just as much a part of the duties of this organization to promote and maintain friendly cooperative relations with the States and render real service, as it is to pass upon the details of plans or projects or to audit vouchers. . . . [T]hose who do not have the qualities of manliness, square dealing, good temper, and ability to get on with people must go.[17]

As MacDonald explained to New Mexico's state highway engineer in 1922, "It is my desire to see a more sympathetic attitude and closer relationships cemented between the Bureau and the States."[18]

Cordial ties, however, were the basis of very detailed BPR oversight of state construction programs. One outside observer claimed that "the highway subsidy is administered more carefully, more thoroughly, and more completely than any other form of federal aid." Some engineering matters, MacDonald made clear, could not be compromised, and he told district engineers that they should never hesitate to demand compliance with the rules.[19] Moreover, he expected district engineers to be more than inspectors of state plans. In reply to a query from the BPR's general inspector on the west coast, regarding the extent to which district engineers should help the states design federal-aid highways, MacDonald answered that the BPR should be involved with state projects from the beginning and not wait to respond to state initiatives. The bureau could have influenced design choices by using its authority to withhold funds until state plans conformed to federal ideas. But the same outside commentator explained that while the BPR occasionally had to threaten to cut off funds "to force upon [a state] a sense of its responsibility . . . such instances are rare."[20] Instead, MacDonald adopted a nonconfrontational manner that avoided adversarial positions. This nonthreatening style led another later observer to comment that

> in the highway field at least, formal regulations are not the substance of intergovernmental relations. . . . There are compulsory aspects . . . but the most striking thing about them is that they do not play as important a part in the day-to-day operations of highway agencies as a reading of the laws and administrative regulations would lead one to believe. Not only is less stress put on these matters . . . but more stress is put on voluntary techniques.[21]

MacDonald's noncoercive administrative style was vital to maintaining the BPR's influence in the federal-aid program, but just as important was his ability to develop associative structures that gave the states a voice in policy matters without upsetting the basic goals of the BPR. The essential cooperative mechanism for the federal-aid program was the American Association of State Highway Officials, which not only provided a forum that admitted the states to policy discussions but also gave the BPR a means of transmitting its technical ideas to the states. In short, the existence of the AASHO eliminated the need for the BPR to dictate policy to the states. MacDonald strengthened the linkages between the BPR and the AASHO first built by Page and then refined by himself after 1919, and after 1921 he gradually made the AASHO the spokesperson for state road builders even while he remained the arbiter of highway policy.

MacDonald's method was illustrated by the bureau's revision of federal-aid rules and regulations after November 1921. The basic procedures were not altered; states still had to submit routes, plans, and estimates for inspection before they could receive federal-aid funds. Only the maintenance provision and a few other clauses were tightened. After personally reviewing these changes, MacDonald presented them to the Executive Committee of the AASHO "to discuss in detail." The rules as issued incorporated every suggestion the state officials made. This was not a minor concession, as the law gave the secretary of agriculture the right to make the rules. But review by the AASHO backed MacDonald's claim that federal action was not arbitrary. After 1921, all modifications in BPR rules went through this procedure: District engineers canvassed state officials about minor changes, while the Executive Committee of the AASHO (which included MacDonald) prepared major new rules.[22]

There were occasional complaints about BPR actions, of course, but the AASHO's involvement in policy making defused most charges of federal interference in state affairs. In fact, MacDonald did "everything possible to strengthen the policymaking role of the American Association of State Highway Officials and its numerous technical committees."[23] Largely through his efforts, the AASHO became a forum for establishing standards and specifications, a subject significant enough to be treated in a separate chapter. The AASHO also entered other policy areas after 1920, when it formed technical committees that began to shape policies affecting every aspect of road building, maintenance, administration, and finance. The BPR made this work possible, for until 1953 every AASHO committee secretary was a BPR engineer. Most state officials could not squeeze committee work into their busy schedules, but MacDonald viewed committee assignments as a natural extension of the BPR's position as the technical leader of the field. Furthermore, several BPR engineers chaired AASHO committees; some served for decades. Balancing this federal role were the review and approval procedures to which all policies were subjected before they emerged from the committees, a process that gave state engineers a voice in all technical decisions. As one AASHO official explained, "After a great many discussions, the committee often arrives at consensus. Sometimes consensus is tantamount to policy." But the essential role of BPR engineers in preparing the proposals that served as the basis of discussion indicated where the technical strength of AASHO lay.[24]

Once he had built this cooperative arrangement with the AASHO, MacDonald used it to solve special problems, such as creating a national scheme for numbering interstate routes on the federal-aid system. In 1918, Wisconsin introduced numbered highways to help motorists, and by 1920 Illinois and Michigan had coordinated their connecting highways with Wisconsin's. But few other states followed this example, and by 1922, calls for marking the main national routes were strengthened by the inconsistency of the signs for the Lincoln Highway and other cross-country trails. The BPR could have imposed a plan, but MacDonald asked the AASHO to develop one. In return, the association appointed MacDonald to chair a committee, with E. W. James, the BPR's leading troubleshooter, as secretary. The committee proposed running odd number routes from north to south, with U.S. 1 on the east coast, and using even numbers on east to west roads, with U.S. 2 near the Canadian border. After a series of state conferences, James prepared the maps from which the committee designated a numbered U.S. system encompassing 50,137 miles of highways. The task was complicated by the inevitable complaints from memorial highways and towns on main highways that were not included in this system, so James deliberately avoided public hearings. By October 1925 the plan, including the design for the familiar shield to designate federal highway numbers and standard shapes and colors for highway signs, was complete.[25]

Determining highway numbers and signage was truly a cooperative venture. The AASHO's sanction shielded state highway officials from local attempts to influence route selection, while the BPR avoided complaints about federal interference. This arrangement explained the AASHO's technical significance to MacDonald, but it did not exhaust the reasons for BPR support of the AASHO. One important role that the AASHO assumed after 1921 was preparation of the biennial federal-aid bill for Congress. The association prided itself on this function—until 1943 every proposal was accepted—arguing that it "has never lobbied for an action, but has only expressed opinions and given information upon invitation or request, which has been frequent. The Association, through this process, has enjoyed immense prestige and respect on Capitol Hill and many times the executive secretary has been asked to submit legislative language for consideration." In reality, the AASHO drafted the bills with MacDonald, who was in constant touch with W. C. Markham, the AASHO's executive secretary. Thus the AASHO, while portraying itself as a group of apolitical experts, actually became a conduit for BPR legislative proposals to Congress.[26]

A good example of this procedure was the effort to limit the construction of private toll bridges on federal-aid roads. Widespread antipathy for toll roads had existed since the nineteenth century, and the 1916 legislation prohibited tolls on federal-aid highways. But bridges at state lines were a problem, as speculators controlled most existing or potential sites. In 1923 MacDonald set out to limit toll bridges by offering advice to congressmen and drafting bills when requested. Then, after the AASHO passed a supporting resolution in 1926, toll bridge restrictions appeared in the federal-aid appropriation bills. After several states acted on their own, in the mid-1930s Congress finally required state ownership of toll bridges on federal-aid routes and permitted the use of federal-aid funds for their purchase. The victory belonged to the AASHO, but the idea had been MacDonald's.[27] Nor was this an isolated example of the congruence between the legislative positions of the AASHO and the BPR. Especially on the issues of federal funding, the limitation of funds per mile, the use of federal-aid funds for urban roads, and the prevention of the diversion of highway-user revenues, the AASHO led the legislative efforts.

Perhaps as important as its legislative role, however, was the social aspect of the AASHO, for it provided a point of informal contact between state and federal engineers. A steady interchange of personnel between the states and the BPR took place, as many federal officials

Figures 1 and 2 (facing page). A 3,500-foot object-lesson road before and after construction by the OPRI, at Chattanooga, Tennessee, November 13–16, 1901. Top view shows the road's original rutted surface and lack of drainage ditches. The bottom view, taken after the OPRI was finished, shows a widened macadam surface with a crown to drain water into ditches. Note steam roller in background. (Source: OPRI, *Road Conventions in the Southern States and Object-Lesson Roads*, Bulletin No. 23 [Washington, D.C., 1902], pl. VII, figs. 1 and 2.)

Figure 3. The Good Roads Train sponsored by the Southern Railway from October 1901 through April 1902 at a cost of $82,000. The equipment, which was donated by various machinery builders, is visible on the flatcars. (Source: OPRI, *Road Conventions in the Southern States and Object-Lesson Roads*, Bulletin No. 23 [Washingon, D.C., 1902], pl. I.)

Figure 4. Stone crushing plant in operation on the Southern Railway's Good Roads Train, Winston-Salem, North Carolina, October 30–November 3, 1901. Note the crowd attracted by the object-lesson project. (Source: OPRI, *Road Conventions in the Southern States and Object-Lesson Roads*, Bulletin No. 23 [Washington, D.C., 1902], pl. III.)

Figure 5. Logan W. Page, a civil engineer with degrees from Virginia Polytechnic Institute and Harvard, developed the OPRI's testing laboratory from 1900 through 1905, and then served as director of the OPR until his death in 1918. He became the leader of both the technical efforts, especially research, and the political activities related to highways. (Source: "Logan Waller Page," *Public Roads* 1 [December 1918]: 3.)

Figure 6. A classic view of the dilemma facing motorists who ventured into the country in the early years of the century. Roads that turned into quagmires after a rainfall were frequently encountered in every part of the country and caused many people to support the good roads movement. These intrepid adventurers were in California's Sacramento Canyon. (Source: U.S., Department of Transportation, Federal Highway Administration, *America's Highways, 1776–1976: A History of the Federal-Aid Program* [Washington, D.C., c. 1977], p. 52; originally from California Division of Highways.)

Figure 8 (facing page, below). Federal highway engineers conducting the initial studies of the effect of vehicle impact on highways in 1919. The truck was being driven off a two-inch incline on to an apparatus that enabled the investigators to calculate the impact force. This project inaugurated the BPR's twenty-year examination of vehicle impact that featured the increasing use of complex equipment. (Source: Original drawing by Carl Rakeman, in Albert C. Rose, *Historic American Roads: From Frontier Trails to Superhighways* [New York, 1976] p. 95; reproduced courtesy of Crown Publishers, Inc.)

Figure 7. This highway in Pennsylvania, with a relatively narrow concrete surface, gravel shoulders, and wire-rope guard fences, exhibited the initial construction standards adopted for the federal-aid highway system, which grew to 160,000 miles by the end of the 1920s. By the late 1930s, however, this design was obsolete, and the shortcomings of such highways contributed to the traffic problems that helped to destroy MacDonald's highway policy consensus. (Source: *Public Roads* 3 [September 1920]: front cover.)

Figure 10. After 1922, the BPR began to encourage and help the states to conduct origin and destination surveys in an attempt to identify the reasons and the length of automobile journeys. The BPR believed the states could use such data to plan future highway construction programs, and federal funds for traffic surveys were made available in 1936. This information also provided the basis for the BPR's 1939 report *Toll Roads and Free Roads* and the National Interregional Highway Committee's 1944 report on a postwar highway program. (Source: U.S., Department of Transportation, Federal Highway Administration, *America's Highways, 1776–1976: A History of the Federal-Aid Program* [Washington, D.C., 1977], p. 156.)

Figure 9 (facing page). Thomas H. MacDonald, who graduated from the Iowa State Agriculture and Mechanical College in 1904 and served as the engineer-director of the Iowa State Highway Commission from that year until early 1919, headed the BPR until 1953. Known as "the Chief," he was the highly respected leader of the highway community throughout this period, an ardent proponent of highway research, a firm believer in the cooperative federal-aid program, and the architect of many features of American highway policy. (Source: Courtesy of Texas A&M University Archives, College Station, Texas.)

J. MacGilchrist
DEL

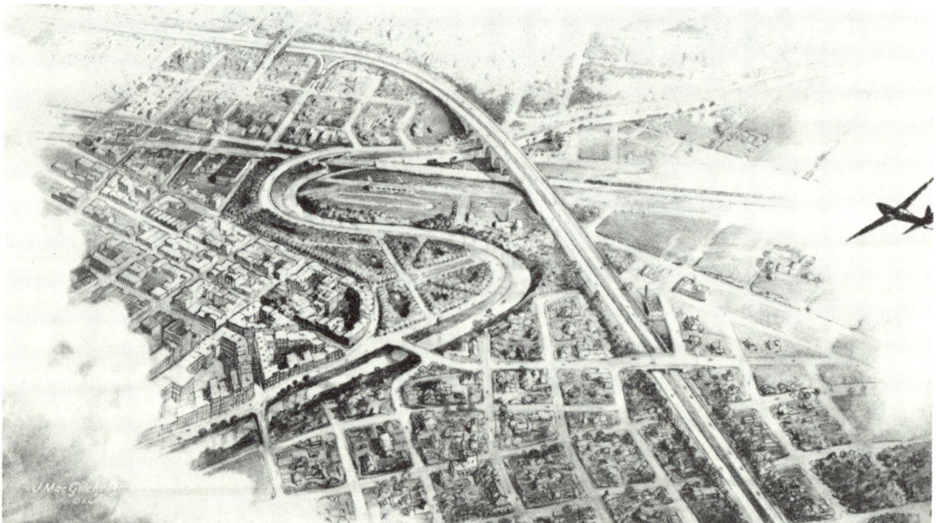

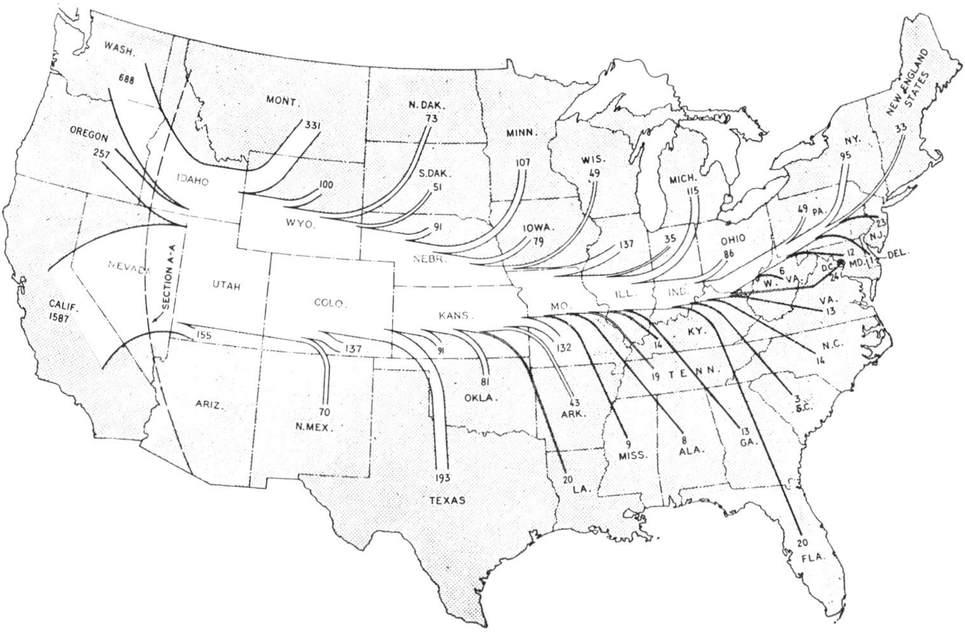

Figure 13. Map from *Toll Roads and Free Roads* depicting the number of cars traveling daily from the West Coast to all states east of the Sierra Nevada Mountains. Relying upon data from the state highway planning surveys, the BPR used such maps to back its contention that there was little need for a transcontinental network of high-speed toll highways. (Source: BPR, *Toll Roads and Free Roads*, 76th Cong., 1st sess., 1939, H. Doc. 272, pl. 1.)

Figure 11 (facing page, above). Design for a section of a rural toll road, which appeared in the BPR's 1939 report *Toll Roads and Free Roads.* This study derided the prospects for a national system of high-speed toll roads and proposed instead a network of free roads built on the existing federal-aid system. The BPR's plan adopted many design ideas from toll roads, especially the vision of limited-access adopted on parkways in New York, Connecticut, and many cities in the 1930s. (Source: BPR, *Toll Roads and Free Roads*, 76th Cong., 1st sess., 1939, H. Doc. 272, pl. 33.)

Figure 12 (facing page, below). This plan for a four-lane toll road completely separated from local urban streets appeared in *Toll Roads and Free Roads* in 1939, and reflected the BPR's attempt to shift the focus of American road-building efforts away from rural primary roads and toward multilane urban arterials. (Source: BPR, *Toll Roads and Free Roads*, 76th Cong., 1st sess., 1939, H. Doc. 272, pl. 30.)

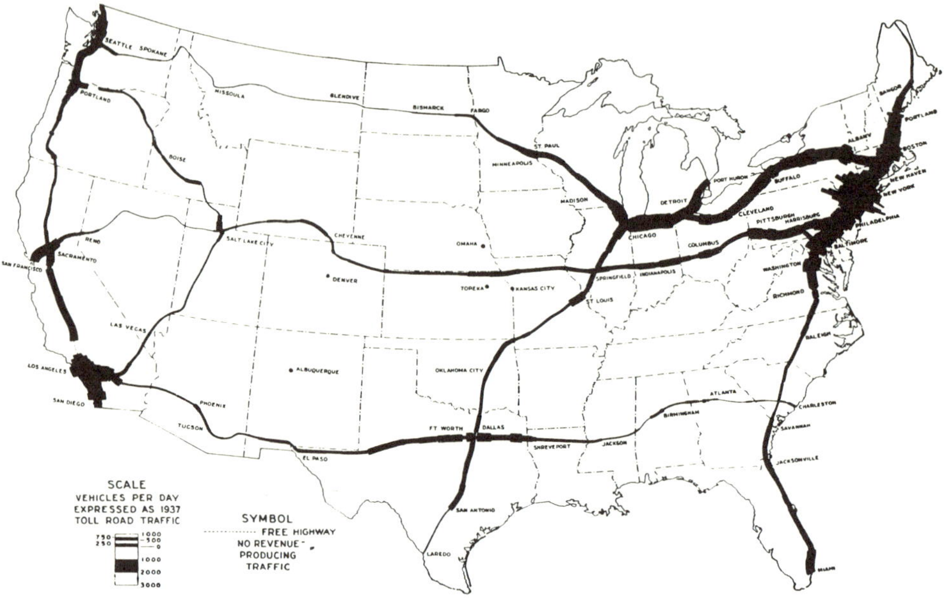

Figure 14. In 1939, the BPR used this graphic representation of projected traffic volume on several proposed toll highway routes to back its claim that only in the northeast would traffic generate sufficient toll revenue to pay off the construction bonds issued to build the roads. (Source: BPR, *Toll Roads and Free Roads*, 76th Cong., 1st sess., 1939, H. Doc. 272, pl. 11.)

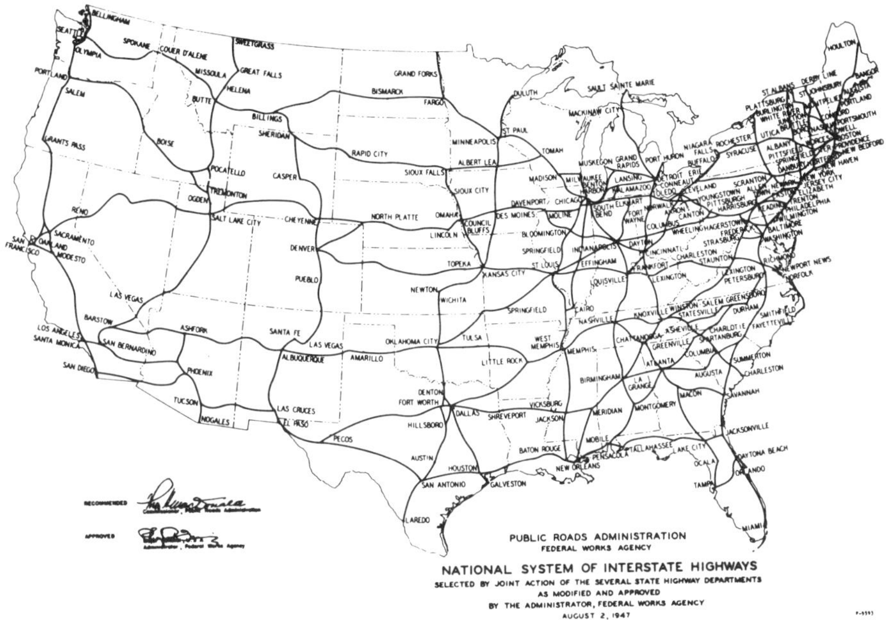

Figure 15. This map shows the original system of 37,681 miles of Interstate highways recommended by Thomas MacDonald and approved by the Federal Works Administrator on August 2, 1947. With the sole exception of the almost 2,300 miles reserved for additional urban routes that were not shown because they were not designated until 1955, the routes shown on this map became the National System of Interstate and Defense Highways in 1956. The only significant alterations to the original plan have been expansions provided by Congress—1,000 miles in 1956 and another 1,500 miles in 1968. (Source: U.S., Department of Transportation, Federal Highway Administration, *America's Highways, 1976–1976: A History of the Federal-Aid Program* [Washington, D.C., c. 1977], p. 158.)

began their careers in state highway departments, often in the districts they later oversaw. According to one commentator, the AASHO provided a means of "encouraging and continuing these intimate relationships," which "creates a sort of camaraderie which, when coupled with a common interest in the civil engineering profession, makes easier the contacts between the two levels."[28] Because the AASHO made these contacts easier, the rigid federalism described in the highway legislation of 1916 and 1921 was much weaker in practice, as the BPR, not the states, controlled the federal-aid program.

Federal Aid in Michigan

The BPR's associative relationship with the AASHO, when combined with the general cooperative style MacDonald used in dealing directly with the states, produced harmonious state-federal relations even as the BPR continued to guide road construction. The extent of BPR influence becomes apparent in an examination of the federal-aid construction program in Michigan. Because financial and traffic conditions varied so widely across the country, there was no typical state, but Michigan offers a good case study. Its highway department, founded in 1905, had struggled against a constitutional restriction prohibiting internal improvements, but it quickly grew into a road-building leader during the 1920s after the lure of $8 million in federal funds overcame doubts about direct state construction. After a slow start—only sixteen projects totaling fifty-three miles by 1921—federal-aid construction quickly increased, and by 1930, 1,605 miles of the state's 5,238-mile federal-aid system were paved. The Michigan road department was one of the best in the country, yet the BPR's presence was constantly felt, confirming the pattern of limited power but extensive influence that cooperation and expertise gave federal engineers.[29]

Michigan conformed to another national pattern in that the level of funding was not an indicator of federal influence in the state. Between 1920 and 1946 the state received only 15.6 percent of its highway funds from federal sources, yet the BPR's influence was always visible.[30] For example, the BPR's district engineer played a major role in helping Michigan meet the requirements for obtaining federal aid after 1916. The selection of Michigan's 7 percent system of federal-aid highways also illustrated the paradox of influence and cooperation. The state's highway department submitted its tentative federal-aid system to district engineer J. T. Voshell in January 1922, and by October E. W. James and the BPR's chief engineer had met with state officials to approve "as much of the system as is possible." In March 1923, MacDonald sent the revised map of the Michigan system to the secretary of agriculture, but the BPR then shifted several secondary roads to the primary system and the final map was not drawn until February 1925.[31]

This sequence demonstrated how the BPR used the inspection process, the only legal basis of federal influence in the states, to guide state actions, for every plan, drawing, estimate, and payment voucher received this close attention in both the district and Washington offices of the bureau. Over the next twenty years, the form and content of Voshell's comments on Michigan's plans changed very little. Few proposals escaped without some BPR comment, ranging from the trivial, such as the suggestion that a highway fill be built parallel to an adjacent railroad embankment for aesthetic reasons, to the important, such as Voshell's refusal to approve projects he considered extravagant. BPR engineers frequently redrew state plans to improve alignments and eliminate curves. However, their comments were not restricted to the review of plans. The district engineer paid equally close attention to contracts, and BPR construction inspections also showed the same style, as everything from cement stored in the open to poor protection of newly poured pavements was under scrutiny.

The credibility of federal engineers in district offices was increased by their seniority. Voshell, for example, remained in District 7 from 1917 until the early 1950s, whereas state engineers were replaced at every transfer of state political power. Thus continuity in the federal-aid effort came from the BPR, a point Voshell made clear to Michigan's highway department in 1933, when a Democratic engineer named Murray Van Wagoner replaced Republican Grover Dillman. The department's new business manager (Van Wagoner's campaign manager) remembered the first day of Van Wagoner's tenure as the only time Voshell visited Lansing. State Engineer G. Donald Kennedy explained that during the visit Voshell "outlined the federal-aid program and what the states had to do and what the procedures were. This was to our newly organized staff and he made it very apparent that he wasn't going to let these new people get out of hand." Moreover, the political rhetoric of campaign speeches was dampened by Voshell as well. In 1943, Republicans regained control of Michigan's highway department after their candidate for highway commissioner, Charles M. Ziegler, campaigned against Democratic corruption, favoritism, and inefficiency. It was difficult, however, to discern any significant alteration in highway policy after his victory, because the BPR had established the basic parameters of road building in the state.[32]

Nonetheless, MacDonald's desire to allow the states with sound engineering staffs to make their own decisions whenever possible showed when the BPR's Washington office occasionally overruled Voshell. This basic attitude of reasonable cooperation thus eliminated many state-federal conflicts before they emerged. As Voshell himself explained,

> For several years we have realized that we cannot actually control construction work to the extent that we could if the work were actually done by Government contract and on which we had inspectors employed by the Bureau all of the time. We must depend on the integrity of the States, and our view is that our inspections are more for the purpose of seeing that the State had competent men in charge of the work than to actually control the construction.[33]

State engineers generally responded with a similar attitude that was most evident in the widespread practice of asking for informal federal inspections of preliminary drawings "in order that we might have the benefit of criticisms or suggestions prior to the completion of final plans."[34] BPR engineers had succeeded in creating a harmonious federal-aid partnership with the states.

The annual effort to help Michigan obligate its share of federal highway funds demonstrated the cooperative strategy at work. Given the federal-aid procedures, one might expect that systematic attempts were made to complete the 7 percent federal-aid roads in each state, but progress was actually haphazard in order to accommodate the specific circumstances in the various states. Rather than laying out each year's program in advance, Michigan forwarded proposals to the district office three or four times a year. Voshell then shifted funds between projects, moved funds from one year's account to the next, delayed approvals, and withdrew or added projects to meet state needs. Thus the state's federal-aid system was in constant flux, but the complicated juggling act dictated by the three-year limit on each year's federal-aid appropriation insured that Michigan spent every penny of its federal aid. The best indication of the smooth state-federal relationship in Michigan is the fact that the commitment to construct a maximum number of miles took precedence over efficient long-range planning.

Departures from this spirit of cooperation with the states during the 1920s was rare. One example of disharmony occurred in 1925, when Montana withdrew from the federal-aid highway program on the pretext that rigid federal standards needlessly raised the cost of construction; there were occasional political attacks as well. But Montana state highway engineers, who had opposed their state's withdrawal, led the state back into the system in 1926.[35] This move reflected the highway community's almost unanimous support for the federal-aid program. A survey of state highway departments in the mid-1920s found that most considered the BPR's role beneficial. One engineer noted, "I cannot recall a single instance in which the federal government forced us to do anything that did not ultimately turn out to be for the best interests of everybody."[36] Of the forty-one states polled, thirty-six felt that charges of federal meddling were "a product of the imagination," three saw only occasional interference, and just two complained of intrusions.[37] Michigan's highway chief and AASHO president Frank Rogers spoke for the great majority when he told Congress in 1925,

> I wish to be emphatic in saying that there is no loss of initiative when the state and Federal engineers jointly strive for the best there is in highway practice. . . . Neither group feels it has lost anything by contact with the other and I cannot see wherein any state's rights or local home rule principle is violated in this kind of partnership which is producing at the rate of between 11,000 and 12,000 miles a year of the best roads . . . ever built.[38]

When one remembers the entirely typical level of BPR involvement seen in Michigan, Roger's defense of federal aid was amazing.

This discussion of the BPR's administration of federal aid might seem to suggest that MacDonald's political ability to keep people happy provided the basis of the bureau's influence in the states. Yet while the BPR's chief was a consummate politician, cooperation alone could not explain his success. MacDonald was above all a stereotypical engineer: dour, plain-spoken, never given to small talk or backslapping, and totally wrapped up in the work that was his life. It was this engineering image, the picture of the expert, that let MacDonald implement the BPR's ideas. Styled as a rational expert, he could make the hard political choices if he believed good roads were threatened, as he did in 1926 when he temporarily withheld federal aid from Arkansas for the second time in an effort to force apolitical control of highways and higher technical standards.[39] Indeed, the entire basis of MacDonald's cooperative structure was a tool unavailable to the shrewdest politician—the image of apolitical technical expertise. This becomes clearer as we examine the BPR's involvement in state road building *outside* the federal-aid system. Because work on such roads was free from the political overtones introduced by the presence of federal money, the role of expertise in state-federal relations is more easily discerned in this area.

The BPR and State Highway Systems: The Role of Expertise

BPR assistance on nonfederal-aid highway projects after 1921 differed little from the help rendered by the OPR before 1916. The federal presence in exclusively state projects was not undergirded by the coercive authority that enabled MacDonald's staff to veto any state plans for federal-aid construction. In fact, the bureau could assist a state highway department on such projects only when the states asked for help. These requests demonstrated that state respect for the BPR's technical superiority had not declined since Page's day, because such calls for help were not motivated by a desire for money. In some respects, BPR influence outside the federal-aid system during the 1920s was even more important in shaping the nation's road-building program than was its federal-aid work, for the BPR struggled hard to create sound state highway administrations.

The BPR's clear leadership in technical aspects of highway construction continued through the 1920s and was the reason that state requests for assistance on all types of road questions continued to flow to Washington. This should not be interpreted to mean that state highway engineers were incompetent, for most were not. But even in states with the most advanced highway systems, the BPR continued to serve as the technical adviser because of its research and standardization programs. Federal engineers were especially active, for example, in encouraging the states to adopt the newest mechanized construction technologies, presenting a series of articles on such equipment in the agency's monthly research magazine, *Public Roads*. BPR engineers thus continued to focus on practical questions, relying on the expertise they had gained while building roads in the National Forests after 1916 and in the National Parks after 1925. By 1929, they had built 4,091 miles of improved roads, mostly in the National Forests, at a cost of $77 million. They spent another $93 million on such projects during the 1930s, not counting funds for work relief; part of this sum enabled federal engineers to complete 1,577 roads in or leading to forty-three national parks and monuments. Moreover, the BPR designed and constructed several national parkways, including the Mt. Vernon Memorial Parkway from Arlington, Virginia to George Washington's residence, from 1928 to 1932, the Skyline Drive and Blue Ridge Parkway, along the Appalachian Mountain ridge, beginning in 1930, and the Natchez Trace Parkway in Tennessee and Mississippi, starting in 1934. With the exceptions of the parkways, which the BPR used as experiments in building limited-access highways, BPR construction rarely involved high-type pavements, such as concrete or asphalt. But projects in mountainous western terrain presented some of the most challenging grading problems and let the BPR work with the machinery that had the greatest technological potential to alter road building.[40]

This experience kept the BPR on top of road-building technology and explained the state's respect for MacDonald. Michigan highway engineer

G. Donald Kennedy later recalled "the chief's" tremendous integrity, but it was his expertise that Kennedy remembered most vividly: "The appearance of Thomas MacDonald on the first afternoon of the AASHO meeting was always the highlight of the convention, and his addresses were historic and very, very influential. They were timely . . . and laid down the Federal position very clearly." Kennedy's esteem extended to other BPR engineers as well, including E. W. James, L. I. Hewes on the west coast, and MacDonald's deputy, H. S. Fairbank: "I can't emphasize the influence of those men as individuals. This group . . . also would attend AASHO meetings and work with the states and they were powerful influences, by sheer weight of their own ability. . . . Just powerful men. Brilliant."[41]

This technical stature was attested by hundreds of inquiries that the BPR continued to receive from state and local authorities. Even states with advanced highway departments sent the bureau enough queries to justify the retention of the Special Advice and Inspections unit, and by the mid-1930s, federal engineers were making repeat visits to some counties. Most problems, however, were solved with a letter. Typical of the problems that the BPR handled were the requests from officials in New Bethlehem, Pennsylvania, for information about Kentucky rock asphalt and the Delaware River Bridge Commission's call for advice on paving materials. Numerous city officials thought and acted like one local engineer, who in a letter to MacDonald explained, "I wrote your bureau as the only place I knew of where I could get some absolutely disinterested advice on the material question."[42]

But the most important contribution that the bureau made to highway programs outside the federal-aid framework was guiding the formation of state highway departments. It is not an exaggeration to say that the federal-aid legislation underlay every important improvement in state highway organizations in the early 1920s. Even the best state departments had been given permission to build roads only after their legislatures had extended federal requirements to state systems, and some states acted only to meet federal requirements. The BPR demanded only that state highway officials be able to exercise direct control of construction on federal-aid projects, but this usually led to state road departments being given similar control over state highways. Always, the BPR insisted that giving engineers control of state highway construction organizations defined an acceptable highway department. As G. Donald Kennedy explained, "I don't know whether you realize how strong that motivation was in having a state highway engineer run the department. . . . Believe you me, during all those formative years the Bureau of Public Roads was stressing the state highway engineer theory."[43] So persuasive was the BPR that at least two states, Indiana and South Carolina, hired BPR engineers to run the new state agencies.

Some states, however, proved unable to build organizations without more direct federal assistance. Generally, eastern and midwestern states with older highway departments required little BPR intervention, but even the most advanced departments continued to rely on the BPR for technical and organizational advice. Western states, which had lagged in organizing departments before 1916, solved some problems by defining state systems that were synonymous with their federal-aid networks. The bureau's attempt to improve state administrative abilities was most evident in southern highway departments, which had been organized only after the federal-aid legislation demanded them. They were given only minimal authority by legislators, who were influenced by county officials jealous of centralized authority, so into the 1920s, most southern highway departments relied on counties to build, maintain, fund, and even select federal-aid projects. The BPR's regulations eventually altered this situation, and a survey of southern states by noted highway engineer H. G. Shirley in 1925 showed a clear correlation between federal legislation and changes in the states. Still, resistance to central authority remained as some southern states made only minimal reforms, so federal supervision of southern states was more rigid. This included assigning resident federal engineers to every struggling highway department to introduce professional standards of conduct. Generally, they produced noticeable improvements, as occurred in Florida, where H. J. Morison was the BPR representative from 1920 into the 1950s. The state's highway department later gave him "major credit for the high caliber of the work and the standards which the State Road Department insists on today."[44]

Few state engineers objected to this federal scrutiny, for they were able to use it not only as a crutch but also as a bludgeon against opponents. Southern highway engineers, for example, used BPR demands for minimal standards on federal-aid work to impose similar standards on roads built by counties, thereby deflecting local criticism. This prompted some southern politicians to rail against federal interference, but as they rarely wished to forgo federal money, the changes slowly took root. This approach obviously required considerable patience from the BPR, for the weak state departments often only bordered on the edge of acceptability. It was also the clearest reminder of the federalist structure of the road program. But MacDonald believed that by assisting with slow changes, such as the building of capable highway organizations without making political waves, the BPR insured better state-federal relations in the long run. Most state engineers appreciated the federal help, and their complaints against federal standards were strictly for home consumption. Over time, as the editor of the *Engineering News-Record* observed in 1925, BPR assistance improved the southern highway agencies: "Nowhere had local control of roads offered a more sturdy resistance to abolition. Nowhere has federal-aid road legislation been a greater force for the betterment of highway practices."[45]

Southern state highway departments were joined by older and stronger departments in all parts of the country in calling for federal help to solve strictly state problems. MacDonald noted that by the mid-1920s most states had adopted some version of federal-aid construction specifications for county and state projects. Moreover, both weak and strong departments continued to request BPR help in research.[46]

From the BPR's perspective, these years were its golden age: Its control of highway building was never so firm yet so readily accepted. Yet while the nation's highway efforts in the 1920s showed both enormous energy and great diversity, they were also far from uniform. Indeed, the clearest proof that MacDonald operated cooperatively was that some states continued to be cursed with awful roads. This was the price of a federalist system that did not allow federal engineers to order state compliance with the best engineering practice. Nonetheless, the BPR's influence was still considerable and reached far beyond its narrowly prescribed role as the inspector of federal-aid projects. What unity existed among road builders in both state programs and the federal-aid system was a product of the influence of the BPR. The level and type of its assistance varied from state to state; its tone ranged from motherly cajolery to fatherly reprimands and outright punishment. But one cannot examine highway design and construction in the 1920s without finding the BPR.

Moreover, the pattern of state-federal cooperation, the influence of expertise, and the general boom in road building was largely unaffected by the Great Depression, which historians have traditionally seen as introducing a sharply different conception of government. Herbert Hoover's denunciations of Franklin Roosevelt for shifting responsibility for relief from local government and private charity to federal authorities helped establish this view. Yet while Roosevelt more readily accepted the idea that government had a responsibility for individual citizens, the New Deal's intervention in the economy and the lives of Americans was foreshadowed by Hoover's own associative programs developed to deal with the Depression.[47] The highway building program provided clear indications of continuity between the policies of the two presidents, with Thomas MacDonald's cooperative road-building program one of the bridges connecting the two decades.

Roads and Relief, 1929–36

About the only elements of American society that are usually acclaimed for having had a "golden age" in the 1930s are radio, cinema, and streamlined trains, but road building also continued to merit that description. The highway boom of the 1920s continued despite the Depression, as road builders faced a sharp drop in funds only during 1932 and 1933. By 1934 both state disbursements and highway user-fee receipts were above 1929 levels and continued to increase through the end of the decade. Almost no other area of the economy "recovered" so quickly. By 1940 surfaced highway mileage had increased from 694,000 miles in 1930 to 1,367,000 miles. Yet the Depression eventually had an impact on road work, although the changes were not felt until later in the decade and form the subjects of later chapters. Broadly speaking, however, potentially disruptive structural changes threatened MacDonald's consensus about roads after 1936. Although the total funds for highways grew as they had in the 1920s, the state's ability to match federal aid was greatly diminished. Even more damaged were construction budgets for local and municipal roads. Thus state highway expenditures in 1940 surpassed 1931 levels, but outlays by county and municipal governments remained below those of 1929. Filling this gap were larger federal-aid appropriations and a variety of federal relief programs for the jobless that increased federal spending both relatively and absolutely.[48] However, the disruptive effects of this shift were felt only later, and through the mid-1930s, the BPR continued to dominate the expanding highway program with its expertise and cooperative administration rather than its control of funds. The best proof of the BPR's continued leadership was the perpetuation of the cooperative federal-aid structure even though Congress and the president had considered experimenting with different systems of highway funding and administration.

The increase in the pace of road building was almost unchecked during the 1930s because highways became the largest public works program undertaken by the federal government. Even Herbert Hoover, who opposed large-scale relief projects, supported federal-aid highway construction as a jobs program because the money went to an existing program that produced needed physical improvements, not make-work. Hence in April 1930 he signed a bill that added $50 million to the regular $75 million federal-aid appropriation for highways for fiscal 1931 and provided $125 million each for 1932 and 1933. Six months later, he requested another $150 million for federal construction projects, assigning $80 million to the states as advances to be deducted from future federal-aid funds over five years beginning in 1933. These funds were important because several states could not match the increased federal-aid sums.[49] But others, led by Senator Robert Wagner, fought steadily for more extensive public works efforts, and they also favored roads. In a December 1930 interview, Thomas MacDonald embraced both arguments, arguing that "Common labor forms a large proportion of the forces used for road work and for supplying the materials used. As there are roads to be built in practically every county in every state, the relief afforded is not confined to a few centers of population." He noted that safeguards in the federal-aid program prevented waste and abuse.[50]

Reasoning such as MacDonald's can explain the emphasis given to roads in a bill that was the first step toward Wagner's goal of more public works spending, the Emergency Relief and Construction Act of 1932. This legislation provided another $120 million in loans for federal-aid roads, with repayment scheduled over ten years starting in 1938. Thus under Hoover, highway construction received $200 million in advances even as regular federal-aid increased from $75 million to $125 million annually.[51] Although these amounts paled beside the relief efforts of the New Deal, they did firmly establish highway work as the leading solution to unemployment. In March 1931, almost 200,000 people had road-related jobs, and this rose to 365,000 by June. After a decline during the winter of 1931, over 500,000 men were engaged in road work by June 1932.[52]

Road construction fit Hoover's associative approach to the crisis, for packaging highway funds as loans against future appropriations preserved the federal-aid mechanism. But by 1932, MacDonald accepted the congressional directive to give jobs the highest priority in distributing the $120 million provided by the Emergency Relief and Construction Act, even though Hoover balked at this idea. The BPR limited workers to thirty hours per week and specified hand labor rather than machines for certain types of work. Contractors protested these guidelines, and the BPR itself was not convinced that they were necessary. Its statistics showed that, counting the labor involved in preparing materials, 91 percent of highway funds reached workers. But by 1932, MacDonald, unlike Hoover, agreed that jobs came first.[53]

As with other government programs, the BPR's place under Franklin Roosevelt was initially unclear as the new administration groped to develop a policy on public works. On March 21, 1933, FDR halted the letting of federal construction contracts even as he received letters from almost every governor urging that highways be the foundation of any public works program. Only the National Industrial Recovery Act (NIRA), passed in June 1933, ended the uncertainty about the BPR's position in the administration. This bill allotted $400 million of the $3.2

billion total for road construction; seven-eighths were distributed to the states on the basis of the federal-aid formula and one-eighth on the basis of population. Although features of the usual cooperative mechanisms were evident in the act, for the first time federal highway funds were outright grants, not advances or loans to the states. Moreover, for the first time since Page's death, a social goal was given top billing at the bureau, as the shift toward maximizing employment, which was begun by Congress in 1932, became the justification for federal road building.[54]

The Public Works Administration (PWA) turned these funds over to the BPR for distribution, and director Harold L. Ickes singled out MacDonald's administration for praise: "The rapidity with which PWA was able to start the roads program was due to the efficient organization of the Bureau of Public Roads."[55] Within two weeks of the NIRA's passage, the BPR announced procedures for obtaining and utilizing this money, but as with other PWA operations, work was delayed as the states prepared and submitted plans. By mid-September, MacDonald had authorized contracts worth only $34 million and employing 20,558 men, but two months later, the BPR had released $232.6 million for 4,128 road projects employing 276,000 men.[56] MacDonald cautioned the state highway agencies about the purpose of those funds, warning that "there will be no sympathy with any reluctance of administrative officers to make the principle of maximum employment the first consideration. . . . There will be no temporizing with evasions or attempts to defeat the high and practical idealism behind this measure."[57] The PWA rules followed those developed in 1932 at Hoover's Reconstruction Finance Corporation (RFC) and continued to give preference to hand labor. They permitted MacDonald to order the use of force-account labor, day workers employed directly by the highway departments. He even allowed work to proceed without plans if a competent engineer was in charge. These provisions provoked more angry protests from contractors, but MacDonald noted that "wholly unreasonable" bids on many jobs disqualified their complaints.[58]

Through 1936, solving the problem of unemployment guided the BPR's efforts as relief funds from various agencies enabled the states to continue building roads. From March 1933 through September 1936, PWA funds built 60,361 miles of roads and 2,641 grade-crossing viaducts. The Federal Emergency Relief Agency (FERA; existed from May 1933 to 1935), Harry Hopkins's Civil Works Administration (CWA; existed from fall 1933 to March 1934), and the Works Progress Administration (WPA; later the Works Projects Administration, existed from 1935 to 1943); all provided additional highway funds, and in fact surpassed the

PWA's concern for jobs. Together, federal relief programs kept the golden age of highways alive, for between 35 percent and 45 percent of all workers on federal relief built roads. From 1933 through 1938, more than $1 billion in federal-aid and PWA funds went to roads and grade-crossings, while 38 percent of WPA expenditures, $3.69 billion, went to city streets, grade-crossings, and local rural roads. Translated into physical results, the CWA repaired 255,000 miles of roads, while the WPA built 572,000 miles of highways, 67,000 miles of city streets, and 78,000 bridges. The WPA road programs alone provided 7.25 billion hours of work, 82 percent of it for unskilled labor. As Ickes explained, "Dollar for dollar, more money was spent for direct labor in road building than in any other kind of work."[59]

The effect of this federal money on state highway programs is shown in Tables 1 and 2. Relieved of the need to match federal-aid funds, state highway departments channeled more of their shrinking budgets to rural secondary projects during 1932 and 1933. But after 1935, federal funds

Table 1

Highway Expenditures during the Depression, 1929–40 (in Millions of Dollars)

Year	State Highway System Funds	Rural Roads: State Funds	Rural Roads: Local Funds	Rural Roads: Total Funds	Urban Roads: State Funds	Urban Roads: Local Funds	Urban Roads: Total Funds
1929	866	154	636	790	14	847	861
1930	1,039	296	623	819	12	899	911
1931	1,127	209	574	784	20	716	741
1932	939	208	441	649	17	516	538
1933	897	222	311	558	18	386	438
1934	953	223	277	654	25	366	568
1935	845	215	308	618	23	352	490
1936	1,110	235	318	894	31	367	373
1937	1,083	302	327	869	56	359	615
1938	1,120	289	332	1,022	53	407	837
1939	1,049	308	319	1,000	62	401	708
1940	1,100	318	309	937	62	397	639

Source: U.S., Bureau of the Census, *Historical Statistics of the United States, Colonial Times To 1955* (Washington, D.C., 1960), pp. 458–60.

Table 2

Federal Highway Expenditures during the Depression, 1929–40 (in Millions of Dollars)

Year	Federal-Aid Appropri-ations	Federal-Aid Funds Awarded	State Funds for Matching Federal-Aid Funds	Federal Funds for County and Rural Roads*	Federal Funds for Urban Highways*
1929	75	76	117	—	—
1930	75	94	137	—	—
1931	125 + 80†	218	97	1	—
1932	125 + 120†	95	110	>0.5	—
1933	200 + 50†	223	41	25	29
1934	200 + 24†	311	47	154	172
1935	200 + 200†	218	24	95	103
1936	125	225	13	341	264
1937	125	348	173	240	203
1938	125	183	125	401	367
1939	125	176	130	373	233
1940	125	150	119	309	168

Source: U.S., Bureau of the Census, *Historical Statistics of the United States, Colonial Times To 1955* (Washington, D.C., 1960), pp. 458–60.

*Most of these funds were from the WPA.

†The additional sums listed for 1931–35 were provided as follows: 1931—Emergency Construction Act, $80 million advance on future federal-aid funds; 1932—Emergency Relief and Construction Act, $120 million advance; 1933—National Industrial Recovery Act, $50 million for roads in National Forests and Parks; 1934—Hayden-Cartwright Act, $200 million of unmatched federal-aid and $24 million for National Parks and Forests roads; 1935—Emergency Relief Appropriation Act, $200 million for grade-crossing elimination.

overtook state expenditures even in this area, and assumed paramount importance in urban street construction as well. After the middle of the decade, federal funds raised state and local road-building activity above pre-Depression levels everywhere except in the cities. The Michigan State Highway Department simply stated, "Practically all road construction in Michigan would have stopped during the depths of the Depression had it not been for substantial increases in federal aid." In 1934, federal aid accounted for 69 percent of the state's trunkline construction, with the average being 34 percent for 1929 to 1934 and 37 percent for 1935 to 1940. Federal funds comprised a steadily increasing percentage of the state's total road expenditures, peaking at 53 percent in 1939. Significantly, 61 percent, or $229 million, of the federal funds for highway building received by Michigan between 1910 and 1946 came in the 1934–42 period.[60]

Federal grants of this magnitude obviously reinforced the BPR's position as leader of the highway community. In addition, federal aid was extended to roads previously excluded from the 7 percent system—to city streets in 1936 and to secondary roads in 1938. After 1936, Congress also permitted the states to use federal-aid funds for highway planning surveys. These extensions of federal activity (which will be discussed in Chapter Seven) were only partially related to the New Deal, but they do suggest an apparent increase in federal domination. After surveying the situation in early 1934, the *Engineering News-Record* concluded that "a revolution, in truth, is taking place."[61] Another writer added, "By its large financial contributions federal highway authority has assumed virtual dictatorship of public road development."[62] A BPR engineer's prediction of increased centralization following the introduction of highway planning seemed to confirm the *Engineering News-Record*'s fears.[63]

None of these observations proved accurate, however, because the basic direction of highway policy remained unchanged. First, most relief funds, including PWA and FERA money, were distributed cooperatively. More importantly, as Harold Ickes pointed out in 1935, the $400 million in NIRA funds "did not represent an entirely new attitude toward roads. Ever since 1916, the federal policy had been to supplant the chaos of independent and unconnected road systems in the States with uniform highways integrated on a national scale, although until 1930 the states had been required to match each federal dollar." In spite of the concern for jobs, Ickes stressed the need to apply sound engineering to all road work and supported MacDonald's attempts to focus construction where it was most needed. Thus the secondary goal of all relief work was to fill gaps in the federal-aid system or to eliminate traffic bottlenecks or engineering hazards. In other words, the focus on efficiency, as defined in engineering terms, remained intact. Ickes therefore concluded that federal highway policy had not changed, although later events challenged his claim that there would be no unanticipated side effects.[64] Through the mid-1930s, however, Ickes was largely correct about the Depression's effect on highways.

Numerous indications showed that a revolution in highway policy was *not* underway. Most importantly, the BPR's administrative style was unchanged despite the transfusion of federal money. As before, MacDonald used the AASHO to inform the states about relief programs. Another mark of continuity was the unchanged nature of the bureau's review of the use of federal-aid and relief funds in the states. The BPR believed that its responsibility to insure efficient use of federal funds increased with unmatched grants, but the tone of its dealings with Michigan's highway department, especially the inspection of plans, was not altered.[65]

As a result, the sense of a state-federal partnership also remained intact, in spite of potentially intrusive changes imposed by the 1934 federal-aid bill. This legislation, discussed in more detail below, contained a provision to penalize states that diverted highway-user taxes (gas taxes, license fees, etc.) to purposes other than road building. User fees, which nationally totaled $882 million in 1931, fell only in the following two years, to a low of $816 million in 1933, before climbing to $1.06 billion in 1936 and $1.3 billion in 1940. Many states could not resist diverting this money to their general revenue fund, and one might have expected the reduction in the federal aid to some states to provoke an outcry against federal interference, especially after Maryland, Massachusetts, and New Jersey were penalized in 1937. Yet the strongest supporters of these penalties were state highway officials. The AASHO had urged such action in its legislative drafts since the late 1920s, while officials in Michigan welcomed BPR help in a 1938 campaign to install a constitutional amendment prohibiting such diversion of funds.[66]

In fact, the only complaint of BPR interference made in the 1930s came from Governor Eugene Talmadge of Georgia, after MacDonald had withdrawn certification of the state's highway department in 1935 and cut off $19 million. Talmadge had used political criteria to appoint the highway commission, a move that MacDonald saw as preventing the efficient administration of federal aid. When Talmadge restored engineering control of the department, federal funds again flowed to the state. Yet the grounds for MacDonald's actions, his belief that apolitical expertise should guide highway work, precluded their appearance as dictatorial. Interestingly, BPR warnings about a similar use of politics in South Carolina's highway department in 1936 aroused no state resentment.[67]

The final indication that the basic highway policy of the nation was not altered in the early 1930s was MacDonald's ability to protect the federal-aid principle of cooperative road construction. After 1930, Congress considered several plans for bypassing the system of shared power during the emergency, usually by providing unmatched grants to the states or by having the BPR construct roads directly. Had MacDonald advocated such programs, his record suggests that they might have won congressional acceptance. The BPR chose, however, to defend the existing shared power arrangment by repeatedly asserting that Congress should maintain federal-aid procedures, for "any other plan would result in waste and inefficiency."[68]

MacDonald's commitment to federal aid was visible in Hoover's decision to loan federal-aid money to the states in 1930 and 1932, but MacDonald's role in forming the New Deal's road programs in early 1933 seems to have been limited to providing figures for determining the size of the relief appropriations. This was understandable, given his status as a holdover from the Hoover administration, but it is significant that when MacDonald's influence was reduced, both FDR and Congress temporarily retreated from federal aid, passing instead the NIRA with outright grants for road building. Yet only a year later, Congress reinstated federal aid rather than providing road funds as relief grants, thus accepting MacDonald's claim that no other satisfactory way to build roads existed. Congress ignored a plan to give the PWA another $400 million for road-building grants, and the Hayden-Cartwright Act of 1934 returned to federal aid.[69] Two years later, MacDonald remained convinced of the importance of federal aid:

> My judgment is that the Federal-aid appropriations are important, particularly important to some of the States of small populations and vast areas, but that the now demonstrated effectiveness of these funds to establish and to keep an administrative system of uniform methods and objectives in all of the States, is vastly more important than the money. . . . [Moreover, a national highway network has been built.] It has not in an arbitrary manner determined the States' policies or infringed on States' rights, but it has given direction and has formed a nucleus around which the States could gather their resources and their efforts to produce coordinated systems of roads."[70]

Another congressional decision in 1934 reinforced MacDonald's victory. Even as Congress restored federal aid for highways, it was debating the bill creating the WPA, which provided $800 million for roads and railroad grade-crossings. As passed, the measure used the federal-aid formula, not any criterion of need, to allocate these funds and imposed federal-aid standards on all work. Senator Carl Hayden, chairperson of the Roads Committee, was behind this move, which led one contemporary observer to claim that the senator "had come to occupy an almost unique position as a combined expert, advocate, and arbiter for the federal-aid system." Yet Hayden had only based his position on the arguments of Thomas MacDonald, with whom he was in constant contact. Indeed, the above description of Hayden more truthfully belonged to the chief of the BPR.[71]

The key to both these decisions remained the BPR's reputation for expertise. This showed in MacDonald's dealings with Congress and in his influence with men such as Hayden. During a 1934 discussion regarding the allocation of funds to various types of roads, one senator captured the prevailing belief that the BPR continued to embody apolitical expertise when he suggested, as the solution to a thorny problem, " 'Why not leave it in the hands of Mr. MacDonald's office to decide.' "[72] Hayden consistently adopted a similar attitude, frequently asking the BPR to prepare the language of amendments his committee sought, often at MacDonald's suggestion. Hayden's successor as chairperson, Kenneth McKellar, also followed MacDonald's lead on all major highway matters. Indeed, McKellar seemed more interested in the state patronage inherent in his Senate Committee assignment, as this kept him in contact with Tennessee's highway department. Likewise, McKellar's counterpart in the House, Wilburn Cartwright, worked closely with MacDonald on all road legislation.[73]

Similarly, the office's technical skill eventually enabled MacDonald to develop a good relationship with the administrators of the New Deal. The praise of Harold Ickes has been noted above, but MacDonald worked even more closely with Rexford G. Tugwell, a bright young New Dealer who later developed the greenbelt town concept. In his Agriculture Department post, Tugwell examined all highway contracts before they went to the secretary of agriculture for final approval. Tugwell later explained that this contact inspired his interest in setting road construction priorities, for "Thomas MacDonald . . . was assiduous in educating me, and through me, Roosevelt. We supported MacDonald and in so doing made some changes that could begin within sight." Tugwell came to respect many bureau chiefs in the Department of Agriculture, but especially MacDonald, because he "was there both before and after we, shaping, through one Presidency after another, and one Congress after another, the nation's highway policy."[74]

Highway Building, 1921–36

An examination of state and federal-aid road building in the 1920s and 1930s reveals a pattern of slow, uneven, but steady improvement in roads and technology, larger programs, and basic agreement on national and state policy. Other facets of road construction during this period, including the construction of urban highways and secondary roads and the development of the concept of interregional express highways, will be discussed later in this study. But the building of the state and federal-aid systems by themselves constituted a significant accomplishment. By the mid-1930s, the country had a national highway network, although many roads were already obsolete due to larger vehicles and rapidly increasing traffic. As a result, in 1941 *Fortune* condemned American roads, while a more recent account blamed the engineers for not showing more imagination in solving the problems of the nation's highways.[75] Some of the unevenness of the states' efforts involved the lack of funds, but this outcome could more generally be traced to the federalist system rather than to short-sighted designers. Perhaps the BPR could have *demanded* uniform administrative and construction practices across the country had they not been "hindered" by the cooperative structure of federal aid. Yet thinking along this line misses the basic point of this chapter. The structure of the national highway program was in accord with this country's traditional constitutional concerns for states' rights, a cornerstone of the founders' sense of democratic values. Indeed, the entire purpose of federal aid would have been understood by any American from the Revolutionary period onward who had learned that strong central governmental institutions were to be feared, for federal aid was intended precisely to limit the BPR's authority. The significant point is that in spite of federalist restrictions, BPR engineers played a truly decisive role in transforming the nation's highways. Not only did they guide the improvement of the federal-aid system, so that by 1930 few people were further than ten miles from an improved road on the U.S. numbered system, but they also insured that most states had competent engineering staffs in their highway agencies. The BPR's ability to promote these changes demonstrated a significant change in the nature of federalism, as engineers used the cloak of expertise to extend their influence beyond the traditional boundaries separating the states and the federal government.

Contemporary commentators, however, reacted only to the improvements visible in the nation's roads, not to the changes that expertise caused in American federalism. Thus in 1926 *Engineering News-Record* identified Thomas MacDonald and the BPR as "a mighty coordinator of the 48 great state highway improvement agencies."[76] Another observer from Texas credited the bureau with the success of that state's highway department, concluding that "no other feature of state and federal cooperation had produced such remarkable results in securing uniformity, efficiency, and freedom from political meddling in all the states."[77] The key to this limited recognition of the changes that the BPR was introducing to the nation's road system was that Thomas MacDonald's BPR acted within the framework of the associative ideal of cooperating rather than openly interfering with the states and business. Even when new features of federal aid and larger amounts of money became available in the 1930s, they existed within the cooperative structure of federal aid.

Yet it is also clear that MacDonald's far-reaching role directly contradicted the ideal of cooperation to an extent, however much he appealed to the image of unbiased, apolitical expertise. He created a situation in the BPR that matched Robert Cuff's description of the War Industries Board in World War I, for both shared "the paradox of an agency exemplifying aspects of both dictatorship and cooperation."[78] MacDonald skirted the dictatorial aspects both by relying on expertise to create the image of the BPR as disinterested specialist and by using the AASHO as an associative forum. Yet equally important in masking the paradoxes of federal aid was the justification for highways that he had developed in the 1920s. MacDonald avoided espousing the moral and social improvement goals that had driven Logan Page and turned instead to the colder, dispassionate, efficiency arguments characteristic of Herbert Hoover. Clearly, the federalist structure of American highway policy limited and handicapped the implementation of such a policy in some states, as *Fortune* made clear in 1941. Nonetheless, the BPR developed a widely adopted rationale for road building based on the savings to the wear and tear on vehicles, the expense of freight shipments, and the time spent in travel. Highways were meant to accommodate traffic, and the slogan of this efficiency-oriented vision was "we pay for good roads whether we have them or not."

In other words, engineering and traffic considerations should determine the routes and priorities of road construction, so by 1921 the BPR no longer talked about getting farmers out of the mud. The elevation of employment to first priority from 1932 through 1936 only temporarily restored a social goal for road work. The more lasting result of the Depression was the boost it gave to highway planning, which will be discussed in Chapter Seven. This effort fit into the BPR's demands for engineering efficiency and further extended the technical side of the Progressive ideology of expertise. The result was the assumption that roads should be built by experts to serve cars. Even the social engineers of the New Deal never challenged this formula; even Tugwell's Greenbelt towns, which sought to replace crowded urban housing with suburban-style house-and-park settings, relied on automotive transportation, although those roads served social goals in addition to moving traffic.[79]

This apolitical approach to road building meshed smoothly with the associative view, but as a recent commentator has noted, " 'apoliticalism' was itself a political strategy designed to strengthen a political position by insisting that it was a nonnegotiable technological necessity."[80] Yet MacDonald was not deliberately manipulative, nor was he concerned solely with the technical aspects of road building. Contained within his basically economic justification was a vision of highways that retained the Progressive crusader's goals, an outlook that was apparent in his first memo to his staff in 1919:

> Through enthusiastic endeavor we will try to justify the confidence that has been reposed in us by Congress. . . . There is no work more worthwhile. I have yet to know any man who has devoted a considerable period of his life to the building of roads who is not conscious of having accomplished results whose returns to the public in service can scarcely be measured.[81]

Moreover, he attempted to instill this approach to public service in the states as well:

> Here is an opportunity to do a big, basic work, such as comes to few in the course of a life-time. The individual who fails to vision the importance of the task has no moral right to hold a position of authority in its performance. . . . We have in the short period of five years visioned our more important highways extended and interconnected to form a vast network, serving local, state, and national traffic, only limited by the confines of the United States. This is the conception which has been written into the law, and which . . . lifts the importance of . . . the selection of the Federal-Aid system, above any other principle or duty therein announced.[82]

MacDonald believed that roads would be the avenue to progress if engineers and other experts could guide their development in the most efficient way. This attitude explained his efforts to prevent political corruption and his demands for minimum standards of performance in highway work. By combining this outlook with his belief in cooperation, symbolized by the federalist structure of the federal-aid law, MacDonald continued to find a deeper justification for road construction. Although there was wide variation in the conditions of roads and in the capabilities of state highway departments, he was able to form the state engineers into a broadly coherent group by accepting slow change as the price of harmony, all the while avoiding coercion. From 1921 through the mid-1930s, the BPR combined cooperation with unrivaled expertise to hold its place as the arbiter of American highway building. This chapter has discussed the importance of the BPR's status as technical expert in administering the construction of federal aid and influencing the work on state roads. Two other areas of its activities—its highway research program and the standards that emerged from this research—more clearly demonstrate that the foundation of this influence was expertise. A close examination of these BPR activities, which vary in both their degree of cooperation and their results, provides an opportunity to see not only how expertise supported the cooperative edifice of federal-aid road building but also how federalism imposed limits on the experts.

Chapter Five

Highway Research and the Cooperative Ethos, 1918–40

Scattered through the many institutions of learning and the laboratories of the smaller industrial works there are innumerable investigators . . . . [T]hese men collectively have enormous potencies for research, if their several gifts and skills can be brought into broad cooperation.

Henry M. Howe, 1918[1]

In considering the BPR's ability to transcend the limits imposed by a federal road-building structure, emphasis has fallen on the cooperative aspects of Thomas MacDonald's administration, but there have also been clear indications that cooperation was possible because the states respected the bureau's technical expertise. Throughout the interwar period, the BPR's highway research program was the symbol and basis of this expertise, with few questions escaping the attention of bureau investigators. Moreover, the BPR also attempted to transform highway engineering into a precise science by introducing the scientific style of research that had spread rapidly through American industry after 1920. Even though it had little effect on highway construction practice, this effort won the BPR further praise from the engineering community. Yet despite the BPR's dominance of highway research, no activity better illustrated the bureau's operation within the associative pattern of the 1920s, with the Highway Research Board (HRB) of the National Research Council providing the institutional mechanism of cooperation. In research as elsewhere, the bureau claimed that federal aid had not altered the pattern of voluntary cooperation pioneered by the Office of Public Roads before 1916. The arrangement of highway research in this country lent more credibility to this claim than any other aspect of road building under federalism.

A paradox, however, surrounds the BPR's vision of cooperative highway research, for although it demonstrated the BPR's technical superiority, it also failed to improve the quality of American roads significantly. The problem lay with the cooperative style itself, for in creating a genuinely cooperative research program, the BPR deliberately downplayed its leadership in research more than in other areas of road building and chose not to superintend research efforts. This decision, however, left the coordination of highway research to the HRB, which could not perform this task because it lacked any real authority. Thus highway research exhibited the ultimate irony of cooperation in the associative style, failing to live up to expectations because it lacked clear leadership. This chapter will develop these points first by examining the BPR's dominant expertise as reflected through highway research, including the attempt to make such research more scientific. Then it will consider MacDonald's deliberate attempt to pursue research cooperatively, especially through the HRB, before highlighting the pitfalls of the associative ideal.

The BPR and Highway Research

The development of a formal highway research program after 1920 was an important expansion of the mission that had justified the creation of the ORI in 1893—to collect and disseminate information about highway materials and construction methods. This effort had accelerated after 1900 under Logan Page's direction of the testing laboratory and later of the entire OPR, which under his guidance used field investigations and testing to buttress its position as the unrivaled source of information. Moreover, Page introduced the first economic and statistical studies of American highways. With a certain degree of smugness, by 1916 the OPR believed that it had resolved the crucial technical questions facing the highway system, with financial and management problems remaining as the primary shortcomings of American roads. *Scientific American* conveyed this attitude in 1918, when it explained that in the BPR laboratory, "samples of rock and other building material are made to tell their story—tell it so well and completely that once the test is finished, those who are to make roads know to a certainty just what they can expect of the material, just how it will wear, just what value it has as a foundation or surfacing material."[2]

Thomas MacDonald shared Page's belief that technical studies were essential to the BPR's mission, equating the significance of the research gains of 1921 and the establishment of the 7 percent federal-aid system. "The adoption of the federal-aid system," he explained, "and the significant researches of the past year constitute the greatest forward steps that have ever been made [for American highways]."[3] For the next twenty years, the BPR's leadership of the highway community rested in large part on this research program, which annually tackled about fifty projects that spanned the spectrum of road construction and management topics.

The strength of the BPR's research program was its focus on practical problems, which had been apparent before the war in its interest in the performance of specific highway materials, such as blast furnace slag, asphalt, and road oil. This work continued after the war, as did the BPR's free public testing program, which was halted only in 1924. By that time, the number of samples tested had risen from more than 1,000 a year in 1910 to 3,000, and local governments and contractors were permitted to utilize the laboratory's services after 1924.[4] Most postwar materials projects in the BPR, however, slowly began to consider materials as part of the whole road, reflecting a concern brought about by the highway failures caused by the truck convoys of 1918. This was most apparent in a battery of studies of concrete roads that included examinations of the strength of sand, the quality of cement, the optimal placement of reinforcing, and different types of joint fillers. After 1930, attention turned to the effect of aggregate quality on concrete and the ability of aggregate testing methods to predict satisfactory highway performance. In addition, the BPR studied a variety of production problems related to concrete pavements, including the effect of mixing time on quality, the value of different curing procedures, the effect of vibration during casting, and the merit of delayed finishing processes.

The other significant BPR materials project was a long-running study of secondary road construction from 1925 to 1940. Working with the Asphalt Association, bureau investigators studied low-cost bituminous surface treatments, the performance of different binders, pavement life, hot-mix application versus emulsions, the feasibility of cotton-fabric reinforcement, and comparisons of plant-mix and road-mix asphalt construction. Many other federal projects during this period also examined highway materials, but even this short list suggests the breadth of the BPR's research activity. The concrete investigation alone insured that the BPR maintained its position as the leading source of information in this area. The bureau not only sponsored the research leading to the first mathematical model of concrete slab behavior but also disseminated much of the information on improved practice through articles in *Public Roads* and other journals and numerous technical treatises.[5]

Clearly, these projects exhibited the BPR's orientation toward practical problem solving that MacDonald stressed in 1931: "The Bureau has never looked upon itself as an agency of research in the field of pure science. The central object, underlying all its researches, is the devising of ways by means of which highway transportation may be made more efficient."[6] This goal also explained the numerous smaller tasks that federal engineers tackled, such as testing bridge decks for the Benjamin Franklin Bridge in Philadelphia in 1926 and the George Washington

Bridge in New York City in 1929. Several BPR investigators working with the AASHO and the National Bureau of Standards examined the visibility of highway signs, while the BPR as a whole devoted much attention to the efficient utilization of mechanized construction equipment.

Other federal engineers during these years applied the same approach to economic problems. Some studies concerned vehicles, as did efforts to encourage the use of motor trucks for milk and fruit delivery and examinations of the feasibility of six-wheeled trucks and the operation of buses and other common carriers. But the crucial economic question facing road builders remained the matter of paying for highways. At the start of the 1920s, BPR economists focused on bond issue questions, but gasoline taxes soon assumed priority. Information from the BPR consistently buttressed the argument that highway user fees should be applied exclusively to road construction. With the onset of the Depression, BPR economists worked to prevent the diversion of this revenue, launching a study of highway taxation in Illinois, Wisconsin, and Michigan in an effort to undermine legislative sentiment favoring the use of motor vehicle taxes for nonhighway purposes.[7]

This roster of research activities is far from comprehensive, but it does demonstrate the wide variety of questions the BPR examined. No other agency rivaled the bureau in the scope of this research; indeed, until the late 1910s, few centers were capable of the kind of work that the OPR performed. Few commercial laboratories existed, and only California, Illinois, New York, Pennsylvania, Michigan, Maryland, Maine, and Kentucky had created rudimantary highway department laboratories by 1916.[8] It was not surprising, then, that one-third of all highway research funds for 1920, or $150,000, came from the BPR, with another $175,000 coming from twenty-one state highway departments and the final $150,000 from twenty-two universities and experiment stations. More interesting, however, was the survival of this pattern of funding throughout the interwar period even as the number of research centers increased, for by 1930 half of all research funds came from the bureau. The staff of eighty-one full-time and fifty-six part-time researchers at the BPR in 1928 was the largest in the country.[9] They worked hard to circulate their findings, using both the BPR's journal, *Public Roads*, which was the leading highway digest, and a number of technical treatises and journal articles.[10] The Michigan highway department accurately described the bureau's position in highway research in 1930, noting that the "investigational work by the Bureau of Public Roads . . . is followed closely, at all times, through technical publications and also by the departmental representatives on the research committees of the Highway Research Board[,]

. . . of AASHO, and of the American Road Builders' Association." The flow of requests from state highway engineers, trade association researchers, and even commercial testing laboratories for information, advice, and general pleas for help from the Washington offered equally eloquent testimony of the BPR's stature.[11] The construction of a new research station at Abingdon, Virginia, in the late 1930s expanded the BPR's facilities, and its impressive, colonial-style buildings symbolized the bureau's place in the highway research field.[12]

The final indication of the BPR's leadership of highway research was its effort to move the field into conformity with the increasingly scientific and mathematical style of research that flourished after World War I. The OPR's routine technical studies of individual road materials and its field tests of materials and construction methods were systematic but not scientific. These early investigations were intended to find the best materials, not new theories. Testing and field projects were not consolidated under Director of Tests A. T. Goldbeck until July 1915, and the OPR only identified such work as "research" after 1916. The OPR thus worked within a tradition of empirical engineering, an approach that enabled it to provide reliable information to American road builders. Moreover, like other bureaus in the Department of Agriculture, including those of Chemistry, Animal Industry, Entomology, and Plant Industry, the OPR had developed a national reputation for expertise.[13]

More than any other event, World War I forced the BPR to shift the focus of its research in more scientific directions. The war, in fact, altered the entire place of research in this country. Despite the apparent lack of sophistication in the OPR's program, its technical studies were far in advance of those of most industries, for only a few large chemical and electrical corporations had established scientific laboratories by 1905.[14] The conflict in Europe, however, led Woodrow Wilson to accept the argument of Willis R. Whitney, head of General Electric's research laboratory, that research was "a necessity to any people who are ever to become a leading nation or world power." In 1916, he created the National Research Council in the National Academy of Sciences to coordinate scientific efforts for military uses, and researchers became the heroes of an otherwise unheroic war, assisting in the development of sonar, radio, anti-gas measures, and aviation.[15] The results were increased public prestige for scientists and the rapid spread of scientific research into the economy during the 1920s.[16]

Given this change, it was not surprising that the BPR, as the leader of the highway field, moved to make its technical activities more scientific. The increased interest in research was also an extension of the Progressive enthusiasm for experts, as researchers offered the hope of solving problems ranging from inefficiency to disease. Moreover, the massive road failures during the military truck convoys in early 1918 had demolished the BPR's confident belief that highways were well understood. This shift in research strategies was further promoted by the congressional attack on the federal-aid concept after 1918, for the highway failures had led some to claim that the OPR was incompetent. A final motive for moving in new directions was MacDonald's desire to elevate the prestige of engineers by having them emulate scientists, who had emerged from the war with much greater public esteem.[17] Thus the BPR embarked on a highly visible campaign to introduce scientific research to an industry far removed from the competitive demands and technical changes that encouraged sophisticated manufacturing industries to support such study. By the standards of the time, roads were a "low-tech" field relying on empirically developed rules. The BPR's advocacy of science was only partially successful, but to most observers the attempt confirmed the bureau's continued technical leadership.

The scientific research style appeared first in the BPR's effort to explain the wartime pavement failures. BPR engineers began with studies that resembled the OPR's usual field investigations and quickly found that the problems had been caused by water freezing in the subgrade (soil foundation) and heavy trucks. Capillary action carried water into the road foundations, and when these frozen, waterlogged foundations thawed, heavily loaded trucks, especially those with solid rubber tires, shattered the pavements. These findings altered basic assumptions about drainage practices that had been established by the nineteenth-century British engineers Thomas Telford and John Macadam.[18] The results shook the BPR out of its complacency, and by 1919 a research methodology was emerging that emphasized numerical results and the development of theories.

These new concerns were hallmarks of science rather than engineering, and in a 1919 explanation of the bureau's research aims, the head of the BPR's Tests Division wrote that "with the rational design of road surfaces as a goal, the Bureau of Public Roads has begun experiments to find out something of the fundamentals affecting road design."[19] Repeated statements of concern for fundamentals marked the shift toward scientific research through the early 1920s as the BPR worked to develop theoretical explanations rather than simple solutions to highway problems.

The BPR pursued two questions with this intent—the effect of trucks on pavements and the behavior of soils supporting highways. Both differed significantly from previous federal field studies. The truck studies focused on determining the force with which the wheels of trucks struck the pavement, especially at expansion joints. After simple tests with trucks provided rough estimates, researchers examined the ability of different pavements to withstand such pounding by using a machine to apply impact forces. This use of apparatus introduced an enduring pattern, as complicated devices replaced trucks as the source of impact. This choice, which reflected a desire to generate unequivocal mathematical data, also required simplifications in the research design. In 1933 the researchers moved their tests completely into the laboratory by using accelerometers and a pendulum-type device to apply impact to miniature slabs. Although every step limited the number of variables and provided better numerical data on the force of impact, more artificiality was also introduced. However, chief researcher A. T. Goldbeck claimed that this approach gave the BPR an opportunity "to establish the art of roadbuilding as an exact science."[20]

Concurrent with the impact studies, BPR engineers conducted a series of tests on soil behavior, which the field investigations and impact tests both suggested held the key to pavement life. After 1919 the goal of this work also reflected the new scientific orientation as researchers attempted to develop laboratory tests that measured soil qualities regarding moisture retention and strength that could lead to a theory of soil behavior. Again Goldbeck explained in 1921, "The day is past for building roads of designs arrived at by rule-of-thumb. Our future designs [for foundations] must be based on sound, scientific, fundamental data."[21] But soils were little studied before 1920, as only two or three Europeans and the Agriculture Department's Bureau of Soils had examined soils. Therefore, the BPR was only slowly able to give meaning to its laboratory tests by correlating results with soils that satisfactorily supported highways. The BPR's 1926 report noted that "those who are working in this important field are urged less by the hope of finding a solution than by the passion of the scientist for facts."[22] By the early 1930's, thanks to assistance from the German-born engineer Charles Terzaghi, the bureau had developed means of soil classification that formed the basis for soil mechanics, the engineering modification of soil.[23]

These projects were recognizably different from both the prewar studies of the OPR and many of the practical postwar efforts because of their scientific component. This was evident in the use of machines to reproduce identical impact forces, the focus on obtaining numerical data, and the concern for expressing results in theoretical terms. But unfortunately for the bureau, both of these studies resisted theory building because of the enormously variable conditions road builders faced. The BPR simply proved unable to meet its goal of replacing the practical experience of engineers with simple formulas or tests. These unrealistic goals prevented both the impact studies and the soil research from having more than a limited effect on road-building practice.

The impact investigation showed this problem more clearly, for in spite of pressure to explain the road failures of 1918, suggestions for building better roads came slowly from the research. By 1920, the bureau had issued only common sense suggestions, urging the states to build thicker pavements and avoid poor ground. Over the next fifteen years, the goal of developing a rational design for pavement slabs—that is, a formula to guide designers—eluded BPR researchers. Ironically, the first significant design change to prevent truck-related highway failures, a slab with thickened edges, emerged from studies conducted by the Illinois Highway Department, which built a test road consisting of several different designs on various soil conditions.[24] The durability of each was tested by running trucks around the circular road for two summers, beginning in 1921. Largely because this arrangement re-created real conditions more closely than the BPR's experiments, the Illinois Bates Test Road was more helpful to road designers.[25]

The bureau's soil studies eventually proved valuable, but they, too, failed to provide the simple scientific guidance the BPR had hoped to find. Before its soil classification theories could be adopted, they required significant modification, and soil mechanics was widely accepted only after World War II. Moreover, the goal of using laboratory tests to replace the field experience of engineers has never been achieved. Indeed, soil science is still considered more art than science by its practitioners.[26]

In spite of these difficulties, the BPR's research program received significant praise from the profession. T. D. Mylrea, a civil engineering professor at the University of Delaware, considered the twenty-year impact project a model of engineering research, writing that "the care with which these tests are being conducted has impressed me very much, and I have come to the conclusion that it would do our Senior students a great deal of good to witness this test procedure."[27] Charles Terzaghi, the father of soil mechanics, called the BPR's early work "one of the prominent landmarks in the Field of Civil Engineering."[28] This approval stemmed from the desire of engineering profession to move in the same research direction as the BPR and repeat the successful adoption of scientific techniques by electrical and chemical engineers, especially the use of the powerful tools of mathematical analysis.[29] The National Research Council's Engineering Division, established in 1919, became a powerful champion of such a goal. Alfred D. Flinn, a moving force behind the division, who believed that a scientific style was essential for the advancement of engineering, found much support between the wars. By 1935, Earle B. Norris, Dean of Engineering at Virginia Polytechnic Institute, explained with approval that "the scientist is pretty much of an engineer and the engineer must be a fair sort of scientist."[30]

Thus the BPR was not alone in trying to bring science into engineering, nor were its soil and impact projects the only attempts to do so. Other federal engineers examined the stresses in concrete slabs, the design of bridges, and the construction of curves and intersections with an eye to developing theories. In one project, conducted with the ASCE, the Yadkin River Bridge in North Carolina was destroyed to compare theoretical predictions against actual behavior. This and several other projects proved more successful than the impact work.[31]

Nontechnical studies, especially on the economic value of good roads, also reflected the new mathematical style of the BPR's research. Such work had begun in the 1910s, but after 1920 these studies became more sophisticated in their attempt to measure the benefits of good roads. As the BPR explained in 1924, "Realizing that road improvement is justified only if it facilitates or reduces the cost of necessary and legitimate highway transportation, we have been endeavoring to establish a scientific and businesslike administration of highway improvement. . . . Methods have been devised by which it can be determined whether or not the improvement of a road is desirable and the form which the proposed improvement should take."[32] The key to these early cost-benefit studies were traffic surveys that counted the number of vehicles using a road, their destination, and their purpose. Through these and other projects, the bureau was demonstrating its belief that traffic was the primary determinant of highway construction. Cooperative ventures with state or local authorities, beginning with a study of the Boston Post Road in Connecticut in 1922, eventually grew into the statewide planning surveys mandated by the 1934 Hayden-Cartwright Act, and by the late 1930s, economic studies consumed fully two-thirds of the BPR's research budget.[33]

Significantly, whatever the results of these efforts to bring science into highway research, the goal itself won unanimous applause from leading engineering figures, including the head of the HRB, Roy W. Crum. In 1935, he argued for more highway research, defining it as "scientific study rather than experimentation. Much of our present day practice is based upon the experimental method applied in the field, but from now on we must go much deeper into fundamentals."[34] As usual, the BPR led the way. It is ironic, however, that the simple investigations of bridge decks, road machinery, and materials probably were more beneficial to road builders than the scientific studies praised by the engineering profession.

The BPR and Cooperative Research: The Highway Research Board

Despite the BPR's commanding authority in highway research, Thomas MacDonald never deviated from his goal of creating a genuinely cooperative highway program. Research especially lent itself to the associative framework of the 1920s. The war had thrown together teams of scientists and engineers, and they remembered the cooperative atmosphere. The growing number of trade association laboratories created to benefit entire industries preserved this environment. The National Research Council, which after the war continued as a strictly informal coordinator for scientific research, provided another beacon for the cooperative approach. In 1922 one investigator adopted Progressive language to stress the importance of cooperation, arguing that "against the background of research and a common effort to determine the facts, acrimonious disputes are quieted, mere opinions are held in abeyance under the benign influence of research for the truth."[35]

Like its effort to develop a scientific approach to research, the BPR's desire to operate cooperatively followed national trends. Many of the bureau's research projects just discussed, for example, were joint investigations with trade groups such as the Asphalt Institute or professional organizations such as the American Society for Testing Materials and the American Society of Civil Engineers. From this basic approach, which Logan Page had pioneered after 1900, grew cooperative projects with almost every member of the highway community. Indeed the BPR devoted much of its budget to joint projects and was not guilty of overstatement when it observed in 1922 that "to some extent, the bureau has been associated with practically all the major investigations that have been in progress."[36]

Importantly, the BPR worked steadily to make both state highway departments and universities reliable research partners. MacDonald wanted every highway department to maintain a testing facility capable of conducting serious research. The BPR provided equipment, observers, and financial assistance to the first large-scale state research projects, the Bates and Pittsburg test roads in Illinois and California, respectively, in 1921–22, and many other cooperative projects followed, especially on soil-related topics. The AASHO conducted research censuses in 1928 and 1933 that documented the extent of state-federal research cooperation, finding that fifteen states were involved in joint soil projects with the BPR each year and that in 1933 eight of these had not appeared on the earlier list. The BPR's also had extensive ties to engineering schools and experiment stations, including those at Purdue, Iowa State, the Massachusetts Institute of Technology, Maryland, Ohio State, and Georgia. Its closest contacts, however, were probably with Illinois, where joint concrete studies were conducted throughout the 1920s and 1930s.[37]

As a result of these efforts, the number of agencies engaged in research climbed steadily. By 1923, twenty-eight state, twenty-four university, and thirty-five commercial laboratories were in operation. BPR personnel visited many of these, especially the state facilities, to introduce new tests and check older procedures.[38] Even so, few laboratories, particularly in state highway departments, reached beyond routine materials testing operations; only Pennsylvania, Illinois, Iowa, and California had truly professional facilities that pursued genuine research questions. The situation had been even worse immediately after the war, when all the rules of road building had been destroyed by the failures of 1918. This confusion created a strong demand in the states for a formal highway research program, and not surprisingly, the state highway departments turned to the BPR for leadership. The president of the AASHO in 1919, A. R. Hirst of Wisconsin, wrote that " it is to my mind certain that unless the United States Bureau of Public Roads builds up a good and extensive organization to prosecute these inquiries, that nothing of value can or will be done in any other existing organization."[39]

The BPR's clear technical leadership and its research journal, *Public Roads*, established in 1918, were among the reasons the states looked to the bureau to organize highway research. Moreover, only the BPR had the resources to conduct in-depth research. Yet in answering the call of the states, MacDonald insisted on following the cooperative philosophy. His reticence about BPR coordination of a program of highway research did not stem from doubts about the value of research. Rather, he knew that a coordinator might have to set tasks for individual states, a chore that would inevitably attract complaints from unhappy state engineers and other researchers. Therefore he declined to establish the bureau as the official highway research center. Instead, he answered the pleas of Hirst and others by acting as midwife for a new agency within the National Research Council that mirrored the cooperative principles of federal aid—the Advisory Board for Highway Research (later the HRB).

From the start, MacDonald envisioned a highway research group in the associative mold, functioning as a clearinghouse for research results rather than as a rigid, centralized coordinator. The official history of the HRB explained his thinking.

> Decentralization could enlist the interest and aid of more people, it could encourage more individual initiative, it could provide for the important investigations of peculiar local problems, it could institute cooperative re-

> search activities in its own locale, it could spread the efforts necessary on nationwide problems, it could profit by blending the viewpoints from many regions and institutes, and finally it could and would use the findings from its own efforts in its own operations.[40]

Importantly, this goal meshed with the desires of several scientists, including the leading force behind the National Research Council, George Ellery Hale, who hoped to use the council to turn science toward practical problems but who also realized the importance of cooperation.[41]

The man who proposed the idea of a highway research group in the National Academy of Sciences's National Research Council was Anson Marston, MacDonald's mentor at Iowa State College and a member of the Engineering Division of the National Research Council. In August 1919 MacDonald agreed with Marston's proposal and worked with other leading engineers in the council to prepare a plan of action. After Marston presented their ideas to the AASHO's 1919 annual meeting, MacDonald and BPR staffers developed an outline of research problems, membership rules (which favored trade associations as liaisons to industry), and a funding policy that avoided direct corporate financing to preserve objectivity. At the organizational meeting for the Advisory Board for Highway Research in November 1920, all observers agreed that MacDonald had been its "prime mover and catalyst."[42]

Over the next years, the HRB worked to become the clearinghouse and encourager of highway research. Its first director, W. K. Hatt of Purdue University, traveled widely and published and delivered many papers to introduce the board to state highway departments and university engineers. He retained the Progressive theme of letting expertise control policy, arguing in a 1922 lecture that "there are engineering problems to be solved . . . upon the basis of scientifically determined information, with politics and special interests excluded from the council."[43] But always he stressed that the board only *recommended* areas for study. To emphasize this point, the executive committee removed the words *coordination* of research from the board's statement of purpose and replaced it with the less threatening *correlation* of research. Significantly, the number of trade association members grew from thirteen to twenty-three during the board's first year and steadily expanded.

Under its next two directors, Charles Upham (from 1924 to 1927) and Roy Crum (from 1927 to 1951), the HRB slowly grew, adding more technical committees and launching its own technical investigations. Both men were highly respected engineers. Upham had been a state highway engineer in Delaware and North Carolina, and would later make the ARBA the spokesperson for the commercial highway community.

Crum, another Anson Marston student from Iowa State, had taught at the college and worked at its engineering experiment station before becoming testing engineer for the Iowa State Highway Commission, where he won national recognition. He carried the HRB through the Depression unscathed, and in fact actually expanded its research program. By 1935, twenty-three project committees with nine subcommittees were at work in the HRB; by 1940, fifty-two project committees with nineteen subcommittees were active.[44]

Because the board relied heavily on outside agencies to conduct the research that its committees considered, it was a model of the associative style. State highway department laboratories that the BPR had prodded into existence in all but eight states by 1930, became important partners of the HRB. Except in smaller states in the West and South, thirty-six of these laboratories were conducting more than just routine materials tests by 1932. This enabled the AASHO to carry out joint projects with the board after 1930, even though only fifteen highway department laboratories employed ten or more workers.[45] The HRB also relied on investigators in trade association laboratories, including those of the Portland Cement and American Paving Brick Manufacturers Associations, the Asphalt Association, the Crushed Stone Manufacturers Association (the latter two headed by former BPR engineers), and the National Sand and Gravel Association. Finally, a number of universities and engineering experiment stations continued a tradition of highway research.[46] In short, highway research activity by these various agencies increased steadily through the late 1920s, rising from the 525 projects and $775,000 budget identified by a 1928 census to the $1,000,000 budget in 1931. After research budgets dropped to $838,000 in 1932, they grew steadily, and in 1936 another research census found that 1,500 investigations had been conducted since 1920.[47]

The apparent success of this cooperative research is deceptive, however, because the Bureau of Public Roads remained the paramount influence in highway research. Although MacDonald had refused to let the bureau become the national research coordinator in 1919, even a low profile could not hide its position in the community of highway researchers. The surest indication of the continuing federal preeminence was simply the scope of BPR activity. Its research budget dwarfed those of all other highway research centers, and its share of the national total rose steadily. In 1920, the bureau's budget accounted for one-third of the total; by 1928, the bureau contributed $400,000 of the $775,000 available. Only three states devoted more than $25,000 to research. In 1931, $611,000 came from the bureau and $287,000 from all other sources. This ratio held through the 1930s, as the BPR's research budget rose to

$950,000.[48] The bureau lessened this dominance, however, by sharing its resources through cooperative projects with state highway departments, trade associations, and professional groups.

More importantly, for twenty-five years MacDonald nurtured the Highway Research Board as a sign of his support for cooperative research. Having played the principle role in creating the board, MacDonald accepted responsibility for its survival. As the board's first director explained in his 1922 annual report, "Without the sympathetic and active support of the Bureau of Public Roads, through Thomas H. MacDonald, Chief, the Advisory Board would not have functioned."[49] The most obvious BPR assistance was financial. The pattern was established in 1922, when MacDonald contributed $12,000 of the HRB's $14,500 budget after the states were unable to provide money. By 1945, the BPR had given the board more than $2 million, or 62 percent of its budget. Annual support from any other group rarely exceeded $1,000, and even appeals to industry for funds for special projects after 1925 rarely produced significant amounts. In this area also the BPR's role was apparent, as one of its many grants to the HRB totaled over $79,000 for an investigation of speed and drivers from 1936–40.[50]

Probably as important to the operation of the HRB was the large number of BPR engineers who conducted research as members of the board's committees (see Table 3). Chief BPR researcher A. T. Goldbeck became the first chairperson of the committee on structural design and another BPR engineer was the board's secretary. Thus, as they did with the AASHO, federal engineers made the research of the board an extension of BPR projects. Thus when Goldbeck's committee on structural design reported on seven projects in 1924, six were BPR investiga-

Table 3
BPR Representatives on HRB Committees

Year	Total Committees	Chairpersons from BPR	Total Committee Assignments	Members from BPR
1922	6	2	50	9
1934	36	11	253	40
1944	50	15	431	73

Source: National Academy of Sciences, National Research Council, HRB, *Ideas and Actions: A History of the Highway Research Board, 1920–1970* (Washington, D.C., 1971), p. 139.

tions, and the bureau had funded the other at Illinois.[51] In several instances, the HRB established new committees just as the BPR began to take an interest in a particular subject or was ready to publicize its research findings in areas such soil investigations and pavement joints (1935), sight distances (1936), and highway accounting procedures (1937).[52]

Finally, Thomas MacDonald himself played a commanding role in the HRB. He frequently addressed its annual meetings, and his suggestions on the purposes and approaches to research were the bases for the board's activities. He was a permanent fixture on its executive committee, as well as chairperson of the committee on finance and administration from 1934 to 1943. But his greatest influence may have been as adviser to the directors of the HRB, all three of whom were personal acquaintences. His ties with Roy Crum were especially strong since both had been students of Anson Marston at Iowa State. W. K. Hatt best summarized MacDonald's role when he told Marston in 1921 that, "as you know, Mr. MacDonald is in the research work heart and soul and provided its main support. I have been depending upon him very largely for direction from time to time."[53] Others on the board also relied on "the chief," including A. J. Brousseau, the National Automobile Chamber of Commerce's representative and a member of the HRB's executive committee in the 1920s. Before the board's 1922 annual meeting, he sent MacDonald the speech he was to deliver with this request:

> I will appreciate it very much if you will tell me what you think of it and whether it agrees with your ideas on the subject. I am attempting to adjust my ideas of the highway problem to what I conceive to be your ideas. If I have not succeeded in reflecting your thoughts as to principles or important details, do not hesitate to comment freely.[54]

It was thus only fitting that MacDonald received the HRB's first George A. Bartlett Award for outstanding contributions to highway progress in 1931.

The Limits of Cooperation

Superficially, it might seem that the BPR had little need for the HRB after the crisis over pavement design eased in 1921. But several reasons explained the bureau's continuing support. One was MacDonald's desire to raise the prestige of engineers. He saw research as crucial to this, and by locating the HRB inside the National Academy of Sciences, he indicated how important he believed it was for engineers to be identified with science and research. More specifically, MacDonald's cooperative philosophy demanded that some private agency be able to provide the broadest possible auspices for research. Without the cooperative mechanism of the HRB, there surely would have been complaints and resentment over BPR dictation of policy. But by using the HRB, MacDonald could mold a community of researchers from every area of the field by providing a sense of common purpose. Although everyone knew the BPR's position, by working through the HRB, the bureau tempered its dominance and acted as if it was only another member of the highway community. Just as the AASHO enabled the federal-aid program to unite the state highway departments and the BPR, so the HRB genuinely created cooperation among researchers. As MacDonald explained to the states in 1929, "I think it important that the research worker himself be provided a place to give voice to his work and progress . . . and gain such inspiration as he may from men engaged in a like field."[55] Significantly, this unity reinforced the general harmony within the federal-aid structure.

Yet in spite of its success in creating a cooperative atmosphere, the HRB failed as the coordinator of highway research activity, a situation that highlighted the limits of cooperation. Although the board functioned as the unifying forum for the field, it largely left the task of preventing duplication of research to the BPR, which informally correlated research through its funding of cooperative ventures. The only other purpose filled by the HRB was to provide a forum for disseminating research results, a point that Alfred D. Flinn of the National Research Council had stressed in 1921 when he noted that "the information already assembled and the new knowledge being gained ought to be put within reach of everyone who should be using it."[56] But turning information into new procedures or methods proved to be difficult. The HRB's annual meetings in Washington were intended to serve this purpose, but the *Engineering News-Record* complained in 1927 that "very few of the committee reports were presented in such form that more than fragmentary understanding was possible. Brief and aimless discussion, except in a few notable instances, was the result."[57] A year later it added that

> the many-sided problems of the highway . . . make it peculiarly important that the men who must solve them should verify and enliven their thinking through exchange of views with their colleagues. For this they have the most excellent opportunity at the Highway Research Board meetings. But how fairly these gatherings are entitled to the use of the word "research" is less certain. . . . The high ability of the several hundred men . . . and their intense personal concern with all the complex factors of road problems, represent a potentiality which deserves to be utilized to the full.[58]

The board's ability to respond to these difficulties, however, was limited by the associative framework itself. Without a single definitive head, coordination of research from conceptualization through implementation was impossible. The BPR had refused this task, while the HRB had dropped the word *coordinate* from its statement of purpose. The clearest proof of the board's shortcomings were the continuous difficulties in translating its research findings into practice. The HRB started several publication programs to supplement its annual convention and published proceedings in trying to implement this work, beginning in 1931 with *Highway Research Abstracts*. But the hope of encouraging the adoption of research findings has never been realized. After World War II, the number of board publications grew rapidly, but this information explosion itself became a complication. One effort to bring some order to research efforts was made in 1945, when the states received authority to utilize 1.5 percent of their federal-aid allocations for research. The AASHO asked the HRB to coordinate this fund by creating the Highway Research Correlation Service, which according to the HRB would have its own "staff of competent engineers and other technologists who can work in the various areas of interest, travel about the country . . . . They will carry information from one state to another, or to other research agencies. . . . A series of publications, called Research Correlation Circulars will be used for transmitting miscellaneous information." The board added, "Where there are so many technical agencies throughout the country interested in highway technology and its development, there should be some central organization that is thoroughly familiar with all the research facilities available."[59]

The problem of implementation, however, has continued to bedevil highway researchers. In 1962, the AASHO and the BPR organized the National Cooperative Highway Research Program as "a problem solving . . . [device] for the state highway departments."[60] Three years later, the BPR itself organized a National Program of Research and Development for Highway Transportation, arguing that "research programming and management could no longer be a loose collection of a number of separate, isolated, totally unrelated studies. . . . Comprehensive, integrated and balanced research was the only answer." Yet even these efforts fell short of expectations, necessitating further program changes in the 1970s, and as recently as 1984 a study argued for the careful targeting of highway research and the establishment of priorities and central control.[61]

Clearly the organization of highway research, while harmonious, has not completely satisfied the needs of road builders. That this occurred in the most truly cooperative program in a field that prided itself on noncoercive state-federal relations illuminates a basic paradox in the associative philosophy. In earlier chapters, it was noted that actual highway construction under federal-aid auspices exhibited a contradiction between cooperation and the influence of experts, as the BPR used the AASHO to transmit federal initiatives to the states. Nonetheless, this structure provided a clear avenue for state involvement and produced generally satisfactory roads and highway building organizations. But the research program raised the opposite problem: How can a genuinely voluntary cooperative effort achieve its goals without a leader? To a certain extent, the BPR always remained the recognized center for research, exercising an authoritative, but never authoritarian, influence. If for no other reason than the example it set, the BPR established the tempo of most highway research. Yet in a reversal of its role in the development of highway legislation and construction policy, the bureau refused to act as the leader.

Whether MacDonald could have acted in any other way is, of course, unanswerable, although the difficulty in carrying research into practice has not been limited to highways and the HRB. The more important issue is the contrast between the BPR's direct involvement in highway construction under federal aid and its reticence about leading highway research. This matter is of some import in understanding the actual operation of the associative philosophy, but it is easier to determine which pattern was more common—genuine cooperation or the reliance on expertise to bring influence—if another example of the BPR's cooperative style is considered, specifically the introduction of standards for road building. Because here both research and construction intersected, a study of the BPR's work in standardization can help clarify the nature of associative relationships.

Chapter Six

The Paradox of Cooperation and Expertise: Consensus Highway Standards, 1921–39

Standardization can be made effective only by dictatorship or by common agreement and cooperation. In American industry dictatorship is inconceivable.

E. C. Crittenden, 1952[1]

MacDonald was unceasing, though without fanfare, in his efforts to bring about an elevation of the standards of engineering and administrative proficiency at all the state highway departments.

H. S. Fairbank, ca. 1953[2]

The Bureau of Public Roads' role in encouraging and funding highway research clearly maintained its reputation as the primary information source for state and local road builders, regardless of the difficulties in translating findings into practice. At the same time, the bureau's cooperative style bolstered the federal-aid approach to road building after 1921 and contributed to the durability of the consensus regarding American highway policy during its "golden age." Further facilitating the basic agreement on highway development from 1921 through the mid-1930s was the manner in which the BPR presided over the most important and successful associative effort to translate new knowledge into practice—the establishment of material standards and construction specifications. This work was not new, for under Logan Page the OPR had become the center for highway standards after 1900. But standards assumed special importance during the 1920s, when law and custom dictated that they be developed cooperatively. On the one hand, federal-aid legislation allowed the states to determine their own specifications subject only to BPR approval, an arrangement that made the American Association of State Highway Officials the cooperative arena for state-federal interaction. At the same time, American industry had evolved a pattern of setting standards through voluntary agreements between producers and consumers, using groups like the American Society for Testing Materials.

Thus highway standardization, like research, was achieved through the associative mechanisms of trade groups and professional societies. But in keeping with the pattern seen elsewhere in road building, standard setting exhibited the paradox of expertise overpowering cooperative structures. Unlike research, however, the effort to establish standards proved relatively successful and materially benefited road builders. The crucial difference was the attitude of the Bureau of Public Roads, which was more willing to play the leadership role in standardization that it had avoided in research. It therefore dominated standardization attempts even while using the same cooperative mechanisms that maintained harmony in the highway community. This contrast suggests that expertise was a central component of the associative philosophy. Indeed, the introduction of standards is an even better example than the BPR's oversight of federal-aid construction of the bureau's use of expertise to lead the highway community. This will become clear through an examination of how federal engineers worked with state highway officials and the materials industry in forums provided by trade and professional groups to develop highway standards.

The Crusade for Standards in Industry

Industrial standardization is a seldom-studied aspect of technological history, but the ease with which Americans can replace light bulbs should convince anyone of the value of uniform sizes. Almost without exception, American standards have been set by consensus, meaning, in Herbert Hoover's words, that "only a few . . . have been imposed by law. The vast numbers of them have been the result of spontaneous, voluntary, yet organized cooperation within highly individualized industry."[3] But not everyone has an equal role in this process, for corporate size as well as technical expertise can influence these decisions. The widespread compatibility of personal computers with IBM machines since 1982 says more about IBM's size than its computers' technical superiority. American standards, therefore, are compromises representing the broadest acceptable practice, not attempts to encourage the best available technology.[4] This consensus approach to national standards, in which government played little role, began in the nineteenth century as large corporations sought order amidst rapid technical change and the confusion of competitive national markets. First the railroads, and then machine tool makers and manufacturers of technologically sophisticated electrical equipment needed interchangeability between firms. By 1890, the electrical and mechanical engineering societies furnished forums for setting intercompany standards, and after 1902, the ASTM led standardization efforts.[5]

By 1900, interest in standards had emerged among highway builders, with the Office of Public Roads under Logan Page the leading promoter. Page built ties to the ASTM after 1904 and used its sanctioning authority to enhance the acceptability of the first highway materials standards released by the OPR about 1910. The adoption of these technical guidelines by eastern and midwestern road builders was a clear mark of the Progressive concern for efficiency. Indeed, the primary justification for establishing state highway departments was to install statewide construction standards for local roads. At the same time, the OPR's typical specifications appeared just as national interest in standards accelerated. Mass-produced automobiles, for example, spurred the Society of Automotive Engineers to coordinate the development of intercompany dimensional standards for spark plugs, tires, wheels, and many other parts. Similarly, Progressive concerns for efficiency and safety merged with corporate efforts to limit competition to produce the American Society of Mechanical Engineer's widely adopted boiler code of 1914.[6] World War I, however, provided the most important boost for uniformity, as material shortages led the War Industries Board to reduce the numbers of styles and sizes of many products. The results were surprising, as tire sizes dropped from 187 to 32 and shoe colors from 81 to 6 during the war. And as it had for research, wartime success with standardization created enormous enthusiasm for this activity in American industry. In 1918, electrical engineer Comfort A. Adams created the American Engineering Standards Committee (later the American Standards Association) to prevent the duplication of efforts by the various engineering societies. Another impetus to standardization came from the Federated American Engineering Societies under the presidency of Herbert Hoover, which undertook a study that Adams called "Waste in Industry." This study, with its recommendations for uniformity, commanded much attention when released in 1921 and signaled Hoover's enthusiastic support as secretary of commerce (1921–29) for product simplification. By 1927, 778 trade associations had accepted sixty standards prepared by the National Bureau of Standards Division of Simplified Practice, and 100 others were under investigation, including plans to reduce varieties of paving brick from sixty-six to four and sizes of milk bottles from forty-nine to nine.[7] Hoover later dubbed this enthusiasm a "crusade for standards," and it continued into the Depression, although with lower visibility. By 1935, twenty-four different national standardizing bodies had appeared, while half of the National Recovery Administration codes of industrial conduct contained standards clauses. World

War II only reinforced the lessons of the Great War, and the importance of standards certainly has not changed in our "high-tech" world.[8]

The basic method of developing standards also has not changed since the 1880s. Both the American Standards Association and Hoover valued voluntary consensus standards, and they relied on trade associations to communicate information among industrial firms as well as between business and government. These groups were essential to Hoover's conception of cooperative capitalism, and their numbers grew quickly, with standardization activities providing perhaps their best chance to meet associative expectations. One observer claimed that "these associations have become so important that to a large extent they dominate the entire standardization movement."[9]

The highway industry's standardization efforts after 1920 followed these national trends. Just as the interest in highway standards before World War I typified Progressive concerns for order, the promotion of uniformity in road building after 1921 matched the associative pattern of activity. Two separate groups had stakes in this—state highway departments and materials suppliers. The BPR exercised the same measure of influence over both groups after 1921 that the OPR had held before 1916, although the cooperation that remained the hallmark of highway standardization masked the BPR's dominant expertise.

State Highway Specifications: The BPR and the AASHO

The establishment of highway construction standards for the federal-aid system nicely demonstrated both the BPR's predominance in and the cooperative mechanisms for achieving standardization. The Federal-Aid Road Act of 1916 required each state highway department to establish construction standards, but few could do so independently since they lacked research or testing facilities. The states could turn for help to the American Society of Civil Engineers, the American Road Builders' Association, and a few other groups. The Mississippi Valley Conference of State Highway Officials, for example, pooled their information and issued regional specifications for concrete highways.[10] The BPR, however, retained a paramount influence in state decisions on standards despite the authority given the states by the 1916 legislation.

The BPR's position rested in part on Page's early work and in part on the legislative requirement that the bureau approve all state standards for federal-aid roads. But Thomas MacDonald chose to avoid using such inspections to dictate to the states, adopting instead the cooperative style that led him to claim that "the Bureau has been a clearinghouse for the collection and dissemination of the best ideas of all the State forces, and through its contact with all the States it has been able to raise the standard of construction in many States by directing attention to improved methods in use in other States."[11] Although this statement implied that the BPR's role was passive, in reality its examinations were quite thorough, as federal engineers compared state standards to their own minimum guidelines. This procedure was identical to that used to approve the roads selected by the states for inclusion in the federal-aid system, when the BPR had compared state maps of such roads with its own. In the case of standards, federal engineers relied on the standards and testing methods that the BPR had circulated widely after 1910. Moreover, because the bureau released more standards *after* the 1916 act than before and because most states lacked the knowledge and engineering staffs for this work, the initial state proposals usually resembled the bureau's ideas.

The bureau, therefore, played a much more direct role in the development of state highway standards after 1916 than MacDonald's statement suggested. Yet this federal involvement aroused no complaints, in part because a basic argument for the first federal-aid act had been the promise that federal review would insure higher state standards. Thus a survey of state highway engineers in 1926 found that the majority regarded the federal insistence on high standards as the BPR's most significant contribution. One eastern highway engineer observed that "ten years ago [1916] when we first began to cooperate with the federal authorities, we thought they were being too exacting. . . . But we were wrong. Our roads would not have stood up under the strain of greatly increased traffic. The federal engineers saw ten years ahead."[12]

But more important in keeping the BPR's intimate involvement in state highway standards from arousing antagonism were its associative procedures. After MacDonald's arrival at the bureau in 1919, he established as official policy, in the words of his deputy E. W. James, that "no standard specifications should be prepared [by the BPR] but comprehensive provisions for preparing specifications should be included in all the rules and regulations."[13] The statement that the BPR had not prepared standards thereafter routinely appeared in replies to public or business requests for standard specifications. As MacDonald explained to Senator James E. Watson in 1922, "The Federal Highway Act recognizes the right of each State to pursue such administrative policies as it may care to follow, and this Department would certainly adopt no regulation which would force upon the consideration of any State plans or specifications, particularly of a patented type, which the State of its own volition was unwilling to consider."[14]

The bureau made this strategy work by again using the AASHO, especially its technical committees, as a conduit for passing federal initiatives on standards to the states. In fact, the AASHO became the agency that promulgated highway standards for the federal-aid system, a role it acquired because it was congenial to both state highway officials and the BPR. The engineers who had founded the AASHO in 1914 had hoped that the organization could prepare uniform standards for materials and thereby ease the difficulty of choosing the proper construction materials. But the OPR's near-monopoly on materials information concerned state highway officials, including Thomas MacDonald, then head of Iowa's road department. In 1916 he argued that the AASHO should form its own committee with "a personnel that will not be dominated by the Office of Public Roads, but that will be willing to work with that Department and coordinate the results."[15] The BPR, on the other hand, also found the concept of an AASHO standards committee appealing because cooperation on standards conformed to the letter of the federal-aid law as well as to the associative principles that were popular at the time. Through AASHO, the states had a voice in all decisions, thereby forestalling charges of federal dictation of policy. Thus the Federal Highway Administration later argued that "the Government let the States set standards by agreement among themselves through the American Association of State Highway Officials and then made adherence to these standards a condition for receiving Federal aid."[16]

But the highway standards released by the AASHO were actually developed quite differently, although still within a cooperative structure. In 1917, Logan Page had encouraged the AASHO's attempt to form a standards committee by asking the association to prepare the basic rules governing federal-aid construction. From the start, however, the BPR dominated the committee's work, for most of the data used to define the minimum federal-aid standards were provided by the bureau's representative on the committee. Ironically, MacDonald institutionalized this arrangement in 1919 when he encouraged the AASHO to form a permanent standards committee to consider all changes and additions to these federal-aid guidelines. A history of the federal-aid program explained that MacDonald, "in recognition of the sensitivity of State officials, if he could he preferred working through the AASHO committees in developing standards and thereby avoiding the appearance of a Bureau crusade." But through its expertise, the BPR in fact controlled this process.[17]

The key to BPR's dominance of the AASHO's committee structure was the choice of BPR engineers as chairpersons of every technical committee. MacDonald himself headed the standards committee from its founding in 1920 until 1924, presiding over three other federal engineers and fifteen state highway officials. E. W. James of the bureau then replaced MacDonald, serving until 1943, and another BPR engineer held the post for the next decade. Other federal highway engineers chaired the AASHO's committee on bridges and structures from 1922 through 1953, while MacDonald ran a committee on planning and design policies from 1937 to 1945. The BPR also provided what one state official called the "strong secretaries" who established committee agendas;[18] many were the most talented men in their fields. But in addition to expertise, practical considerations encouraged the appointment of federal engineers because most state engineers faced time constraints that kept them from accepting committee chairs. Even in 1944, when many states had good research facilities, the AASHO's executive committee noted that although some members believed "it might not be desirable to have such posts filled to too large an extent by employees of the Public Roads Administration [the name of BPR after 1938]," most agreed that state officials could only rarely accept committee chairs.[19]

This situation meant that almost every major standard released by the AASHO originated in the BPR. Until 1931, the association considered the updated versions of the standard specifications and testing procedures first prepared by the BPR in 1910 as its own. The AASHO adopted almost without revision the BPR's bituminous materials specifications released in the early 1920s, while the bridge specifications the association released in 1924 had been prepared and published by the bureau. The AASHO committee report on uniform accounting procedures in 1936–37, as well as the design policies on geometric pavement standards, alignment and grade, intersections, grade separations, sight distances, and passing zones that AASHO released from 1938 through 1943 were the work of federal engineers. G. Donald Kennedy, AASHO president in the early 1940s, noted without criticism that MacDonald and his chief of research, H. S. Fairbank, "determined geometric and structural design standards of the highway system of America, believe you me, and nobody else."[20] Yet as always, the bureau was uncomfortable with any suggestion of dominance, so the BPR developed a standard litany for responding to requests for standard specifications during the 1920s and 1930s. A typical example was found in a 1933 letter from the BPR's chief engineer, who wrote, "we are enclosing a copy of the specifications for concrete paving of the American Association of State Highway Officials in the preparation of which this Bureau assisted."[21]

The advantage of having the AASHO promulgate standards become apparent in the BPR's periodic reexamination of state standards after 1921. Changes in construction technology and vehicle size demanded that design standards be upgraded and new information included, a potentially tricky test of state-federal relations. The BPR's handling of its inspection of Michigan's standards after 1921 highlighted the bureau's use of the associative framework of the AASHO to eliminate friction. Approximately every two years, Michigan updated the booklet of highway construction rules that it issued to contractors, and this revision was reviewed by the federal district engineer, J. T. Voshell. The MacDonald style of cooperation was always visible as Voshell prevented the state from lowering standards below the BPR's acceptable minimums. At various times, Voshell opposed state proposals to alter the ratio of sand, gravel, and cement in concrete highways without making compensating changes in slab thickness, to build lower guardrails, and to introduce more liberal tolerances for pavement thickness. Always, however, Voshell presented his ideas as requests, not demands, usually asking the state to change the next version of the manual. Moreover, such suggestions were always discussed in conferences, and the BPR often compromised on questions about which the states felt strongly.[22]

The value of the AASHO's role as the promulgator of highway standards became apparent when the BPR wanted Michigan to accept higher requirements, For example, when federal engineers found several deficient components in Michigan's bituminous materials requirements in 1923, they had to cite BPR publications as the sources of the preferred guidelines. But in 1927, federal engineers could request that the state bring its cement testing methods and its sand and concrete specifications into conformity with the typical requirements issued by the AASHO. This advice sounded less threatening and thus removed the appearance of policy dictation by the BPR.[23]

The BPR could adopt this cooperative tone because it always had a second opportunity to insure that state projects passed federal muster. Specifically, federal engineers could require higher standards as any project went through the usual approval process. From the very beginning, this authority to check every project and withhold federal funds meant that the BPR could compromise on the general cooperative standards without losing control of construction practices on individual projects. At the same time, federal engineers hoped that this inspection process would discourage the states from submitting projects "that are below the standard sought by the Office of Public Roads."[24]

Overall, however, the development of highway standards must be considered a cooperative process, for the BPR was consistently flexible in dealing with the states and the AASHO. To be sure, the bureau's chief engineer, P. St. J. Wilson, hoped "to carry standardization as far as expedient toward the ultimate objective—the standardization of practice itself."[25] Major steps were taken toward this goal in the area of testing procedures, and the bureau occasionally distributed desirable standards to the states, as it did with guidelines on tree planting, guardrails, and signs in 1926. Through its periodic inspections, the BPR slowly achieved some uniformity in state specification books for contractors. Nonetheless, every state border still announced itself by changes in road construction, with the variations only partially explainable by differences in traffic demands and natural conditions. The BPR made clear, in fact, that "we do not insist upon uniformity of language or methods, but we do expect specifications to represent average good practice or better."[26] The principle of compromise was clear in the use of the word *average*, and represented another concession to federalism. The goal of perfect uniformity may have been unrealistic in any event, so the BPR was guided in its efforts to improve a state's practice by the state's capabilities. As the chief federal engineer explained, "If any large number of States cannot, on account of lack of funds or other reasons beyond their control, meet with the requirements, it may be better to make the requirements such that they can be met by most of the States."[27]

This was an ideal statement of consensus standard setting, and with this approach, however weak it sounded, the bureau raised the quality of most state construction without encountering serious opposition. For example, most state highway departments quickly adopted the federal-aid standards for all other primary highways on state systems. By 1921, forty states had adopted "our standard plans," and, as one state engineer explained in 1926, "I believe federal requirements have changed almost completely the standards of highway engineering in most states of the Union."[28] Even after state highway agencies had acquired the research and testing ability necessary for setting standards, state engineers continued to turn to Washington for guidance when developing new requirements or revising old specifications.

The BPR's reputation, based upon its research and accumulated experience, attracted these requests for help, but the cooperative atmosphere also played a part. Michigan was typical in this regard, asking for BPR specifications on structural paint, temperature ranges for testing bituminous materials, pavement thickness, concrete culvert pipes, galvanized wire, corrugated metal culverts, and steel reinforcement as well as for information on testing procedures. In only one of many such instances, Michigan and the bureau's testing engineers worked for six years to improve the state's cement testing program.[29] These contacts were totally voluntary, and reflected the state's sincere respect for federal expertise. This situation not only explained the success of the consensus process but also the harmonious nature of state-federal relations.

Consensus Standards for Highway Materials: The BPR and the ASTM

While the AASHO committees and BPR inspections slowly established uniform highway construction specifications, these efforts comprised only half the standardization story. The other aspect, the introduction of standards for materials, was more complicated, because these requirements had to be hammered out with manufacturers in genuinely cooperative forums. The BPR possessed no implicit threats to intimidate manufacturers; it could not, for example, withhold funds from balky companies. Even the slightest suggestion that the government was imposing standards provoked howls of protest from industry, as the director of the National Bureau of Standards discovered in 1919 when he suggested that the government test and certify all commercial cement. The *Engineering News-Record* responded with a stinging editorial, asking, "Do we want an official control of materials in this country? If cement, why not steel, and if steel, why not canned beans and motor tires? Some definite restriction must be made, some limit fixed to the functions of a Government research and standardization body."[30] Even while still in Iowa, MacDonald had learned that highway builders had to "take into consideration the views of the men who must furnish the material."[31] Once in Washington, he moved quickly to establish agreements between users and manufacturers: "I have more confidence in the specifications which are produced by agreement after suitable tests and investigations have been made than in those which assume to set arbitrary standards without regard to some of the difficulties or conditions which may be encountered in the actual manufacture."[32] Led by this attitude, the BPR gained as influential a voice in shaping tests and the physical or chemical requirements for highway materials as it had in guiding state construction and design specifications.

The bureau's leadership was evident in three forums that issued highway material standards—the American Society for Testing Materials, the Federal Specifications Board, and the AASHO. In every instance, the BPR's partners included trade associations, the indispensable element of the associative style. This was especially true in the materials industry, which consisted of small local producers, few of which felt the pressures that had encouraged uniformity in automobile parts or electrical equipment. For example, in 1902, 5,700 stone quarries supplied the basic material of road construction, but 96 percent had annual sales under $100,000; only three firms had annual sales over $500,000. The brick industry was another localized business, with 2,200 firms in 1914.[33] Because no large corporations existed to force uniformity on these industries, trade associations were essential in communicating and encouraging national standards. The BPR, however, performed the crucial tasks of conducting and organizing the research and drafting the resulting standards, regardless of the forum in which the standards were issued. Because of the fragmented nature of the highway materials industry, the importance of the BPR was increased, for often it was both the only source of expertise and the only promoter of standards.

The BPR's efforts in bringing a "backward" industry into the new age of uniformity through cooperation embodied the noncoercive, advisory role envisioned for government in the associative ideals of the 1920s. Interestingly, the Highway Research Board, itself a cooperative mechanism, was not a factor in this process, despite its possession of the research base obviously needed to set standards. The National Research Council was extremely wary of performing work that industry could do itself, so the HRB's first director told committee chairpersons in 1923 that "research committees do not formulate standards or methods of tests, or approve practices of construction."[34] Instead, the leading forum for developing standards for highway materials was the American Society for Testing Materials. Trade associations were in turn central to its efforts, but its connections with the BPR were even more important. These had been firmly established by Logan Page, who had chaired its Road Materials Committee from 1904 until 1918.

The number of federal engineers working on ASTM projects was never as great as those serving on the AASHO's standards committee, but they were found on all committees related to road materials. BPR research director A. T. Goldbeck chaired seven subcommittees of three research committees in 1920, and other federal engineers were as active in the ASTM. Because few sources of technical information rivaled the bureau, the contribution of BPR engineers was out of proportion to their numbers, as they prepared numerous papers and committee reports.[35] In this way, they funneled their research into the ASTM's approval process, so that the research and standards topics that interested the BPR during the 1920s and 1930s, such as concrete and oil and asphalt, constituted the agenda of the ASTM. This congruity meant that almost every standard specification and testing method for road materials adopted by the ASTM emerged from BPR researchers. By 1924, when the bureau issued its last set of typical specifications not promulgated by the AASHO, twenty-nine of its tests had already been sanctioned by the ASTM because that approval further certified these procedures for industry.[36]

The development of standards for crushed stone typified the process by which the BPR turned its proposals into ASTM standards. Federal engineers issued their first standard stone sizes for road building purposes in 1915, but the war encouraged efforts to reduce the number of sizes of crushed stone on the market. Following one such attempt by several state highway departments, the BPR proposed a system of uniform stone sizes after a study in 1917 and 1918 that also encouraged the formation of a national trade association. In fact, R. W. Scherer, the head of the Wisconsin Crushed Stone Association, believed that "the movement to standardize commercial crushed stone sizes, inaugurated by engineers of the Bureau of Public Roads, should be and will be heartily welcomed by crusher operators." Yet the BPR chose not to act unilaterally and instead let the ASTM formally release these guidelines as national standards in 1920, after the BPR had presented its research for discussion.[37]

The BPR thus relied on the testing society in part because the consensus process was deeply ingrained in American industry. ASTM approval meant wider acceptance of BPR initiatives. One authority on standards noted that "the breadth of acceptance depends to an extent on the scope of influence and authority of the standardizing agency."[38] In the ASTM, the bureau found the broadest authority. By 1918, the society's membership of 2,261 was the largest of any standardizing agency. Its nationally recognized road materials committee had sixty members on forty subcommittees in 1922, and they approved the bulk of the testing methods and material specifications used in this country. Many of these had their origin, if not their final wording, in the BPR.[39]

Over the years, however, other research groups were admitted to the ASTM's road materials committee, and the BPR's overpowering dominance faded. By 1930, the state highway department and university laboratories that the BPR had encouraged with cooperative projects began to conduct research for the ASTM. Moreover, former BPR researchers played similarly prominent roles in developing trade association laboratories for the Asphalt Association and the National Crushed Stone Manufacturers' Association, groups that worked with the BPR and with the ASTM. As a result, the membership of the ASTM's road materials committee reached 100 in 1935 and 129 in 1940. With such broad representation, the BPR's efforts with the ASTM may have been its most truly cooperative activity. Yet the bureau remained a visible participant in ASTM activities. Symbiotically, the BPR's involvement with the ASTM insured the acceptance of the bureau's ideas on standards and materials quality, just as the BPR's high-quality studies helped make the ASTM, according to David Hemenway, the "dominant standards-writing body in America." Hemenway explained the procedures, which have not changed since 1900, that account for the ASTM's prestige:

> The ASTM actually encourages the various conflicting interests in standardization questions to represent those interests. The ASTM operates under stricter procedures than do most other standardizing bodies. Committee membership, for example, is balanced between producers, buyers, and general interest groups, with a producer-oriented voting strength not to exceed 50 percent. Technical committee chairmanship is restricted to a member of the non-producing sector. These and other ASTM procedures make it *more* likely, in the words of one district court judge, "that results reached by them will be scientifically sound and will represent the general interest."[40]

In fact, *Civil Engineering* reported that "no published ASTM standard has ever been declared faulty for substantive reasons."[41] Thus the BPR-ASTM interaction benefited both groups as well as it produced the consensus standards for highway materials.

The ASTM was not, however, the only forum in which BPR researchers worked for road material standards. The bureau used the AASHO to sanction certain standards that the BPR had developed through cooperative research with well-organized trade groups that the bureau had worked with before, such as the Vitrified Clay Products Association, the National Paving Brick Manufacturers Association, and the Asphalt Association. In cases in which an industry was deeply fragmented and lacked any trade group, it was again easier to use the AASHO as the sanctioning body.[42] In very infrequent instances, the BPR worked just with the Asphalt Association, as in an early 1930s project to develop standards for secondary road work. In 1930, when the BPR wanted to propose standards for road tars, MacDonald worked with a former BPR researcher at the Asphalt Association to create a trade group for a cooperative study. He explained, "As a Federal bureau, we cannot conduct a research and standardizing program in cooperation with individual companies. Coordination would be impossible under such a plan, and there would be other equally grave objections which I need not particularize.[43]

Clearly, the BPR always used a cooperative approach in setting standards. The one instance that violated this rule involved the Federal Specification Board, and demonstrated the pitfalls of unilateral standardization by government. Herbert Hoover created the Federal Specifications Board in 1921 as part of his multipronged effort to develop national standards, reasoning that, given the size of federal contracts, the introduction of uniform federal purchase requirements would encourage manufacturers to standardize products. Among the board's programs in the mid-1920s was an effort to introduce binding specifications for highway materials.[44] Several BPR engineers were involved in this work, which by 1930 had produced U.S. Government Master Standards for slag, road oil, asphalt, tars, nonbituminous materials, premolded expansion joints, asphalt emulsions, aggregate for concrete, sieve sizes, and other materials. In every possible instance, these standards had already been approved by the AASHO or the ASTM.

Despite this cautious approach, the board encountered difficulties because of the absence of industrial representatives. Instead, proposals were usually circulated to manufacturers only after tentative requirements had been set. When adopting ASTM or AASHO standards, this created few obstacles for the committee. But in 1924–25 the standards for asphalt ran into opposition from both natural asphalt producers and oil refiners who manufactured asphalt as a by-product, because each wanted specifications that would exclude the other from the market.[45] Under BPR research chief A. T. Goldbeck, the board's technical committee tried to skirt this issue by using ASTM guidelines that established separate specifications for each type of asphalt, since the argument had prevented the adoption of standards based on chemical composition. Yet when the board proposed this dual standard, the Asphalt Association protested and delayed the acceptance of the specifications for two years. The mere appearance of arbitrary federal action had been enough to provoke complaints.

The Contradiction of the Associative Philosophy

By the end of the 1930s, almost every aspect of the standardization of highway materials had been considered in some cooperative forum. As was typical of the consensus process, perfection was not achieved. One BPR engineer, in requesting further investigations of culvert standards in 1926, noted that because ASTM specifications were not based on tests that reflected actual conditions, "the results have only served more or less to befog the issue. . . . Manufacturers of all brands of culvert metal are apparently able to derive from the ASTM reports results favorable to each particular brand."[46] Yet the BPR's efforts produced greater uniformity despite the obvious contrasts between highways in different states. A 1933 book on secondary roads, then outside BPR jurisdiction, noted "a definite trend toward standardization."[47] MacDonald himself observed in 1937 that "there are a considerable number of lingering sighs for a return to the time when the contractors operating were not so closely determined by rules and regulations. That time is pretty definitely past."[48]

The associative model of business-government relations provided the framework for introducing these standards, just as the BPR relied on structures similar to those it had erected for guiding highway research and overseeing construction of federal-aid highways. Importantly, this approach to problem solving did not end with the Depression. Yet as closely as highway building conformed to the cooperative ideal, a basic contradiction existed between the BPR's administrative style and the goal of limiting federal involvement in business and the states. No matter how well the bureau masked its role in the cooperative administrative mechanisms, federal engineers remained the leading source of information on highway matters. There was no doubt as to who was in ultimate charge of federal-aid construction, for federal engineers took their inspection responsibilities seriously. In the case of standards, the BPR dominated every forum devoted to uniformity. Only in research did the bureau really play the cooperative game, despite its actual preeminence, and this effort alone failed to meet expectations. Generally, highway work improved significantly in this period; there was little fraud in the federal-aid program, and considering the point from which road construction started in 1921, the results were good. Still, the contradiction between an ideal of limited government and the actuality of central authority in the BPR remains, and is etched more deeply by the pattern that suggests the most successful areas of federal aid were those in which BPR expertise was given the freest reign and most obviously overstepped the boundaries of federalism and the associative ideal.

Switching the focus from this apparent contradiction to the relationship of cooperation and expertise opens a more useful means of understanding the nature of the associative process. The construction of highways was not the only project pursued through cooperative mechanisms in the 1920s. Significantly, the other success stories also exhibit the pattern whereby agencies that commanded superior expertise exercised extensive influence by using associative procedures. Thus the oil industry developed a variety of efforts for achieving order in this period, such as standardization and information collection. The key role was played by the American Petroleum Institute, the recognized center of expertise in the industry. Because it also was seen as a center for technical information, Hoover's Department of Commerce helped stabilize the new aviation and radio industries.[49] By contrast, associative efforts in industries that lacked such a seat of expertise seemed to fail. Thus Hoover struggled unsuccessfully to solve the problems of housing and coal, but these chaotic, localized industries resisted his touch. Others have pointed to the shortcomings of the associative philosophy during the Depression, noting that the absence of coercive power caused the failure of Hoover's conferences and pleas for cooperative action.[50] Again, that Hoover's government was not seen as the country's economic expert seemed to account for his inability to persuade corporate leaders to follow his lead. It is this difference in results that adds significance to Robert Cuff's observation that the War Industries Board, itself a associative agency, was both cooperative and dictatorial.[51] If cooperation was to succeed, someone, it seems, had to exercise authority. In the case of highways and other matters, technical expertise was the key to such legitimacy.

This suggests that the contradiction between cooperation and expertise was in fact a mark of success in this associative world. It mattered little whether government, corporations, or trade associations acted as leaders; the crucial point was that *some* group was recognized as the expert. Thus when the BPR deliberately limited its role in research, it all but guaranteed that the field would flounder, while federal-aid construction proceeded steadily under its direct gaze. The bureau's success in standardization, which grew from both its voluntary cooperation with industry and its more rigid control under federal-aid requirements, fell between these two poles. Yet expertise alone could not have produced the harmony that characterized road building during the 1920s and 1930s. Cooperative mechanisms proved vital for this administrative style, offering the arena within which expertise could function. Cooperative ties defined the field and set the limits on participants, while giving everyone inside a place to be heard. No one was forced to join, but as reasonable members of a reasonable process, acceptance of superior expertise was usually forthcoming. Thus the AASHO, the HRB, and the ASTM provided the essential forums for uniting road builders behind the programs plotted by the BPR.

It might appear that simple political power was the key to MacDonald's administrative success and that he used the image of apolitical experts to advance his ends within the framework of cooperation. Certainly cooperation within the highway policy was, by its very nature, a political matter related to both the federalist structure of the program and the associative ideals of the time. Thomas MacDonald undoubtedly understood that cooperation was not absolutely required by the technical goals of highway construction, but rather served important political goals. In fact, there were obvious examples in both research and standardization of cooperation hindering the implementation of the best technical approaches to road building. Yet if this administrative style introduced weakness, it also provided the BPR with a powerful means of compelling state adherence to its ideas in the form of inspections under federal-aid principles; the penalty was the loss of funds. However, these political concerns should not be given too much emphasis, for despite its lack of coercive political authority over materials producers, the BPR was just as influential in setting material standards. Moreover, had the bureau routinely used compulsion, the states might have followed the imposed rules, but the harmony almost certainly would have been absent. Quite simply, technical expertise—not just the rhetoric of such expertise—was the central element of highway policy and road building between the wars.

Significantly, both the image of apolitical expertise and the cooperative approach to problem solving were ideals with Progressive roots. But it is more important to note that the experts themselves created and maintained the cooperative mechanisms within which they functioned. Again, this might seem to have been a self-serving political decision, for these arrangements shielded the intimate control exercised by the experts from public view. Yet the behavior of the BPR again raises questions about such an easy interpretation. MacDonald certainly supported cooperation because he felt it permitted the wider implementation of his agency's findings without having to resort to the dictation that would have antagonized other members of the highway community. But he and other engineers were equally sincere in their efforts to involve as many actors in the decision-making process as possible by raising the technical capabilities required to join this community. The bureau worked as hard to end its monopoly of research and standardization and to create genuine cooperation as it did to resist efforts to end federal aid. The expansion of the HRB and the ASTM after 1930 demonstrated the success of such efforts, as did the emergence of administratively sound state highway departments that the BPR could treat as equal partners. Yet precisely because the BPR worked to expand the field, it retained its influence among the newcomers that assumed important roles in the highway community. These actions hardly fit the image of an organization that was cynically using cooperation as a façade to hide federal influence. In the end, the political framework of cooperation was definitely crucial to the BPR's dominance of highway matters, but the results of this combination of expertise and cooperation came from respect for the BPR's abilities, not from its traditional political power.

The rhetoric of the associative state, in the final analysis, fails to explain how cooperative administration could actually resolve problems. That expertise held the key to power should not be totally surprising, however, if one remembers the cliché about the shortcomings of government by committee. Thus the history of the highway field from 1921 through 1936 illustrates the reality of the associative philosophy, and the Bureau of Public Roads may represent the definitive combination of cooperation and technical superiority, especially because it maintained this stance into the 1930s. But the most important result of the BPR's cooperative world was the bureau's ability to unite the leading participants in road construction—state highway engineers, trade groups, corporate material suppliers, and Congress—around its consensus on highway policy. However, by the mid-1930s, cracks in the harmony were beginning to appear.

PART III

From Rural to Urban Road Building, 1936–56

A common saying holds that all good things must end, and this fate awaited the "golden age" of highway building as well as summer romances. From 1921 through the mid-1930s a unified highway industry worked toward the goals of the BPR—to form engineering organizations, to build roads, and to construct a national highway system. Importantly, this system was limited to roads outside towns of more than 2,500 people. This provision remained from Logan Page's vision of improving rural roads, but it also narrowed the focus of road builders and prevented the diversion of funds or attention to what the BPR considered the peripheral roads in cities or the countryside. The results were not perfect, since this federal approach permitted wide variation in highway administration and construction standards, but the BPR believed that slower change was preferable to dictated policy. In this way, Thomas MacDonald had forged a durable consensus among highway builders. By the mid-1930s, however, this consensus began to unravel as the New Deal expanded the federal-aid system to include rural and urban highways. The different conceptions of highway needs held by the supporters of these and other types of roads and backed by their own experts soon challenged the BPR's consensus. In addition, the continued rapid growth of demand for highways further encouraged the emergence of alternatives to the federal-aid system. By 1940, the unity of the "golden age" had disappeared, and the BPR was no longer able to speak for all road builders.

This situation had an immediate impact on highway policy, for as the highway community began to speak in many tongues for the first time since 1921, congestion and other evidence of inadequate roads attracted the interest of politicans. With the BPR no longer the unrivaled expert, policy making became a more overtly political, pluralistic process; it also became much more confused. MacDonald made strenuous efforts to revise the consensus and restore unity of purpose to the industry through the late 1930s and 1940s, coming close to success in 1939 and 1944. But without the leadership provided by unchallenged technical expertise, installing a new agreement about the goals and approaches to road building proved extraordinarily difficult. When MacDonald retired in 1953, the highway community was still divided, for the enormous postwar traffic increase, rapid economic growth, and the Korean War had further complicated the shaping of a comprehensive policy. Yet ideas developed by MacDonald and the BPR again provided the foundation for the eventual solution, the Federal-Aid Highway Act of 1956.

During this period, the focus of the BPR's efforts to maintain unity shifted considerably. The first chapter in this section examines the expansion of the road community from 1936 to 1938 and the BPR's narrow, technical response to this change as it tried to accommodate some new ideas while discrediting others. For the first time, the BPR's appeal to expertise failed to resolve a problem among road builders. The second chapter follows the BPR's equally unsuccessful efforts to use appeals to planning procedures as the means of reshaping highway policy from 1938 through 1944. Also visible were the first BPR compromises to the realities of a pluralistic political atmosphere. The last chapter traces the full-blown political struggle over highway policy after the war, but finds that, after several basic political concessions, expertise returned to its preeminent position. The history of the entire period shows that the image of apolitical expertise remained vital to the BPR, even when the gap between rhetoric and action was obvious.

Highway policy making from 1936 to 1956 generally conforms to the picture drawn by Louis Galambos in his essay *America at Middle Age*. He noted that many groups had turned to the government for economic security after 1880, but that only the shock of the Depression led them to consider national economic planning, especially after World War II had demonstrated that such coordination might work. The result, said Galambos, was triocratic, or three-sided government, composed of business lobbies and trade groups, government bureaucracies, and congressional committees and staffs. From this balance of forces emerged economic and governmental decisions intended to establish a steadily growing economy.[1] Highway policy making certainly involved these three groups, but power was not equally distributed. Experts continued to enjoy special legitimacy in the political process, thereby skewing the triocracy. The Bureau of Public Roads, in fact, remained the arbiter of American highway policy, even if it no longer exercised the almost unlimited influence of the "golden age." The emergence of alternative conceptions of national policy supported by rival experts splintered MacDonald's unified highway community in the mid-1930s, but the new consensus that was assembled by 1956 generally represented ideas that the bureau had championed after 1939. The new policy shifted priorities from the rural primary system to multilane express highways in American cities while maintaining such longstanding features as the federal-aid mechanism of shared responsibility and the entrusting of highway matters to supposedly apolitical engineers. As always, the BPR's reputation for expertise, challenged though it now was, implemented these approaches to road building in the face of competing ideas.

Chapter Seven

The Failure of Consensus, 1936–38

We have so many parties that are interested in this problem, that if each one of us sets his teeth and starts out to get what he wants at all hazards, we shall get nowhere; out of our own divided counsels we shall be destroyed by the opposition.

Willard T. Chevalier, 1936[1]

From 1920 through the 1930s, the highway construction boom continued, with the mileage of paved roads increasing from 387,000 in 1921 to 1,367,000 in 1940 as highway revenues quadrupled from $430 million to $1.78 billion.[2] The consensus built by Thomas MacDonald united the highway community, but the harmony that characterized the "golden age" of road building began to fade after 1936. New Deal expansion of the federal-aid program was partially responsible for the crumbling of the consensus, but equally to blame was the rising volume of traffic. The BPR's response to these changes was narrowly technical. Some new goals that fit the BPR's reading of highway needs were accommodated, while the bureau, with the full weight of its expertise, attempted to discredit others. But these attacks only alienated some members of the growing highway community, which responded with rival experts supporting alternative visions of highway priorities. By 1939, this situation had destroyed the consensus, ended the BPR's ability to dominate highway policy choices as the unrivaled expert, and began the process of openly politicizing highway policy. This chapter will examine both causes for the collapse of the highway consensus, and then discuss the BPR's technical response to the new ideas about roads.

Road Building in the New Deal

Road building policy and procedures during the early years of the Depression were little affected by federal money because the BPR strove to maintain the basic intergovernmental relationships on which highway builders had come to depend. After 1936, however, New Deal programs that significantly expanded the federal-aid program began to have a decisive impact on highway construction. Until 1933, only state highway departments had received federal aid, and then only for the 7 percent system of roads outside cities. This restriction enabled the BPR's cooperative style of administration to unite the relatively small highway community. The New Deal helped to upset this consensus by providing funds for highways outside the federal-aid system, a move that eventually diversified and expanded the concerns of the federal highway community. This expansion quickly destroyed MacDonald's consensus.

Although these changes was not visible until the late 1930s, the process had begun in 1933, when the National Industrial Recovery Act replaced the usual federal-highway aid appropriation with $400 million in grants and required the BPR to devote part of these funds to roads outside the federal-aid system. To spread the funds evenly across the country, farm-to-market roads in rural areas and railroad grade crossings and feeder roads to the federal-aid networks in cities were targeted for aid.[3] Because local and state funds declined so precipitously, local officials quickly grew dependent on the federal assistance. Congress responded in 1936, making those municipal roads that were extensions of the 7 percent system eligible for regular federal aid. Two years later, Congress created a federal-aid secondary road system. However, these additions brought under the federal umbrella road builders whose concerns were often different from those of the state highway engineers with whom the bureau had long cooperated. Rather quickly, the new groups were engaged in a divisive competition for funds with supporters of the original federal-aid system.

Another new and disruptive member of the highway community was the executive branch. Its interest in roads came from presidential advisers who began tinkering first with central economic planning and later with Keynesian theories of government spending. MacDonald still presented the administration's highway policy, but only after the Bureau of the Budget had assessed the impact of every plan.[4] Previous presidents had set limits on highway funds, but after 1936 the executive's desire to tie government spending to broader economic goals had potentially drastic consequences for highway policy. To give the states time to act, Congress had always approved federal aid for a two-year period. Technically, a separate authorization was required to release the second year's funds, but even without it, the BPR could award contracts that obligated the Treasury. Starting in 1936, however, Franklin Roosevelt argued that such appropriations reduced his ability to use the budget to respond to changing economic conditions. He wanted annual funding, a moratorium on federal aid because of a backlog in funds, and an end to the formula that gave each state a share of federal aid.[5]

Congress ignored this request, but FDR made another attempt to gain more control of highway funds in a special message to Congress in

November 1937. After repeating the argument that annual appropriations allowed him greater freedom to respond to fiscal changes, he asked Congress to spread the $200 million in highway funds for 1938 over two years, to cancel the appropriation for 1939, and to limit future federal-aid bills to $125 million annually.[6] This time, Roosevelt provoked a vehement defense of federal aid from the congressional roads committees. Senator Carl Hayden summarized their sentiment when he retorted that tying the executive's hands was the very intent of the federal-aid procedure. Only the obligatory nature of the funding process, he believed, insured annual continuity. In addition, impassioned editorials in engineering periodicals denounced the president's action as threatening to return roads to the mercy of the pork barrel. Contractors, machinery suppliers, and highway officials flooded Congress and the president with angry mail. In the end, Hayden worked out a compromise with FDR that canceled the 1939 allocation, lowered the 1940 appropriation, and retained the federal-aid procedure, but only after the president had vetoed a larger highway appropriation.[7]

This fight offered an instructive view of the highway community in the mid-1930s. Although New Deal economists saw highway construction serving purposes beyond improving transportation, the response to FDR's changes was automatic and all but unanimous in defense of existing federal-aid procedures.[8] The crucial factor was the manner in which the president's call for executive control seemed to introduce political considerations into the process of building highways. The outcry against his plan demonstrated that the BPR had kept alive an argument based on the Progressive ideology of expertise, namely that highway construction should be conducted outside politics. This image of objectivity was an essential component of the BPR's style, for it enabled the bureau to claim that its motives were different than other bureaucrats and thus gave it a special legitimacy whenever it stepped into the political arena. Although the BPR had preserved this appearance with its strictly nonpartisan behavior, it was clearly not apolitical, nor had roads been removed from politics during the 1920s and 1930s. Thus the president eventually prevailed on the issue of funds, for this decision had always been a presidential, that is, a political, decision. But Congress refused to alter the structure of federal aid, which had become synonymous with MacDonald's administrative style. The unanimity of support for federal aid demonstrated not only that he had built harmony among road builders, but also that respect for the BPR as the central highways expert remained strong. Even so, the argument between Congress and the president was a disturbing indication that expertise could lose its ability to shape highway policy. Indeed, this prospect quickly became a reality as another force was already undermining the respect for the BPR's expertise that was the key to the unity of the highway community.

This second factor was the rising number of automobiles and the resulting congestion on American highways. Registrations were slowed by the Depression and climbed only from 26.7 million in 1930 to 32.5 million in 1940. But as Michigan's highway commissioner explained in 1939, "The automobile changed almost overnight to what 25 years ago we would have regarded as mechanical magic. Its maximum speeds have become higher and higher."[9] The combination of numbers and speed spelled problems of public acceptance for the highway consensus, especially after the annual highway fatality toll exceeded 25,000 in the 1920s. More troubling to many highway engineers was urban congestion, which quickly negated the automobile's growing speed and caused more accidents. By 1925, highway builders faced a serious traffic challenge that in the 1930s led to the first critical attacks on the BPR for failing to provide adequate highways. An article in *Fortune* in 1936, "Unfit for Modern Motor Traffic," was the first of the complaints, arguing that $2.5 billion in federal aid had produced "a patchwork of roads having little or no continuity, leaping without rhyme or reason from dirt to macadam and concrete; from two-lane to three-lane and four-, five- and six-lane breadth; from well-calculated curves to slithering turns that have not the slightest respect for centrifugal force. . . . So conservative an insider as the U.S. Bureau's Chief of the Division of Design, R. E. Toms, pronounces a quarter, perhaps as much as half of all the roads built in the past twenty years unfit for present day high speed traffic."[10] That the automobile was ultimately responsible for these problems became clear when the article chronicled changes on a road in Indiana. Built in 1924 with 7 percent grades and 30 degree curves, it was safe at speeds up to thirty miles per hour. In 1931, it was rebuilt with better sight distances and 10 degree curves for speeds of forty miles per hour, but further improvements were required to set speeds at fifty miles per hour. One highway official observed in 1940 that "in spite of the prodigious efforts and the expenditures of vast sums of money the work has never caught up with the actual requirements and needs of the motoring public."[11]

As the needs of motorists grew more diverse as the car culture spread, MacDonald's consensus was threatened. In 1933, one highway engineer reported that "the general agreement and pronounced enthusiasm for the limited scope of this concentrated highway improvement have been almost unanimous." He concluded, however, that "rural residents have for years supported the state highway program in hopes that they, too, would sooner or later be gotten out of the mud. Many of them are . . . now demanding that we slow up on the main trunk-line improvement and grant them the boon of a low-cost surface."[12] To be sure, there had always been some dissatisfaction with the BPR's policy, and not just in the countryside. But these complaints had usually come from outsiders who were easily refuted by the BPR's experts. After 1936, however, the larger highway community included supporters of ideas that deviated from the BPR's focus on the primary system and that had been legitimized by federal relief programs. By 1938, MacDonald's grand harmony had disintegrated into a discordant scramble for federal funds led by supporters of rural roads, transcontinental toll roads, and urban expressways.

Alternatives to the Federal-Aid System: Rural Roads and Superhighways

The most consistent calls for expanding the federal highway program came from supporters of secondary roads, the farm-to-market and RFD routes initially targeted by Logan Page and the good roads movement. These received much less attention after 1921, although MacDonald argued that the shift in policy meant "more to farmers generally than to any other class of our citizens," because four-sevenths of the mileage was earmarked for roads that would "reach out into the farm lands and draw them closer to the county seat and the railroad."[13] Certainly federal aid provided for the most important roads in rural areas, but MacDonald always stressed that "we look forward to the time not far removed when we shall have a national system of connected roads, each road a link in the national chain, bearing its due proportion of interstate traffic, yet each a local road as well, serving with well placed lateral roads to distribute and collect the traffic of the rural sections."[14] And by 1922, 45 percent of federal aid went to hard-surface roads that comprised only 20 percent of the system, while the earth and gravel roads that comprised the remaining 80 percent of the system received only 41 percent of the funding. A year later, MacDonald was arguing that "the greatest danger we face in completing the Federal aid system or the State systems . . . will be determined efforts to divert the necessary funds to roads of lesser traffic importance, but totaling a much larger aggregate mileage."[15] He carried this argument to its logical conclusion in 1928, by stating that "there can be no intelligent discussion of the ultimately necessary mileage until there is excluded a very large mileage of legally designated public roads which are not actually needed as highways, and which cannot possibly be improved in any foreseen time with the funds now available."[16]

Until 1928–29, when several bills for funding rural roads were introduced in Congress, no one opposed MacDonald's deemphasis of secondary roads, and his recommendation killed the bills in committee. Also in 1928, the ARBA formed a County Highway Officials Division, but it feared that any extension of federal aid to counties "would spread it too thin."[17] But, as the head of the National Automobile Chamber of Commerce noted in 1929, "The demand for farm-to-market roads is a call which cannot be overlooked."[18] North Carolina and Pennsylvania dealt with the problem by giving their state highway departments responsibility for county and even township roads in 1931. The AASHO (and by extension, the BPR) endorsed this approach, but still believed primary roads should receive top priority. The association objected to complaints that "too much money is being spent on the state roads, and that the fellow outside is being allowed to flounder in the mud."[19] Thus the bureau advised congressional committees through 1932 that existing funds could produce no real improvement in rural roads for there were simply too many miles of them.

The Depression, however, upset this position because providing jobs became the key reason for highway spending. Significant amounts of federal relief funds finally reached even the most insignificant rural roads, as first the NIRA provided $100 million to such roads and later the WPA channeled millions more. These programs established a precedent for federal involvement, and rural roads supporters such as the National Rural Letter Carriers' Association and the ARBA's County Highway Officials Division were soon pressing for secondary highways to be included in the regular federal-aid program. By the mid-1930s, the BPR had to face the rural roads issue.[20]

While rural roads supporters urged reviving the original priorities of the good roads movement, advocates of transcontinental superhighways also emerged to challenge the BPR's policy consensus. This superhighway concept had a long history, for visions of "a double stone and sand road with foot paths, shade trees, with all its bridges, tressels [*sic*], through forests and mountains, in a straight line as near as possible" antedated the automobile.[21] Many proposals for a system of federally constructed main roads surfaced during the debates about federal highway building after 1910, but MacDonald's conception of a national system of state roads had deflected them all by 1921. The federal-aid system also terminated most interest in cross-country memorial highways, such as the Lincoln Highway, which had flourished from about 1910 through the early 1920s. But an occasional plan for a coast-to-coast highway still emerged during the 1920s, including a 1925 proposal by the California State Automobile Association for four-lane, high-speed highways. One writer added, "With the rapid strides taken in transcontinental travel, and the constant increase in the efficiency of motor cars, the day of long distance travel by automobile on high speed highways is coming closer each year."[22]

The BPR, however, rejected transcontinental superhighways as the extravagant plans of visionaries, and used its reputation as the leading source of expertise to ridicule them. It never said such roads could not be built. Indeed, the technical challenges had been defined in 1918 by Maryland Highway Commissioner H. G. Shirley—divided lanes on a separated grade; shallow, banked curves; and hills of 3 percent to 5 percent.[23] Over the years, several states, including Delaware, New Jersey, and Massachusetts, developed specific elements of such highways, while the BPR experimented with these features on the Mount Vernon Memorial Highway after 1928. But the BPR opposed a national network of high-speed roads on economic grounds, arguing that the explosion in vehicle registrations and the resulting congestion and accidents remained largely localized, not national problems. Moreover, long-haul trucking and recreational travel, primary uses of cross-country superhighways today, were in their infancy in the 1920s.[24]

The BPR made this argument quite firmly in its replies to several proposals from congressional committees for transcontinental roads in the late 1920s. The BPR's technical judgment that a national network of high-speed roads was a waste of money was accepted by Congress, and none of the proposals emerged from committees. Most state highway officials shared this viewpoint. They recognized the need for multilane highways where traffic demanded, but considered them special roads for special circumstances.[25] If the *Engineering News-Record* reported in 1937 that superhighways encountered popular prejudice as "extravagant speedways, designed to serve the luxurious few," then the 1930 plans for transcontinental superhighways were viewed as the quixotic dreams of wild-eyed promoters.[26]

The advocates of superhighways, like their secondary road colleagues, hoped to use unemployment programs to realize their dreams. Even PWA director Ickes mentioned the possibility of employing a million men to build cross-country roads.[27] But the enormous expense, especially compared to secondary road construction, convinced Congress to accept MacDonald's judgment that a network of superhighways was not needed. But just as it seemed that plans for such roads had reached a dead end, European highway developments during the 1930s, especially the construction of the German autobahns, provided an impetus similar to that given secondary road proponents by the prospect of federal aid.[28] By 1936, the first sections of the 4,300-mile German system were nearing completion, and American engineers examined them closely. After a personal inspection, MacDonald declared that "the highways which have been completed are wonderful examples of the best modern road building."[29] Other American engineers who attended the International Road Congress in The Hague in 1938 took the opportunity to tour more than 800 miles of autobahns. His trip led Michigan highway commissioner Murray D. Van Wagoner to observe in 1939 that "Germany has the roads while we have the traffic. It seems to me that if Germany can build roads of this type, the United States, home of the world's automobile industry, can do the same."[30]

In fact, American engineers were moving closer to building superhighways by the middle of the Depression. Maryland reconstructed Route 40 east of Baltimore to the highest standards, while several other states adopted design speeds of 75 miles-per-hour and median separations. New York unveiled a grandiose plan for 800 miles of four- to eight-lane expressways.[31] But the construction of autobahn-style roads in America was stymied by financial problems, until another idea from the past—toll financing—offered a solution. Such funding had supported the first American road boom after 1790, although its failure in the face of competition from canals and railroads led to an enduring opposition to tolls. But after 1919, several proposals surfaced for a national highway system based on tolls. The Federal Highway Act of 1921 apparently squashed this idea, taking its cue from the BPR's vehement opposition to toll financing, and state highway officials and the automobile industry repeatedly denounced the concept of tolls through the 1920s and 1930s. MacDonald summarized their feelings when he described toll roads as an infringement on the motorists' freedom and a form of double taxation, saying that "there is no substitute for the free public highway."[32]

Nonetheless, toll plans continued to appear on occasion as an answer to the enormous expense of any high-speed highway. One engineer even thought that the public would accept tolls after finding that American tourists using the Italian *autostrade*, which were built as toll roads in the 1920s, believed that tolls "are more than vindicated through the comfort and convenience, the safety of driving at high speed and through the lessened strain on the car."[33] Hence after 1936, as interest in the autobahns grew, several toll highway bills were introduced in Congress. Most called for building three east-west and six north-south cross-country superhighways, relying on tolls to pay for these autobahn-style roads. Either private or quasiprivate corporations were to build this system. An important sign of the growing legitimacy of these plans, which the BPR had always laughed away, came when both congressional roads committees held hearings on these proposals in 1937.[34]

By the mid-1930s, supporters of both transcontinental superhighways and rural roads had thus convinced some members of the highway community that the existing federal-aid focus on primary roads was inadequate. The most pressing threats to the BPR's consensus, however, came from the cities. The Progressive policy shaped by Logan Page had denied federal-aid funds to communities of more than 2,500 people because streets were seen as a local matter. Further, by 1916, most large cities had brick, block, or asphalt surfaces on important thoroughfares; compared to rural America, the cities enjoyed good roads.[35] But after 1920, when a majority of Americans lived in cities for the first time, a geometric increase in vehicle registration left city streets inadequate to the increasing demand. This combined growth in the numbers of people and automobiles presented the cities with an enormous challenge that soon transformed the goals of American roadbuilders.

Urban Highways in the 1920s and 1930s

As the states inaugurated the national highway network between cities in 1921, cities also were beginning to build roads for automobiles. In most cities during the 1920s, a mass-transit system, usually of streetcars, stood ready to compete with the automobile, but by 1940, most cities had chosen to rely on automobiles. This decision formed only one chapter in the long struggle to move people and goods in cities, yet the victory of the automobile was so complete that it requires an effort of will to imagine any other outcome. There were many reasons for this result, although local political, social, and economic conditions shaped each city's struggle for an answer to the urban transportation riddle.[36] Transit operations were in trouble because bad service, crowded and old cars, watered stock, corruption in franchise awards, and general mismanagement produced tight regulation of fares at a time when street railways needed capital for expansion or equipment.[37] Automobiles struck many as an apolitical solution to urban transportation and other problems, such as corruption and the need to ease access to the urban periphery from the squalid inner city. In other words, the auto promised both social and moral reform.[38]

The decision of most urban planners to view the automobile as a technical panacea for municipal problems was reached without even an attempt to gauge the long-term impact of the car, which was still a novelty. Thus in Chicago, mass-transit plans were exhaustively studied, while provisions for the automobile were narrowly conceived, short-term technical fixes. And while residents in some Southern cities were ambivalent about the motorcar's ability to introduce wide-ranging change, most registered their opinion by buying a car as soon as they could afford it.[39]

Optimistic assumptions about the role of the automobile survived even as traffic congestion appeared in the 1920s. Congestion was nothing new in many cities, but the belief that the automobile would reduce it stubbornly survived because a group of experts argued that automotive congestion could be handled with simple technical adjustments. Nationally the most influential figure was Miller McClintock, who argued that timed traffic signals, one-way streets, the prohibition of left turns, right-turn lanes cut out from the curb, lane markings, and the regulation or prohibition of on-street parking could solve most problems. Traffic surveys could show ways to arrange these mechanisms into comprehensive plans for each city. The young planner produced such a report for Chicago in 1926 that, in the words of historian Paul Barrett, "helped to affirm the belief that all traffic problems could be solved by competent engineering." As head of the Albert Russel Erskine Bureau of Street Traffic Research at Harvard after 1928, McClintock promoted this view, developing traffic plans for San Francisco, Boston, Washington, D.C., and Kansas City, Missouri, by 1931.[40]

McClintock dealt with the most immediate result of the pro-automobile tendencies in urban transportation—the threatened strangulation of central business districts. In Los Angeles County, the number of cars increased from 55,217 in 1915 to 110,000 in 1918 and 441,000 in 1924. By 1930, this figure had again doubled as 45,000 new cars had appeared each year after the mid-1920s. The same story was repeated in other metropolitan regions. New York had 125,101 cars in 1918 and 790,173 in 1932; Chicago's registration rose from 48,358 to 423,786 between 1916 and 1931. Even smaller cities felt the problem, as Atlanta's Fulton County experienced a similar percentage increase over the same interval, from 6,301 cars to 66,193 cars.[41]

Traffic control was not the only device that experts proposed for combating this newest form of congestion. Construction, often developed in conjunction with traffic control schemes, was undertaken to widen streets into broad boulevards. A specialized variant of these arterial highways were landscaped parkways. In many cases, by-pass routes or beltways also were designed to carry traffic around the central business districts. By 1924, Thomas MacDonald had envisioned combining parkways with circumferential highways. Rather quickly, then, experts began working on ways to fit cars into cities. An article in the new journal *Nation's Traffic* in 1927 nicely captured the prevailing view with its title, "Traffic Jams Will Be Solved by Engineers."[42]

This pattern of unthinkingly assuming that the automobile was the ideal urban transportation system was visible in several major cities during the 1920s. Detroit had adopted an innovative master plan in 1923 that envisioned a combined system of subways, street railways, and 217 miles of broad automobile boulevards by 1950. Yet the subway system was jettisoned by 1926, and the only aspect of the original plan that moved ahead was the highways. In 1925, the city received legislative authority to work with the surrounding counties on superhighway construction, and by 1932, 140 miles of improved highways were open.[43] Chicago also had decided, almost inadvertently, to favor the automobile by the mid-1920s. In addition to special projects such as implementing McClintock's traffic control plan, double decking Wacker Drive along the Chicago River, and studying a superhighway system akin to Detroit's, the Board of Local Improvement spent $114 million on street widening and another $174 million repaving 2,384 miles of streets and alleys. Even though Chicago's transit system remained profitable in the 1920s, it could not long compete with the kind of treatment that highways were receiving.[44] Nor was the situation different in Los Angeles, where public opposition forced elected officials to shelve a unified system of interurban trains, street railways, and motor buses proposed in 1925. A report for the City Traffic Commission, "A Major Traffic Street Plan for Los Angeles," which stressed improving highways for better automobile service, attracted much more support, especially when the County Regional Planning Commission began to implement the plan. When one route, a multilane highway over Cahuenga Pass, opened in 1929, engineer Wyatt B. Brummitt concluded, "Superhighways such as this will be required in the vicinity of every great city if motor traffic continues to grow; and there is no question on that point." Few in Los Angeles disagreed, because by 1930 the city had decided to rely on the automobile.[45]

New York also showed signs of accommodating itself to the automobile, although its pattern of development was uneven. The city itself, composed of five densely populated boroughs separated by two rivers and Long Island Sound, relied on the largest, most utilized mass-transit system in the country more than on the automobile. Only four bridges and a few ferries linked Brooklyn and Queens with the island of Manhattan, and 238,277 cars used these four bridges each day by 1933. On the west, the Holland Tunnel and the George Washington Bridge dumped traffic into Manhattan that had nowhere to go. The few highway projects begun in the 1920s were tangled in the web of Tammany Hall corruption.[46] On the other hand, the counties adjoining New York City made significant steps toward dealing with the automobile by pioneering the automobile parkway concept in this county, and boasted some of the best urban highways anywhere. Parkways, literally roads in city parks, had appeared as part of the "city beautiful" movement of the late nineteenth century, and Westchester County planned its initial route in this tradition, proposing the Bronx River Parkway in 1907 as a means of restoring the environment around the Bronx Zoo. In 1916 construction began on a landscaped road with gentle curves, controlled access, and separated grade crossings at a cost of a million dollars a mile, and its success led Westchester County to create a park commission to pursue parkways in 1923. Even more important was passage of the legislation creating the Long Island State Park commission in 1924, which also would build parkways. But the traffic volume on the Bronx River route after it opened between 1921 and 1924 all but forced the transformation of parkways into speedways. The Saw Mill River and Hutchinson River parkways that followed were designed as high-speed roads for automobile commuters, although they retained the park atmosphere and limited access. Further ground-breaking steps were taken by Robert Moses, who planned a system of parkways on Long Island, beginning with the Southern State Parkway after 1925. By 1927, 140 miles of parkways were planned outside the city, 85 of these on Long Island. In 1934, the Meadowbrook Causeway to Jones Beach became the first limited-access, high-speed, automobile parkway, finished seven months before the first German autobahn. Moses quickly became the most influential urban road builder in the country.[47]

The parkways of Long Island and Westchester County were exceptional attempts to provide for the use of the automobile in cities during the 1920s. But even a small city such as Atlanta, with only 66,000 cars by 1931, built viaducts over the rail lines that had long split and congested the city only after the appearance of the automobile.[48] Yet this national pattern was not visible to most people in the 1920s. Mark S. Foster's comment that the decline of mass transit in Los Angeles during the 1920s "escaped virtually every local observer at the time and can be thoroughly understood only in retrospect" applied to other cities as well.[49] One factor that obscured the view was the cities' relatively slight progress in building better highways during the 1920s. More efforts resembled Atlanta's limited program than Chicago's street construction projects; major construction on the Long Island model was usually too expensive. Yet congestion was rarely solved by the less costly steps of controlling traffic or widening streets, leading one highway engineer to warn in 1927 that "We are not confronted by a theory but a reality. Traffic congestion is already threatening the progress of many communities."[50]

During the 1930s, more cities were responding to the "reality" that their earlier policies of blind accommodation to the automobile had helped create by developing grander road-building plans. Indicative of this shift was the suggestion made in 1930 by Edward M. Bassett, president of the National Conference on City Planning, that cities ease traffic by building what he called freeways. These roads would have controlled access like a parkway but be open to commercial traffic.[51] This idea, although not totally original, appealed to engineers who saw simpler measures failing to ease congestion. All too often, wider streets were attracting commercial ribbon developments, "the most tawdry type of building construction," that choked the highway with new local traffic and thwarted the gains made with traffic control. By-pass routes were affected as well, convincing many urban planners after 1930 that only limited-access roads would preserve the cars' ability to move.[52] The public applauded this approach as well. When the first section of the Henry Hudson Parkway opened on Manhattan's Upper West Side in 1931, the *Engineering News-Record* commented, "Whether the saving in time justifies the total cost of $16,000,000 is a matter still in dispute, but the popularity with the motoring public of even the short section now in use cannot be denied."[53]

While the popularity of freeways spurred cities to construct similar superhighways, two problems hindered highway builders. First, common law held that property owners had a right of access to adjoining highways. Restricted-access highways, therefore, were largely limited to parks. But even if limiting access had been legal, high-speed roads were expensive, complex engineering projects. For example, the thirteen-mile access road to the Holland Tunnel from Elizabeth, New Jersey, referred to by Thomas MacDonald in 1932 as "the greatest highway project in the United States today," was projected to cost $2 million per mile in 1925, but the final section, a 25,000-foot steel viaduct over the Jersey meadowlands, cost $21 million and brought the total to $40 million.[54]

The impact of these two obstacles showed in the limited progress cities made in the construction of restricted-access expressways after 1930. Chicago's eight-lane Lake Shore Drive, opened in 1933, was one unique example. It not only had controlled access but also employed center curbs that could be raised and lowered to adjust the number of lanes to suit rush-hour traffic. But on other routes, the city never moved beyond the planning stage, although there was significant activity in this area. In 1933 Miller McClintock recommended that Chicago build 160 miles of limited-access highways, much of it elevated and all of it restricted to automobiles, and for the remainder of the decade, the city worked on a superhighway plan, which culminated in a 1939 report that proposed building five expressways at an estimated cost of $60 million. But no construction was begun, a situation mirrored in other cities. Detroit planners proposed to avoid the restrictions of limited access by designing superhighways that used frontage streets rather than wide boulevards to insulate the road from abutting property owners. But lack of funds stymied construction, despite the belief that "public reaction is definitely towards rapid transit on rubber." Detroit did not open its first limited-access road until 1936, and this consisted of one mile of Woodward Avenue. Similarly, Newburyport, Massachusetts, relocated a 6,600-foot section of the Newburyport Turnpike on a 200-foot-wide right of way, while St. Louis began work on a three-mile section of five-lane superhighway with no grade crossings in 1935.[55]

Given the traffic problems facing many cities, two or three miles of superhighway were small comfort, yet just as these expensive engineering solutions became the favored approach, the Depression was crippling the ability of cities to pay for them. Municipalities had provided $899 million for road and street construction in 1929, but by 1935 they could raise only $352 million. Even in 1940, the cities spent only about $400 million on highways. Only at the end of the decade were construction campaigns underway in any significant number of cities. Even Los Angeles did not begin its first real freeway, the Arroyo Seco from Pasadena to downtown, until 1938, and the first 3.7 miles were not in use until August 1940.[56]

The pressure of declining funds and increasing traffic forced the cities to turn to higher levels of government for funds. Even before 1930, urban highway officials had looked covetously at the steadily increasing gasoline taxes and vehicle fees collected by the states. They argued that cities deserved a share of these funds for use on the state routes through the cities, claiming that 60 percent of all motor vehicles were in urban areas. Echoing the proponents of rural road funding, city officials declared that the states should distribute highway funds more equitably.[57] The decline in revenue and increase in congestion during the 1930s only strengthened these demands. Yet there was little pressure, initially, from the cities for federal funds. Sidney D. Waldon, head of Detroit's Rapid Transit Commission, argued in 1930 "that it is doubtful whether even if the law permitted federal aid to be used for trunk-line thoroughfares within cities, it would be wise to do so."[58] This opinion seemed to be widespread. The American Road Builders' Association formed a City Officials Division in 1929, a year after its County Officials group, but only in 1931 did it pass a resolution asking for state and federal help in improving and maintaining the primary highways that ran through the cities. This relatively mild statement was echoed at the American Association of State Highway Officials convention in 1933, which requested an amendment permitting the use of federal aid in the cities.[59] Cities were thus much less aggressive than either rural or transcontinental highway supporters in trying to win federal-aid funds. Yet with only minor fanfare, city streets and highways, along with secondary roads, were made eligible for the

$400 million in highway funds in the 1933 National Industrial Recovery Act.[60]

This decision proved crucial, as noted in Chapter Four, since federal relief funds became the dominant source of revenue for city highway construction after 1936. The importance of these funds was most visible in New York City, the only city to build any significant mileage of express highways during the 1930s. The key figure, again, was Robert Moses, who used federal relief funds to develop the parkways on Long Island that he had planned during the 1920s. The Grand Central Parkway, which made a colossal entry into Manhattan at the Triborough Bridge, was dedicated in 1936; the final pieces of the Henry Hudson Parkway and West Side Highway opened between 1936 and 1938; the Northern-Wantagh Parkway link was completed in 1938; the Whitestone Bridge and Parkway opened in 1939; and the thirty-two-mile Belt Parkway circling Queens and Brooklyn was finished in 1940. Altogether, 153 miles of parkways and five major bridges were built in New York from 1926 to 1941, most under Moses's guidance. By 1936, the Long Island parkways alone carried eighteen million cars, living up to the claim that they represented "one of the most comprehensive networks of automobile arteries in existence."[61] As the *Engineering News-Record* editorialized, "From a national viewpoint, this evolution is important as indicating a type of highway construction that faces every major city."[62] None of this could have happened without federal funds. Largely because Moses could guarantee immediate work for the unemployed since he had started preparing plans in the 1920s, New York received $1.15 billion in WPA, PWA, and CWA funds between 1933 and 1938.[63]

Not every city, however, found the stream of money that Moses had discovered. A survey of seventy-six cities in 1936 had found that fifty-one received no federal relief funds and that the others had received only $14.6 million. Unhappiness over this situation was, however, somewhat mitigated by the Hayden-Cartwright Act of 1934, which opened urban extensions of federal-aid roads to federal funds beginning in 1936. After this change, $367 million was earmarked for city highways in 1938.[64] Thus most city highway officials were pleased at the progress they had made, but some felt they deserved a larger appropriation. This led certain urban road builders to conclude that advocates of rural roads and transcontinental superhighways threatened their federal revenues, and city officials quickly joined the debate over highway priorities. They attacked especially supporters of secondary roads, arguing that rural roads had already been given federal-aid funds and that now it was the city's turn. In a paper presented to the City Officials Division of the ARBA in 1935, a municipal engineer argued that "the problem of highway improvement is shifting from one of rural construction to one in which municipal boundaries have no significance."[65] In short, the cities had become reliant on the federal government for major highway funds by the end of the decade. Thus another voice was added to the debate that was destroying the consensus about highway priorities.

Heading Off the Competition: BPR Responses, 1935–38

While the heavier traffic and structural changes in the highway community created diversity and divisions among road builders, the BPR did not sit idly and watch the disintegration. Indeed, Thomas MacDonald clearly understood the threat to the harmony that had prevailed for more than fifteen years. The dangers were apparent in the 1937 hearing of the House Roads Committee. While two congressmen warmly endorsed transcontinental express roads, the spokesperson for the American Farm Bureau Federation and another representative argued that "'the immediate need, the important need, is really to build the farm-to-market roads before you build the superhighways. . . . I want to assure you that that is the need.'"[66] Obviously, the single-purpose policy of the 1920s had lost its ability to unite road builders. The danger, of course, was that if the highway community could no longer agree on what should be done, then others, especially politicians, might decide. Thus after 1936, MacDonald engaged in a delicate balancing act, trying to form a new consensus on the goals and priorities for highway construction and the means of reaching them. His strategy was to accommodate those changes he believed were needed, while using the BPR's expertise to discredit those he saw as dangerous or wasteful. His goal was to reunite the larger and more diverse highway community of the late 1930s, but his attempt to use the bureau's expertise to overpower opponents backfired, for it led part of that community to lose respect for the BPR and further fractured the highway community.

MacDonald understood the problems long before the difficulties actually became visible. At the ARBA's 1935 convention, he delivered an important address that proposed alterations in the nation's road policy. In "Broadening the Highway Program," he said the "pioneer phase" of road building—laying out the basic primary system—had been successfully completed because of the policy of restricting funds to the most important roads, the federal-aid network:

> No other course would have made it possible to create the main highway system capable of serving, if imperfectly, so large a part of the total of highway traffic. No other course would have so quickly joined with reason-

> ably serviceable highways so many of our towns and cities, or placed a usable road within so short a distance of many of our farms. . . . No other course would, within this relatively short period, have brought us to the point where we are enabled to broaden the highway program soundly and with good effect.

MacDonald then outlined his view of the country's highway needs. Main roads remained atop the list, but he added that the problems were changing; shortcomings, such as railroad crossings at grade, poor alignments, and narrow bridges had to be corrected. He added that because highway-user taxes from urban areas had helped build rural roads, "City residents can now with justice claim more attention." MacDonald concluded that given the need to move "in many directions," the only reasonable course was the "further development of a coherent plan sufficiently broad to encompass the major needs here touched upon."[67]

Others at the convention also saw just how the program should be broadened. The ARBA president, who had preceded MacDonald at the podium, stated that the association advocated "a balanced annual highway program that will permit needed extensions to the present system, the construction of justifiable secondary roads, the improvement of main thoroughfares and bypass routes in and around metropolitan areas, and the modernization of the existing highway system, principally to relieve congestion and insure safety."[68] MacDonald could accept this all-inclusive program, but realized all too well that everyone's needs could not be met. To establish priorities, the BPR promoted highway planning surveys that used vehicle counts and other studies to project future needs. Although work began in the 1920s, a formal planning program was not launched until 1936 and until national results became available, MacDonald had to rely on the traditional source of the BPR's influence—cooperation with all interested parties, which produced respect for the bureau's technical expertise—to prevent the community from dividing.[69]

This strategy worked wonderfully to placate most supporters of urban highways, largely because MacDonald was already sympathetic to the plight of urban road builders. His remarks of the mid-1930s showed his belief in the importance of such roads, a position the BPR had begun to accept as early as 1924. By 1926 MacDonald noted that " 'the greatest necessity is to take care of traffic congestion in and around the congested centers of population.' "[70] The bureau therefore endorsed legislation to include urban portions of state highway systems in the federal-aid network in 1930 and 1932 before it was finally enacted in 1934. This was the first extension of the federal-aid system, and it removed most of the tension between urban road builders and the BPR. Moreover, even before federal construction funds became available, the bureau had played its usual cooperative role as technical expert. Federal engineers were always available with advice for cities struggling to cope with automotive congestion, as when MacDonald encouraged district engineer J. T. Voshell to accept an offer to participate in the mid-1920s highway planning sessions of the Chicago Regional Planning Association. Another BPR engineer, R. E. Toms, won that city's competition for an express highway design in 1927 with a plan for a divided highway with separated grade and limited access using adjacent service roads. In 1934, the BPR issued a model state law describing limited-access freeway construction. When Rhode Island became the first state to approve such legislation, the BPR led the applause. Thus with urban road builders, the BPR retained its traditional position as both expert and spokesperson.[71]

The BPR's cooperative attitude and expert advice was less successful, however, with the other groups clamoring to be admitted to the federal-aid program. The problem was that urban highways were the only expansion of the federal-aid program that MacDonald wholeheartedly favored. This becomes evident when the BPR's efforts on behalf of urban roads are compared with those for rural roads. In the latter case, MacDonald moved hesitantly, for after overseeing secondary road construction with PWA funds starting in 1933 and WPA money starting in 1935, he showed no real joy when Congress authorized the first regular federal-aid funds for rural roads in 1936—$25 million a year for 1938 and 1939.[72] These developments were identical to those affecting urban highways two years earlier. To some observers, just as the original Federal-Aid Act of 1916 had intended to improve the state highway departments, this legislation tried to bring federal expertise to the counties. A consultant to the Portland Cement Association wrote that

> the present chaotic condition which characterizes the expenditure of funds available for secondary roads is generally admitted. By and large, there is no plan, adequate engineering supervision does not exist and even the potential beneficiaries of secondary road improvement have at best a hazy and confused comprehension of where their best interests lie. But this modest Federal appropriation coupled with far reaching administrative policies can bring order out of chaos.[73]

The BPR, however, seemed to have less lofty goals. For the first time, it seemed to use the cooperative procedures of federal aid merely to quiet the demands for rural road construction. The complaints of rural roads supporters had attracted political support, since the issue was easily related to farms. By 1936 even the AASHO had endorsed a federal-aid secondary system.[74] Perhaps MacDonald believed county officials would develop a broader and to his view, more responsible perspective on road problems if they had a place in the federal picture. In any event, his support for this extension of the program was grudging at best, for he continued to give secondary road work a low priority. In the BPR's annual report for 1937 he repeated his conviction that the "improvement of secondary roads is important but such work must not be allowed to impede the necessary further work on the main highways."[75] Importantly, the 1936 federal-aid bill reflected MacDonald's attitude, for the $25 million it provided for secondary roads was only one-eighth of the appropriation for the primary system. Moreover, the bill embodied the BPR's long-standing view that state highway departments, not the counties, should control this work. First each state highway department had to select 10 percent of all rural roads for inclusion in the federal-aid secondary system. According to the bureau, this procedure would insure that the new system encompassed the most important rural roads, as determined by the highway planning surveys, for "any other course must inevitably lead to unbalanced programs and economic loss resulting from inadequate road service. The highway planning surveys are producing the needed facts and the Bureau is cooperating with the States in developing highway programs, giving full attention to the designation of secondary road systems."[76] Once construction funds were approved, the BPR routed them to the counties through the state highway departments, which were given a supervisory role.

There were good reasons for each of these decisions. MacDonald believed that the small appropriation was justified because the program should be concerned with elevating county engineering ability, not constructing roads. Moreover, extensive WPA funds already supplemented the federal-aid dollars. Placing administrative responsibility in state highway departments not only followed the historic pattern of the BPR's operations, but also was justified by the lack of expertise in most county road departments. Only 200 counties maintained engineering staffs. About 700 others had access to engineering advice, but fully 2,100 could not professionally administer federal-aid funds.[77] Finally, both the AASHO and leading members of the ARBA supported the BPR's position. State highway officials feared a large rural system would hurt the primary network, of which, according to its figures, 100,000 miles, or 22 percent, needed to be relocated or rebuilt at an estimated cost greater than all federal aid granted to that point: "Who said our main highways are all properly improved?"[78] The ARBA's agreement with this position showed the persistence of the basic consensus despite some discord. The ARBA's County Officials Division was an important platform for advocates of secondary road construction, yet it always called for "justifiable" secondary roads. This placed the ARBA in agreement with MacDonald as well, even though some its county officials got carried away, according to an ARBA vice-president in 1937: "It is somewhat difficult to get a man who is so preoccupied with his own local problems to see the whole subject broadly from a nation-wide viewpoint."[79]

Yet if MacDonald intended this limited program to quiet the clamor for more rural road construction, his hopes were misplaced. For the first time, the cooperative approach failed, largely because the BPR made little effort to cooperate with secondary road builders. Most federal engineers had little confidence in county road programs, and this bolstered the bureau's decision to cooperate with state highway departments. This attitude was quickly detected by county engineers, however, and poisoned the chances for genuine cooperation. Those county road officials with competent engineering backgrounds resented the suggestion that they could not handle federal aid without state help; others complained that state and federal inspection of county plans introduced unnecessarily higher standards. Otto Hess, a county engineer who had said in 1930 that federal funds should not go to counties, stated in 1937 that "the Bureau has no sympathy with" secondary road construction. Several congressmen also denounced the BPR's hostility to secondary road work, agreeing with Hess that "we can look for little assistance from that source."[80] Including secondary roads in the federal-aid system, therefore, did little to end demands for rural roads and left wounded pride and resentment in its wake. MacDonald would regret this failure of cooperation.

The bureau had better luck deflecting proponents of transcontinental superhighways. The enthusiasm of many engineers for such roads grew with reports of German accomplishments; no topic in highway engineering was hotter in the mid-1930s. By 1936, civil engineers were proposing plans that complemented the visionary legislative schemes that flooded Congress. One elaborate plan was prepared by Frank T. Sheets, formerly chief highway engineer in Illinois and by 1937 a consulting engineer with the Portland Cement Association. He envisioned a national system costing $57.5 million and taking over twenty-five years to build, arguing that the American highway engineer already knew how to construct such roads:

> What he needs is a free reign and a chance to show the American public what really can be done when legal and administrative fetters are removed. The writer has little patience with those who condemn the engineer for a lack of vision. American engineers have done an outstanding job of highway development in spite of hampering restrictions. With a basic plan and

> public support, they can revolutionize highway transport in the next generation. Look at what the German engineers are doing. After generations of ineffective localized highway development, a national plan was conceived. The engineers were turned loose with adequate funds. In a few years Germany will have the world's outstanding system of superhighways.[81]

While echoing the traditional argument that apolitical experts should build highways, Sheets justified his plan on economic grounds, noting that final verification would come with the highway planning survey data. But his and other optimistic presentations about American superhighways rested more on the engineers' enthusiastic response to the technical challenge of a large-scale project than on economics. As another engineer explained in 1938, "Recent news items concerning a federal-built transcontinental highway have rekindled the fires of a dream that many a highway engineer has had."[82]

This enthusiasm made it easy for the BPR to use its technical expertise to discredit the whole idea, even though it was backed by other engineers. The bureau presented counterarguments in reasoned, almost condescending tones that stressed the need to prevent frivolous projects using taxpayers' dollars. MacDonald adopted this line at every congressional hearing, arguing in 1936, for example, that neither the bureau nor the states had any intention of advocating ' "the concentration of vast expenditures on a limited system of transcontinental roads.' "[83] The same expression of level-headed responsibility underlay every reply by the bureau to the Senate and House roads committees about proposed superhighway legislation, which the BPR always denounced as an "extravagent expenditure . . . without justification."[84]

MacDonald reinforced this argument in a series of technical articles about the differences in the German and American highway programs. The autobahns, he noted, were designed to encourage greater use of the motorcar, which had remained a toy for the wealthy in Europe, and to serve military purposes. The United States, however, had no need to encourage automobile use; indeed, highway engineers could not keep up with demand. Moreover, Germany's military motives deliberately ignored the economic justification of road building in America.

> We are proceeding on the principle that the utilization of the highways must directly produce the revenues with which to finance their construction. So long as we adhere to this method of financing, the building of super-highways must be limited to those areas where the present and prospective traffic will justify. . . . Their location will be carefully integrated with the population centers and the layout will not be on the transcontinental basis.[85]

The BPR was far from alone in making this argument, as many civil engineers disliked both the promotional nature and the technical assumptions of superhighway supporters. The *Engineering News-Record* was always cautious about plans for these roads, noting in 1935 that "a task of deep and far-sighted thinking is before the highway engineer in planning the future automobile and truck express roads of the United States." Two years later, this journal editorialized that only planning and not constructing a network of grandiose highways made sense.[86] A more direct rebuttal came from St. Louis consulting engineer Robert Brooks, who argued that engineer John S. Crandell's call for express highways "up and down our land" made no sense at all because these roads could not go "just anywhere." Even supporters of such construction, including engineer John S. Worley, admitted that "difficulty will be encountered in finding a definite and positive justification for the construction of . . . express rural highways."[87]

The disagreement among engineers about the need for superhighways was symbolized by the 1937 debate over a paper by Charles M. Noble, who later helped design the Pennsylvania and New Jersey turnpikes. When he suggested that highway engineers adopt a design speed of 100 miles per hour and other elements of limited-access highways, most respondents criticized the safety problem of high speeds and Noble's failure to consider traffic planning information in the location of his roads. Although some praised his "vision," most expressed some doubts.[88] Even the automobile industry, which favored almost any additional highways, backed the BPR's position. Speaking through the American Automobile Association and the National Highway Users Conference, it opposed any plan for toll superhighways, arguing that "until it can be demonstrated that the traffic is there, there is no warrant in experience to justify a toll-road program [on the German model]."[89]

Finally, Congress also accepted the BPR's reasons for cross-country superhighways. In 1937, both congressional roads committees held hearings on several plans to build networks of border-to-border, high-speed, toll highways. No legislation was forthcoming, as Congress agreed that such roads could not support themselves and thus were unnecessary. According to one proponent of superhighways, MacDonald was responsible for this outcome; T. E. Steiner complained that MacDonald "had been pouring cold water on the proposition for the past six years."[90]

However, toll superhighway plans did not disappear, despite the BPR's steady campaign against them. That the 1937 hearings were even held was an ominous sign, for the BPR had previously prevented the issue from being aired. In fact, the publicity surrounding the hearings actually enhanced the legitimacy of toll superhighways. Thus in 1938, the BPR again had to convince Congress that a national superhighway system was unnecessary. This time, Senator Robert Bulkley pushed forward a typical plan to build a network of three east-west and seven north-south toll highways through a United States Highway Corporation at a cost of up to $2 billion. To help pay for them, he proposed excess condemnation; that is, the corporation could condemn more land than it needed and sell the excess for commercial development. President Roosevelt endorsed this idea in "a long and enthusiastic discussion" with the press. But MacDonald harshly condemned this and all toll roads, and on March 2, 1938, the Senate voted to refer the bill to the Roads Committee, which promptly buried it by a unanimous vote. Significantly, every opponent of the bill had parroted the BPR's arguments; one even requested MacDonald's suggestions about testimony. Once again, the weight of the BPR's expertise destroyed the chances for a visionary highway scheme. The bureau had presented the choice, in the words of a *Better Roads* editorial, as "Transcontinental Superhighways: Traffic Needs or Fairy Tale?"[91] The outcome was predicted by New York's commissioner of public works, who said that "we shall have enough data to knock this idea into a cocked hat."[92]

Yet even as the BPR won this congressional victory, a crack had appeared in the opposition to toll financing in 1937, when the Pennsylvania Turnpike Commission was created to build a toll-supported superhighway across the central portion of that state. Significantly, the state highway department was not given control this project, which it opposed. The turnpike, conceived in large part as an employment measure, utilized the abandoned right of way and several half-finished tunnels of the South Penn Railroad, a company launched by Andrew Carnegie early in the century. A BPR study concluding that toll receipts would not meet costs convinced the banks not to buy the commission's bonds, and the BPR hoped that the project would be stillborn. But a PWA grant of $26.1 million for labor enabled construction to begin, while the Reconstruction Finance Corporation bought $35 million in bonds to keep the project going. Work began in November 1938 on the 160-mile road over the mountains between Harrisburg and the outskirts of Pittsburgh. Toll-road supporters saw this as an opportunity to test the BPR's argument that traffic would not use such roads.[93]

Thus the BPR's efforts to control opposing views of highway priorities were less than successful. While it did bring urban road builders into the federal-aid system, the BPR did little more than preserve the status quo in regard to rural roads, transcontinental superhighways, and toll roads. Worse, the BPR violated MacDonald's own rules about acting cooperatively and even antagonized its partners with an arrogant and condescending attitude. Not surprisingly, neither its limited cooperation nor its brusque technical expertise appeased the demands for a broader highway program, especially when the enthusiasts behind the newest plans refused to accept the BPR's assessment that their roads were less important than the primary system. Ironically, the BPR's behavior during the late 1930s demonstrated just how important the cooperative style of the previous fifteen years had been to maintaining the consensus achieved in 1921. In retrospect, the bureau exhibited little vision, treating the problems in the narrowest possible technical way—as isolated brush fires. Through 1938, it offered only a limited alternative to its traditional emphasis on the federal-aid primary network.

By 1938, it was obvious that the highway community had fractured into several camps and that a different approach was required to restore order. The key to the BPR's attempt to reconcile these competing interests was the highway planning surveys, whose data were beginning to become available. This information, combined with an opportunity provided by President Roosevelt, gave the BPR a chance to develop a broad, new consensus. As usual, expertise was at the core of this effort, in the form of "scientific" data from the planning surveys that the bureau hoped would place its ideas regarding primary and urban highways on an unassailable basis and end the divisive debate. But the fate of this venture and others that would follow was affected by the BPR's occasionally clumsy response to the demands for change after 1936, for its clumsiness eroded respect for the bureau's technical leadership. With the BPR lacking, at least in some people's eyes, the special legitimacy that its expertise had provided, highway policy making began drifting toward the open political arena. From late 1938 through World War II, the BPR attempted to regain its position as the leading highway expert and stop that drift.

Chapter Eight

Planning and Master Plans, 1938–44

> When every state has a planned highway procedure free from politics or the influence of politicians of every caliber, one will be able to say then, in truth, that highway planning is beginning.
>
> *Rex L. Fossnight, 1939*[1]

Between 1936 and 1938, fifteen years of agreement about the purposes of road construction in this country collapsed under the pressure of conflicting demands for different types of roads. The BPR's efforts to resolve the issue on narrow technical grounds through the use of its expertise antagonized every group but urban road builders. As a result, by 1939 the debate about which roads should be built was in danger of becoming more than a disagreement between experts. For the first time since 1918–21, highway policy ran the risk of becoming openly politicized, as indicated by Willard Chevalier's 1936 warning about divided counsels that was quoted at the start of the last chapter.

Thomas MacDonald was determined to prevent the fractures in the community from erupting into a political fight. He still pinned his hopes for success on the use of technical data to establish construction priorities that no member of the highway community would be able to dispute. But if MacDonald hoped to return highway policy discussions to technical channels, it was imperative that the basis for rating the importance of various highways have the aura of scientific objectivity. In other words, the ideal of apolitical expertise continued to retain its appeal to the BPR. Thus it was important to the BPR that it restore its image as the expert respected by all parties. After 1936, the bureau believed that planning was the answer to its dilemma, with the specific mechanism being statewide highway planning surveys. The comment of Rex Fossnight quoted above indicated that the BPR was not alone in seeing highway planning surveys as the answer to political highway debates. Such planning was an especially popular approach to problem solving during the 1930s, but the BPR's goal of generating a new policy consensus with planning data eventually proved hopeless. Twice—first in 1939–40 and again in 1944—the BPR attempted to use such planning to form new policy initiatives, but those placed at a disadvantage by the BPR's data pushed their positions through political avenues. The difficulties encountered by these two efforts showed both the continuing splits among road builders and the limits on experts' roles as policy makers, for the BPR made progress in reuniting road builders only when it bent its image of apolitical expertise and made overtly political concessions.

Highway Planning Surveys and a Master Plan: *Toll Roads and Free Roads*

During the 1930s, the first national planning efforts were made in this country as New Dealers produced such experiments as the Resettlement Administration, the National Resources Planning Board, and various budget and fiscal programs. The ad hoc nature of these programs has inclined recent observers to consider the New Deal as merely a prelude to the truly serious central economic planning that occurred during World War II. Nonetheless, planning was a byword of the Depression, a reaction to the failure of the "free market" in the 1920s. That highway engineers in the BPR turned hopefully to planning to solve the chaotic debate among road builders was one manifestation of this interest in planning.[2]

The BPR, however, was not a newcomer to planning. Almost from the start the federal-aid program had involved planning, as federal engineers sought to introduce rational criteria and procedures for developing and constructing highways. The decision to limit federal aid to a system of main roads after 1921 was one mark of such planning, and as construction began, federal engineers quickly developed a justification for the highway building consensus that also stressed planning. As the BPR explained in 1925, "The traffic importance of the various roads of the highway system provides the only proper index for the distribution of highway funds."[3] Providing for automobile traffic offered a simple, engineering measure of highway utility, and this quickly became the BPR's basic planning criterion. It excluded any vague, difficult-to-measure social goals such as the rural improvement concerns of the good roads movement. Rather, this practical approach determined which roads should be built first, eliminated any visionary schemes, and kept road builders focused on the real problem as defined by the BPR—meeting the needs of motorists. Especially during the 1920s, when people were clamoring for more roads, this narrow approach seemed logical.

The BPR promoted this view of road planning after 1921 by urging the states to institute origin and destination surveys that determined motorists' use of highways and traffic counts. Such information, federal engineers argued, could guide future construction. Moreover, beginning in 1922, the BPR showcased these methods by cooperating on the first regional traffic counting projects in Chicago and Cleveland and on the Boston Post Road in Connecticut. Others followed in Vermont, Ohio, and the District of Columbia, and by 1930 eleven western states had begun similar surveys after prodding by the BPR. Michigan, which introduced a traffic counting program in 1927, was the only state to act without federal help. By 1929, when it joined with the BPR in a more ambitious, statewide study, Michigan had 598 stations for producing quarterly surveys.[4] The Depression temporarily gave employment goals priority over the bureau's traffic criterion, but this hiatus only reinforced the very strong feeling among engineers that comprehensive surveys were needed to prevent WPA-style programs, which highway officials derided as wasteful.[5]

By the 1930s, the BPR was transforming its traffic focus into a somewhat more sophisticated planning determinant that placed priority on roads that could pay for themselves. The bureau had collected good statistics on gasoline taxes and vehicle user fees, and by combining this information with traffic data, it could justify the construction of certain types of roads in economic terms. In fact, these calculations involved nothing more than multiplying traffic by the average tax revenue produced per vehicle per mile. But these figures, when compared to construction costs, could be used to support certain highway improvement projects or to assess the priorities of various projects. These beliefs—that roads should meet current traffic demands plus allow for growth and that those roads that could pay for themselves should receive priority—represented a decisive engineering influence on highway policy and would eventually be adopted by most American road builders.

An important step in promoting this approach to highway planning was taken in 1934, when MacDonald told Senator Carl Hayden that "the pressure to get the present appropriation under contract has meant that those projects which take a longer preliminary study to develop have been forced into the background. Future efficient administration ought to be based on more careful planning and the establishment of priority of improvement."[6] Congress accepted this argument and allowed the states to use 1.5 percent of federal aid for planning, starting in 1936. The statewide traffic surveys that grew out of this action were the ultimate expression of the BPR's logic of basing highway construction on traffic needs. BPR assistant director H. S. Fairbank developed these surveys from conception to implementaton; indeed, Michigan highway engineer G. Donald Kennedy later recalled that these surveys were "pretty much thought through and organized by the time the [1934] act was passed."[7]

The highway surveys began in 1936, with Michigan's serving as the model.[8] Detailed manuals prepared by Fairbank outlined every aspect of the operation; the BPR even inspected recording forms, down to the number of columns. With IBM's help, federal engineers designed an automatic traffic counter; they also approved the location of every counting station. The data was recorded on state highway maps, and these graphic presentations of traffic density had to conform to BPR guidelines for lettering and symbols. All personnel were from the state highway departments but paid with WPA funds, although the BPR's resident manager really ran the survey.

This arrangement prevailed in the forty other states that used their planning funds to create "comprehensive fact-finding surveys" after 1936. After 1938, Congress acceded to BPR requests and *required* every state to conduct a planning survey acceptable to the bureau. In this way, Fairbank intended to introduce order into state road building, making decisions on technical rather than political grounds. For example, the BPR used highway survey data to govern the selection of roads for the federal secondary system in Michigan. Although the state and counties made the initial choices under the traditional federal-aid arrangements, the BPR also prepared its own secondary system map. Significant differences appeared in the two maps when the Michigan highway department used the new mileage as a feeder network for the original 7 percent system, while the BPR created a wholly separate network of secondary roads. Moreover, Michigan engineers distributed the new system equally between the state's Upper and Lower Peninsulas, a decision that reflected political realities, while the BPR used economic criteria to shift much of the mileage to the more populous lower area. These differences were resolved in the BPR's typical cooperative style, for after a two-year debate, state engineers accepted the bureau's position and allotted the Upper Peninsula only 64 percent of the initially proposed.[9] By presenting its argument in reasonable terms, the BPR won its demand that decisions be based on "objective" traffic data.

In addition to creating the system that Rex Fossnight endorsed as the key to apolitical state highway departments, Fairbank hoped to determine broader national priorities for each type of highway by combining state survey information, again relying on the argument that this data represented objective reality.[10] This goal explained the BPR's emphasis on uniformity in the collection and recording of data. It wanted this "scientific" data on highway usage to be the primary weapon in fashioning a new consensus. Only such information, argued west coast BPR director L. I. Hewes, permitted intelligent priorities to be set for rural, primary, and municipal highways, especially during a time of transition. As BPR engineer M. L. Wilson explained to Senator Kenneth McKellar,

> It has been the view of this Department and of practically all State highway officials that any future change in policy, or enlargement of the investment in highway facilities by the Federal and State governments should be based on more complete facts than those heretofore available as to the relative improvement needs of the various parts of the existing highway system and of any additional highways or highways of different character that may be required to serve the real economic and social needs of each State and section.[11]

The opportunity to transform the survey data into a national planning document for highways emerged in 1937, when President Roosevelt asked MacDonald to conduct a study assessing the feasibility of building three east-west and three north-south toll superhighways. This plan was typical of the bills that had appeared in Congress after 1936, and by April 1938, MacDonald had drafted a report that repeated his usual reasons for rejecting such proposals. The BPR agreed that better direct connections between cities were needed, but believed that toll roads would not attract enough traffic to be "self-liquidating." The greater need, argued the report, was "to provide a considerable number of new routes to relieve the congestion in the metropolitan areas and to modernize the standards of existing highways in rural districts." To facilitate these goals, the BPR proposed creating a Federal Land Authority to acquire rights of way, especially for urban highways, using the principle of excess condemnation.[12] FDR liked this "sound policy," which proposed that the government condemn more land than needed for the roads and then lease it back to concessionaires serving the highway or use it to insulate the road from unwanted commercial encroachment. MacDonald had heard about this concept as early as 1928 but dismissed it as impractical; he later changed his mind because it offered a means of solving the financing problems, controlling roadside development, and providing limited access. The BPR based its recommendations, which clearly represented an alteration in its vision of highway needs, on traffic counts from the 1920s and early results from the planning surveys.[13]

Congress also asked the BPR for a report on toll roads in 1938, and the result, entitled *Toll Roads and Free Roads*, repeated the earlier findings in more detail. It attacked a national system of toll superhighways as wasteful, presenting traffic estimates that showed that only 3,346 of the proposed 14,336 miles required more than two lanes. Only 547 miles, it predicted, would return more than 70 percent of the receipts needed to retire the construction bonds, and only the 172 miles from Philadelphia to New Haven might break even. BPR analysts assumed that public resistance to tolls would deter traffic and that limited access would prevent superhighways from serving the local traffic that formed the majority of all trips, even if motorists wished to use them. The BPR simply believed toll roads were unprofitable.[14]

The BPR not only refuted toll road schemes but also outlined an alternative—the construction of 26,700 miles of free highways connecting the nation's cities. The BPR argued that these roads, built to the standards needed to meet traffic and paying for themselves through gas taxes and vehicle fees, would be needed even if a toll system were built, as the toll roads would siphon off only about one-third of the long-distance traffic between any two points. Federal engineers predicted that toll routes would carry 1.2 percent of the nation's traffic, while the larger free network would attract 12.5 percent of the traffic to 1 percent of the road system's mileage. The free road plan, according to the BPR, would do everything the smaller toll network would, yet would cost about the same, $2.9 billion. Again, the bureau recommended a Federal Land Authority be created to implement its plan.

Both Secretary of Agriculture Henry Wallace and Franklin Roosevelt labeled this interregional highway system the central feature of the report. FDR, who had been involved with the Taconic Parkway in New York in the 1920s, was fascinated with the highway plans of the late 1930s. He followed plans for transcontinental toll roads and superhighways for New York closely, and in March 1939, entertained industrial designer Norman Bel Geddes, who was then building a display depicting a visionary world of express highways in 1960 for General Motors' Futurama exhibit at the 1939–40 New York World's Fair. Thus when the Agricultural Department sent a draft of *Toll Roads and Free Roads* to FDR for review in Feburary 1939, the interregional aspect caught Roosevelt's attention. In spite of Wallace's cover letter that stressed urban highway needs, Roosevelt wanted the report to emphasize interregional roads for defense reasons, and after a cabinet meeting on March 29, Wallace noted that FDR "would like Mr. MacDonald's report revised so as to make it less probable that the Cities will be able to get from Congress Federal funds for doing work for which the Cities should pay." The BPR, however, only adjusted the introduction to agree with the President's views. Thus accommodative planning by the BPR meant accommodating not only the automobile but also the president.[15]

The cover letters on the final report restated FDR's ideas, yet the text contained—almost intact—the BPR's view that urban areas most needed better highways. Even when FDR submitted the report to Congress, his letter of submittal recognized that "the need for superhighways . . . exists only where there is congestion on the existing roads, and mainly in metropolitan areas."[16] In its last fifty pages, entitled "A Master Plan for Free Highway Development," the report laid out a way to deal with these problems, using the president's favored interregional system as a means of building urban expressways. Traffic surveys showed the federal-aid primary system was uncongested over most of its length. These roads, if upgraded to higher standards with improved alignments, restricted access, and separated grades, would become the interregional system in most rural areas. Over much of its length, argued the BPR, two-lane roads would adequately serve traffic. Therefore, more attention could be devoted to cities, where, according to the report, "only a major operation will suffice—nothing less than the creation of a depressed or an elevated artery (the former usually to be preferred) that will convey the massed movement pressing into, and through, the heart of the city, under or over the local cross streets without interruption by their conflicting traffic."[17]

This statement was a milestone for American highways, for contrary to earlier BPR arguments, federal engineers found that most traffic moved in and out of, not through, urban centers. Thus the report proposed shifting the nation's highway policy from its traditional rural focus to an urban orientation. The interregional routes remained the central feature of the report, but the actual priorities were arterial highways through city centers, then by-passes around the largest cities, the modernization of the existing primary system, and finally secondary roads.[18] The specific plan for attacking urban traffic problems—expressways that penetrated central business districts and were linked to an improved cross-country system—was the legacy of *Toll Roads and Free Roads*.

The BPR clearly designed this plan as the foundation for a new consensus, for it incorporated ideas of three of the four main groups of road builders—state highway departments concerned with the primary system, the superhighway enthusiasts, and urban road builders. Yet always visible were the BPR's traditional planning goals—to build roads that could meet current traffic needs and pay for themselves. Rural roads, therefore, had the lowest priority, because planning data showed these roads, which comprised 83 percent of the mileage but carried only 13 percent of the traffic, were a poor investment. The BPR billed the report as "the first comprehensive study ever made of the national highway situation,"[19] arguing that it put each type of road into its proper place. As H. S. Fairbank explained,

> We have passed the time when the planning of improvement of the primary rural highways could reasonably be considered apart from the im-

provement of the feeder roads of the rural system, and both of these without regard to the needs and plans for city streets. We have passed the time when each year's construction program on streets or highways could reasonably be prepared without conscious relation to the ultimate objectives of a general long-time plan. For the development of each section of the whole street and highway system there should be a definite prevision of the condition to be achieved within a period of 10 or 20 years. . . . It is such a master plan that is suggested in broadest outline in the report of the Public Roads Administration on "Toll Roads and Free Roads."[20]

The highway community responded enthusiastically to the report with comments indicating that it had the potential to shape a new consensus. ARBA Engineer-Director Charles Upham embraced the BPR's planning philosophy in the ARBA's own proposal for a national highway system.[21] An *Engineering News-Record* editorial called it "one of the most important guide documents of public policy" since the 1921 Federal Highway Act: "Its great contribution is bringing measured fact to bear on the problem and relegating to well-deserved limbo the fantastic schemes for transcontinental superhighway grids that often find voice in Congress." A more detailed factual account in the same issue concluded "in fullness and in accuracy the facts supplied for consideration in the investigation are unmatched by the information elsewhere available."[22] Murray D. Van Wagoner, Michigan's highway commissioner and president of the American Road Builders' Association, called the report "a bible for highway authorities and members of the highway industry as they go about the task of building our 'highways of tomorrow.'"[23] The comments of Paul G. Hoffman, president of Studebaker, should have given the BPR hope that the report would not be disputed:

The significance of the master plan, and its underlying surveys, can not be over-emphasized in any appraisal of the highway problem. It rests upon hard facts, the essential foundation for every intelligent program; facts showing exactly what the traffic is, where it is going, and what facilities are needed to take it there safely and efficiently. As a comprehensive national program, it substitutes for guesswork; it eliminates the cross purposes and confusion which grow out of uncoordinated planning by the individual cities and states; it establishes a blueprint for assisting the national defense . . . and in general puts the highway movement on a unified and orderly basis.[24]

At first glance, then, the BPR apparently had gone a long way toward determining road priorities, for almost everyone accepted statistics compiled and interpreted by the bureau as scientific fact, which fulfilled a central goal of the BPR. Moreover, the BPR also seemed to be regaining its technical prestige. Michigan's Van Wagoner, for example, urged allowing the bureau to direct any superhighway program, "not only because of the long record on highway engineering accomplishment on the part of the Bureau, but also because it has always been a non-political agency of the federal government. The superhighway problem must not become a political football, else it is doomed to failure from the start."[25] Another heartening sign was a burst of interest in urban highways. Plans for expressways in Chicago, Cleveland, and Los Angeles attracted more notice in engineering periodicals, and the ARBA's 1940 convention passed a resolution recommending that Congress implement the BPR's master plan, agreeing that "our primary problem is now within cities." Finally, the BPR joined eight state highway departments in developing plans for a superhighway from Boston to Washington, D.C., the corridor that *Toll Roads and Free Roads* had targeted as the most heavily traveled in the country.[26]

MacDonald sought to use the report's favorable reception to rebuild the cooperative atmosphere among road builders. He highlighted this theme at the twenty-fifth convention of the American Association of State Highways Officials in 1939, tracing the achievement of the federal-aid programs. Even more pointedly, he repeated his call for harmony in a talk entitled "Accord or Discord?" at the ARBA convention in early 1940. Everyone supported roads, he said, since "the chorus remains constant but the verses sung by various groups glorify superhighways, transcontinental routes, farm-to-market roads, city streets, and more lately defense highways." The cause, he argued, was

> the tendency of the democratic process in government to form pressure groups, each demanding a particular service first, [which] introduces discord as to the particular program to be carried forward in prior place. These efforts to secure a favorable place in the sun are not be be criticized. They are natural and healthy and in the long run operate toward a roughly equitable program.

MacDonald felt, however, that this competition obscured "the interdependence of these various interests." *Toll Roads and Free Roads*, he suggested, was a factual national plan for all highways. Because it rested on traffic surveys, he believed the debate should now cease. He told supporters of rural roads that "each of these [local routes] is very important to a very few persons. . . . But let him who would criticize the highway officials answer this question: Shall the roads of the third or fourth class be improved before the real secondary, farm to market roads?" Only "downright selfishness" would explain rejection of the report, he argued, while "the way to accord lies very largely in . . . a willingness of all groups to lend their best cooperative efforts . . . in the development of fair and honestly balanced programs in the use of highway revenues."[27]

MacDonald based this appeal on the concept that had united the highway community under the federal-aid system—that road builders would accept the dictates of superior expertise. The cooperative system, which made sure that every state received some funds, made this approach workable, but it also assumed that highway engineers, unlike politicians, could submerge their narrower, local interests to national needs. It soon became clear, however, that this vision of the highway community no longer held true. In part because of the BPR's treatment of groups outside the federal-aid system and in part because of political considerations over which the BPR had no control, those who felt slighted by the BPR's proposal were unwilling to accept it gracefully. As a result, the controversy soon entered the political arena, where politicans had to choose between the positions of rival experts, and the BPR's first attempt to use planning data to shape a new concensus failed.

This attempt began when Senator Carl Hayden announced in June 1939 that he was preparing legislation to implement *Toll Roads and Free Roads*. His plan proposed that the BPR buy and lease land, which local authorities would pay for with RFC loans over forty years. This arrangement, rather than the creation of a Federal Land Authority, was intended to eliminate a major cause of construction delays. Hearings on the plan began in January 1940, and the BPR was hopeful as MacDonald led off the testimony, stressing that highway user fees could pay for the Master Plan.[28] But the discord over highway policy reappeared as many congressmen chose to avoid any "radical" New Deal plans.[29] This problem was not related to the debate over priorities, but it was important for showing how highway policy had slipped from being a concern of trusted apolitical experts into a matter of partisan politics. On the first day of the hearings, several Senators expressed doubts about federal involvement in land purchases, which they saw as an unwise extension of national authority. When state highway department executives later labeled such a plan unworkable and the AASHO's president demanded that state initiative be preserved, MacDonald backed away from supporting it, leaving an undercurrent of confusion. Only after several months of wrangling about the federal-aid appropriations for fiscal 1942 and 1943 did FDR sign a measure on September 10, 1940, and it contained no significant changes in highway policy. The only feature of *Toll Roads and Free Roads* that had been approved was a watered-down provision permitting RFC funds to be used for land purchases.[30]

But anti-New Deal sentiment alone could not account for the rejection of the report. More important was the unwillingness of those road builders who had been antagonized by the BPR after 1936 and then debunked in *Toll Roads and Free Roads* to defer to the bureau's expertise. Both rural roads and toll highway promoters disputed the BPR's master plan with their own experts, forcing politicans to choose between the rival claims. Complaints about slow progress on the federal-aid secondary system also surfaced, while even J. E. Pennybacker, the former BPR statistician, called for building rural roads first. Toll road supporters also renewed their demands for a cross-country network in several bills.[31]

These challenges were particularly effective because several recent developments suggested that certain technical assertions by the BPR might be incorrect. Coincident with the release of *Toll Roads and Free Roads*, the visionary highway scheme of Norman Bell Geddes opened in the General Motors' Futurama exhibit at the 1939–40 New York World's Fair. The display purported to show the American road system of 1960: twelve-land highways carrying cars at 100 miles per hour on electrically controlled guideways. Although MacDonald considered Bel Geddes, who had no engineering background, and his ideas "bizarre," the exhibit and his 1940 book *Magic Motorways* captured the public's imagination.[32] As highway engineer Frank T. Sheets observed in 1939, "There may arise certain practical questions when this exhibit is viewed. But I venture this opinion, that of every 100 people who are fed through the ramps into that awesome chamber of highway prophecy, 100 emerge from the exit with a very positive conviction that some heroic measures are needed, both in engineering, planning, and financing, if tomorrow's needs are to be met." Certainly the "high-tech" image of highways shown in the Futurama upstaged the conservative approach adopted by the BPR.[33]

More damaging to the BPR's credibility was the success of the first modern toll highways. Connecticut had opened the Merritt Parkway from New Haven to the New York border in 1938, imposing a ten-cent toll on this limited-access parkway to pay for its extension to Hartford. In the first year, almost 5 million cars paid tolls rather than use the adjacent and free, but crowded, Boston Post Road. Spurred by this success, Westchester County temporarily placed a toll on the Hutchinson River Parkway in 1939, with the same results. In a letter to the editor of the *Engineering News-Record* in late 1939, the president of the American Toll Bridge Association claimed these results refuted the BPR's position by demonstrating that "the public is willing to pay for a good facility and that they are not antagonistic to it when its feasibility is properly proved."[34] The Pennsylvania Turnpike only confirmed these words after it opened without fanfare in October 1940. Because it saved trucks five to six hours of travel between Pittsburgh and Philadelphia, it earned $2.6 million in its first eleven months, a sum that defied BPR projections. The chairman of the Pennsylvania Turnpike Commission believed his road opened the way for similar arterials, and within two months, plans were being drawn to extend the turnpike to Chicago. By 1941, five states—New York, Maryland, Maine, Florida, and Illinois—had created toll road commissions.[35]

A final factor in the failure of the BPR's Master Plan in 1940 was the nation's growing preoccupation with defense measures, provoked by the war in Europe. The military utility of roads had been recognized much earlier; MacDonald had worked with the War Department to identify a national system of strategic roads in 1918. Both the military and the BPR had concluded that the army was best served by good civilian highways, and periodic reviews during the 1930s reaffirmed that no special military roads were needed.[36] After France fell in 1940, however, strategic aspects of roads became the bases for determining highway priorities, thus destroying MacDonald's chances to use the master plan to restore a consensus about roads. In the preparedness atmosphere of 1940, BPR arguments about urban highway problems lost their urgency, and the third set of highway priorities in a decade, following first the primary system and then employment, was not the fusion of primary and urban highways sought by MacDonald but roads for wartime.

Yet the new focus on the strategic role of roads did not reintroduce harmony among road builders. Now, instead of road builders fighting over which roads should be funded, Congress and the President fought over how many roads should be built at all. In 1941, a BPR report identified poor access roads to factories and bases as the most significant deficiency in the system, with the 4,000 inadequate bridges on the strategic system ranking second. Further, the report estimated that $220 million was needed for access roads and another $458 million for the strategic system, which after 1939 consisted of the interregional highways described in *Toll Roads and Free Roads*. But the president wanted to keep nonessential programs to a minimum, so he asked for only $100 million for access roads and $25 million for strategic roads. Since most access roads were not on the federal-aid system, MacDonald recommended total federal repayment to states undertaking these projects. Congress, however, wanted to spend more on the strategic system by using the federal-aid formula, which ignored needs in distributing matching funds to the states. FDR again vetoed a highway bill, but finally signed the Defense Highway Act, which appropriated $150 million for access roads and $50 million for the strategic network, on November 21, 1941.[37]

This contest again showed the increasing influence of events outside the control of highway experts, with funding being measured by economic and budgetary criteria unrelated to actual highway needs. Moreover, this struggle caused more friction in the highway community, with state highway officials growing frustrated over the decline in federal-aid funds as they were asked to build roads that met national rather than local needs. The BPR's veto of Oklahoma's construction program in September 1940 for failing to consider defense concerns further strained state-federal relations.[38] Most damaging, however, was the fact that little road work was actually pursued in the year before America entered the war, despite the emphasis on strategic roads. The president refused to spend the funds Congress had provided for major improvements to the strategic system, and the adoption of relatively low material and equipment priorities for all but access road construction in June 1941 reinforced his stand. Finally, the military's decision to use railroads for long-haul transportation ended plans for improving the interregional network for strategic purposes.[39]

Thus despite the BPR's efforts, the highway community was even more divided in 1941 than in 1936. Both rural and urban highways had been brought into the federal-aid system, but complaints continued from those who felt slighted by federal policy. MacDonald's hopes for building a new consensus around *Toll Roads and Free Roads* were derailed by this opposition, to which were added the fears of federal encroachment in state and local affairs and the onset of the war. The planning data that the BPR considered indisputable support for its vision of highway policy failed to silence the critics, who had turned to politics. For the first time since 1921, no viable highway policy was in effect; instead, a temporary focus on military concerns guided road-building efforts. Yet even though no one realized it, the BPR's 1939 report had laid the groundwork for a future consensus on American highway policy. In November 1941, however, that mattered little, for three weeks after the Defense Highway Act was signed, the country entered the war.

The National Interregional Highway Committee and Postwar Planning

After Pearl Harbor, the test of wills over defense road policy and debates over urban, rural, and toll superhighway priorities ceased, as did most road work. Completed federal-aid projects fell from 11,549 miles in 1940 to only 3,035 in 1945, as tar and asphalt supplies on the east coast were restricted and the War Production Board permitted only essential construction. Even the maintenance of damaged roads challenged highway departments as gasoline rationing sharply curtailed tax revenues. Because federal-aid funds went unspent, Congress provided no regular highway appropriation for fiscal 1944 and 1945. Ironically, however, even rationing and the virtual end of civilian car production failed to lessen traffic pressures in many areas.[40] Work on access roads to factories and military bases dominated road building, with 600 projects underway by late 1942. The two largest access-road projects were the type of urban highways called for in *Toll Roads and Free Roads*. The first, begun in October 1941 by BPR urban freeway specialist Joseph Barnett, was designed to move 50,000 workers in and out of the Pentagon. Even grander in scale was the $12 million limited-access link between Henry Ford's Willow Run bomber assembly plant in Ypsilanti and his River Rouge facility in Detroit. The Detroit Industrial Expressway was not completed until March 1945, but the first sections eased congestion at the bomber plant, where almost all workers arrived by car.[41]

In some respects, the Bureau of Public Roads welcomed this construction hiatus as another opportunity for fashioning a consensus about road priorities. One part of this effort was the BPR's strong encouragement of the development of postwar highway plans by the states. This effort was set in motion in February 1941 when FDR urged the preparation of a postwar highway program to cushion future unemployment, which, like most people, he feared would follow the war. The Defense Highway Act of 1941, at MacDonald's urging, had provided $10 million in matching funds to the states for highway planning. However, the loss of engineering staffs to the war meant that state highway departments were limited to identifying basic priorities, and only half of the departments had submitted planning proposals to the BPR by March 1943.[42]

To encourage greater activity, MacDonald and others urged Congress to permit the states to use $171 million in unspent federal-aid funds for 1942 and 1943 for planning, and the Federal-Aid Highway Amendment of July 1943 set aside $50 million of this money for such work. Yet by early 1944 only $5.7 million of the initial $10 million allocation had been approved by the BPR, while most states saved their outstanding federal-aid balances for postwar construction. Nonetheless, a survey by the Special Committee on Post-War Economic Policy and Planning in the fall of 1943 found that "only the Public Roads Administration is about ready for peace. Congress has appropriated money and in addition, suspended funds from the old program will be ready on Armistice Day."[43] Even though fewer engineering drawings were produced during the war than MacDonald wanted, the constant emphasis on planning as defined by the bureau certainly carried the BPR's thinking to the highway community.

Clearly, the BPR hoped that the acceptance of its planning ideas would end the controversy and reunite the highway community around a new policy. In April 1941, a second attempt to develop such an overall policy was made possible with the presidential appointment of the National Interregional Highway Committee. Its members were MacDonald, Michigan highway commissioner and AASHO president G. Donald Kennedy, former Alabama governor Bibb Graves, California highway engineer C. H. Purcell, National Resources Planning Board chairman Frederic A. Delano, city planner Harland Bartholomew, and New York City Planning Commission chairman Rexford Tugwell. They began with the modest goal of recommending an improved system of interregional highways that could provide employment, with FDR asking for the report by October 1941. This deadline passed, and the war slowed the group's progress, yet because of the delay, the report assumed broader significance. Even at the committee's first meeting, a philosophy and plans for executing it had been accepted, and these eventually produced a national blueprint for postwar road work. MacDonald wanted agreement about national highway priorities *before* major construction resumed, and the presence of respected planners on the committee was intended to lend a image of technical objectivity to the results. Yet the new study differed only in detail from the master plan proposed in *Toll Roads and Free Roads* in 1939.

The reason for the similarlity was simple: The bureau played the preeminent role in shaping the committee's report. This was a measure of continued respect for the BPR, which was evident at the very first meeting, when G. Donald Kennedy declined Graves's nomination as chairperson and instead proposed MacDonald, who accepted. The chief appointed H. S. Fairbank, author of the 1939 report, as his secretary and placed the BPR's engineering staff at the committee's disposal. Fairbank then opened the discussion at the first meeting by quickly tracing the BPR's earlier plans for interregional roads—*Toll Roads and Free Roads* and the 1941 report on defense highways. Importantly, the committee accepted, without debate, these documents, the state traffic surveys, and the BPR's planning philosophy of determining which roads could pay for themselves by using traffic counts as the technical foundation for its work. Bartholomew and occasionally Tugwell interjected visions of highways as more than devices for moving traffic, but BPR ideas permeated the committee's deliberations from the beginning.[44]

With the acceptance of the BPR's data and planning approach, everything else fell into place. When Graves and Bartholomew suggested that military needs receive highest priority, Purcell quickly dissented, echoing the bureau's traditional viewpoint that no special military roads were required. Instead, he argued that the highways most in need of improvement were in, near, and through cities. Only Graves resisted what became the underlying assumption of the committee's report, and when Bartholomew listed a dozen objectives that had appeared during initial discussions about interregional roads, eight related to cities. This result was hardly surprising because Tugwell, the man behind the Greenbelt towns of the mid-1930s, had wrestled with transportation planning for New York, and Bartholomew had authored several urban highway plans. In addition, Kennedy and Purcell came from the states leading in the development of urban freeways. The committee also accepted the BPR's anti-toll stance and agreed with MacDonald that the federal-aid procedure should be retained. Moreover, a proposal to increase the federal share of such funding to 75 percent was quickly accepted, again after MacDonald had suggested a temporary adjustment in the formula to hasten the start of postwar construction. Thus every basic decision the committee made reflected the BPR's views.

Using these guidelines and their earlier plans, the BPR's staff prepared plans for three new interregional highway systems with 48,400 miles, 36,000 miles, and 33,920 miles. These proposals dominated the discussion at two committee meetings, which Tugwell and Delano could not attend. Not surprisingly, the questions were resolved in accordance with established BPR positions. For example, after much discussion, the group accepted Grave's suggestion (which reflected MacDonald's view) that highway funding should equal receipts from taxes on tires, motor vehicles, and gasoline. Moreover, they recommended that the entire system be built with restricted access and accepted the standards prepared by the BPR's staff; both decisions echoed the 1939 report. Finally, the committee continued to emphasize urban roads. Wilfred Owen, a noted highway planner who attended the third meeting in Delano's place, commented that the report's focus on interregional routes was "misplaced," not only because the real problems lay in cities but also because National Resources Planning Board studies indicated "a lessened use of highways for long-distance movement in the future, such movement to be made increasingly by air." The committee ignored this opportunity to broaden the inquiry to include alternative transportation systems, and instead assured Owen that the final report would suggest building the road system outward from the cities.[45]

By November 2, 1941, the committee had all but finished its work, despite Tugwell's appointment as governor of Puerto Rico and Bibb Graves's death. But the war slowed the final revisions until 1943, when a clause in the Federal Highway Amendment directed the BPR to study express highways for Congress; one guesses that this action was a mark of MacDonald's desire to release the report. As in 1939, he first submitted *Interregional Highways* to the president on January 5, 1944, and Roosevelt sent it on to Congress.

In every major particular, the final report took its cues from *Toll Roads and Free Roads*.[46] It affirmed the necessity of building roads into cities and located 4,500 miles of interregional routes in cities containing more than 5,000 people. Another 5,000 miles were reserved for undetermined circumferential arterials that swelled the final urban component to almost 9,500 miles of the 39,000-mile network of interregional highways. The increase in total mileage from 26,000 to 39,000 between 1939 and 1941 was the only major difference between the reports. *Interregional Highways* downplayed this fact, presenting a recommended system of only 33,920 miles by not including the 5,000 miles of circumferential roads. This was justified by the claim that not even preliminary routes had been determined; however, the report made it clear that there was nothing tentative about those 5,000 miles. The discrepancy was merely an attempt to package the information according to the president's guidelines.[47] The report also filled in such missing details of *Toll Roads and Free Roads* as the first estimated price tag on the BPR's vision—about $750 million a year for twenty years. It also printed the BPR's model state law for ending restrictions on limited-access construction. Finally, it repeated MacDonald's belief that the federal government should be involved in acquiring rights of way, eliminating grade crossings, and constructing parking facilities near the urban expressways.

Perhaps the most significant feature of the report, however, was the acceptance of the BPR's narrow engineering approach to highway construction by the two regional planners (Tugwell and Bartholomew) and the politician (Graves). Not surprisingly, the three highway engineers (MacDonald, Pursell, and Kennedy) assumed that road construction should be determined almost solely by traffic needs. At various points in the committee deliberations, the planners raised the possibility that highways could revitalize urban business districts, control decentralization, and guide urban land use and growth. The report did mention these points, recognizing that highways in cities had a permanent impact on urban development. But traffic service always ranked first among the purposes of highways. Even unemployment considerations were rejected as a criterion for federal road policy, and the report noted only in passing that the proposed location of the system coincided with areas of predicted unemployment. Any broader social or economic purposes achieved by the new roads were coincidental to the need for express highways. Indicative of this narrow viewpoint was the report's failure to examine the relationship of roads to alternative forms of transportation. Such an outcome only emphasizes that the BPR's viewpoints shaped *Interregional Highways* from start to finish.[48]

Building a New Consensus, 1943–44

One reason the report was not released until early 1944 was the unlikelihood of its receiving serious consideration during the worst days of the war in 1942. A second factor in the delay, however, seems to have been the lesson learned from the effort to turn *Toll Roads and Free Roads* into policy, for this time MacDonald attempted to create favorable conditions before releasing the second document. Therefore, during the war the BPR worked steadily both to promote the report and to include several of its elements in highway legislation before the report was released. It was especially ironic that the major obstacle to developing a new consensus based on the report was a dispute within the AASHO, the BPR's closest ally during the policy debate of the 1930s.

The BPR began to pave the way for the comprehensive highway policy revision proposed in *Interregional Highways* in 1942. Parts of the document were actually released, such as the model state law for limited-access highways in 1943. MacDonald and Fairbank also published a number of articles in technical periodicals hammering at the theme that urban highways should be the highest concern of builders at war's end.[49] Joining them was G. Donald Kennedy, the most prolific advocate of the report. He had lost the 1942 Michigan highway commission election, but immediately became a consulting engineer to the Automotive Safety Foundation, acting, in effect, as a full-time lobbyist for *Interregional Highways*. In numerous articles, he advocated renewed state planning efforts modeled on the 1936–38 highway surveys, believing that such traffic data would automatically put the states to work on the routes selected by the committee.[50] MacDonald added teeth to this effort by informally requiring the states to devote federal planning funds to projects on the urban sections of the highway network proposed in *Toll Roads and Free Roads*. Most state highway departments respected this request.[51]

Moreover, the BPR won acceptance for several elements of the report in legislation drafted by the AASHO during the war. The Federal Highway Amendment of 1943, which provided the states with more planning money, also allowed states to use federal-aid funds for right-of-way purchases. Achieving acceptance of this controversial item without criticism was a major step in the bureau's urban roads program. The bill also requested completion of the interregional report by early 1944.[52] Even more important was the legislative proposal for postwar highway construction released by the AASHO in March 1943, which became the vehicle for implementing *Interregional Highways*. Kennedy, as a former president of the AASHO, and MacDonald, as a member of the association's executive committee, assisted Brady Gentry of Texas, the AASHO's president in 1943, in drafting the bill. The AASHO suggested an annual appropriation of $1 billion, perhaps by adding the traditional federal-aid allocation of $250 million to the $750 million for interregional construction projected by the BPR. Moreover, the draft bill temporarily adjusted the federal/state matching ratio to 80 to 20 until gasoline tax revenues improved, a figure close to the 75 to 25 ratio proposed in the report. The legislation also changed the formula for apportioning funds to the states from equal proportions of population, area, and post-road mileage to one-half based on population and one-fourth each on road mileage and area, thereby providing more funds to those states with the greatest mileage of expensive urban arterials. Finally, federal involvement in right-of-way acquisition was maintained, as was the BPR's off-street parking plan.[53] As Kennedy later told his superiors at the Automotive Safety Foundation, the approval of the draft postwar bill by the AASHO "may be considered the culmination of several months of preliminary work and thinking on the part of the Public Roads Administration."[54]

As the BPR had hoped, the AASHO's comprehensive proposal on postwar roads set the tone for all subsequent policy discussions, and an interregional highway system that passed in and through cities was included in every postwar highway plan, including the ARBA's.[55] But the AASHO's bill also was more generous to other components of the highway community than the BPR's 1939 Master Plan had been, thereby recognizing that funding discussions were inevitably political rather than technical matters. Thus the AASHO balanced its focus on urban road building by calling for more funds for secondary roads. This consensus-building step was important, because even though MacDonald still held the Progressive notion that highway issues should be left to experts, the problems of 1939–40 showed that expertise alone could not resolve funding disputes in the face of conflicting expert opinions. MacDonald, who believed that traffic should determine highway priorities, saw the planning surveys as the indisputable guides for road building programs. The statistics in these surveys supported only one conclusion, MacDonald argued, yet "in spite of the gigantic requirements of highway modernization in and around congested metropolitan areas, force of habit continues to direct our major efforts toward the provision of rural highways."[56] But when nonurban highway supporters refused to accept the engineering logic of basing highway priorities on traffic counts, achieving the new consensus MacDonald sought required a little political bargaining, which the AASHO plan now included.

The acceptance of political reality was, of course, a blow to MacDonald's belief that highway policy should be guided by apolitical experts, but in fact this reality differed little from the aspects of federal aid that had insured harmony over the years. The formula system of distributing funds that FDR had attacked during the 1930s was the most obvious of these, for it satisfied state highway departments by insuring each a share of the total, without regard to need. The BPR's willingness to permit slow compliance with higher construction standards was another political compromise, based on the acceptance of federalism. Thus no great change was behind this attempt to placate opponents to urban highways by providing more funds for other types of roads. The only difference was that the BPR was making a political concession more openly, acting after debate had moved into the political arena. This turn of events, however, was the clearest indication of the gap between the BPR's rhetoric of apolitical expertise and its actions.

The move to accommodate outside demands was, however, especially important in 1943, because diverse opinions regarding federal road priorities remained. New toll road bills appeared in Congress during the war, and although they were killed in the committees, the issue was not defeated. States such as New York had only curtailed planning a toll road network because of the war.[57] More serious challenges came from rural roads supporters, whose irritation at the BPR remained strong. The ARBA convention in 1941 provided a forum for their discontent, with several engineers proposing a Division of Local Road Administration within the BPR. Leading county road builders wanted direct contact with the BPR, not supervision by state highway departments, and the introduction of a bill creating an entirely separate Rural Local Roads Administration in late 1943 grew directly from this discussion. The *Engineering News-Record* called the measure "Needless Duplication," but it did signal deep resentment.[58]

Ironically, the complaints of rural officials stemmed from the very success of the federal-aid program, and had their origins in technical not political concerns. Federal Works Agency solicitor Alan Johnstone remarked that the call for a rural roads agency was "expressive of the cooperative lack of contact, or might I say, the indirectness of the contact between us and city and county governments and recognition of the fact that what we call the "Federal aid" or "State aid" systems do not include all of the highway transportation problems of the people; and that as we perfect these systems, we must have a care for and some relationship to the contributory roads, streets and highways of which they are the arteries."[59] In effect, he was saying the BPR had made a mistake in not trying to work with local road builders. The best county officials were less interested in receiving federal money than in raising engineering capabilities in the same way that the BPR had used federal aid to improve state highway departments. The counties wanted to be a partner in the cooperative procedures that MacDonald himself considered the only way to administer federal highway funds.[60] The chief's instinct had always been to include all concerned parties in the structure, but the thought of working with more than 3,000 counties, as opposed to forty-eight states, caused MacDonald to oppose this extension of cooperation. It was difficult to draw this distinction without appearing to be opposed to secondary road construction, but the AASHO bill attempted to correct the impression that the BPR was totally against all rural roads.

These efforts to deal with opposition to *Interregional Highways* proved even more ironic when the most serious obstacle was a dispute within the AASHO itself. In more than two decades of close cooperation, state officials had rarely disagreed with the BPR. Crucial to this harmony was the allocation formula that determined each state's share of funds. But in 1943, the formula provoked a major argument within the AASHO, with unrest centered in the highway departments in seven industrial, urban states in the Northeast that contributed more in taxes than they received in federal aid. Initially, they wanted more money than the original formula based on post roads, population, and area provided. The AASHO president Brady Gentry acceded to this demand in the 1943 legislative plan by distributing half the funds on a population basis. But this did not satisfy the leader of the rebellion, William Cox of Connecticut, and the dispute quickly caused hard feelings within the organization. Even Robert Moses, the man behind New York City's parkway system, rebuked Cox: "If we can get this [formula revision] through it will be the beginning of fair treatment of the more populous states with the largest amount of automobile traffic. There ought not to be any difference of opinion among people in the more populous states on this issue."[61] Samuel Hadden, who succeeded Gentry as president of the AASHO in 1944, tried to shame them into compromise: "I am told that our Association has never yet failed to secure the enactment of a bill which it has sponsored . . . and I am unwilling to believe that any state highway official would want that record broken."[62]

But for Cox and others, the problem with the report was not only the distribution of money but also the proposal to abandon the policy of matching funds after the war. They saw this basic feature of federalism in highway policy as a bulwark against federal interference. Cox based his argument on principle, for northeastern highway officials had no complaints about BPR administration of the highway program. Moreover, this appeal to principle found support among state engineers in other parts of the country who feared that the sudden increase in federal funds during the 1930s would bring federal domination. Being careful to show that the BPR was not being blamed, New Mexico highway engineer C. Coykendall explained, "On occasion . . . I have felt that Federal rules and regulations are entirely too numerous and restrictive. But in all candor, the fault is as likely to be as traceable to our own door as to that of the P.R.A."[63] But even with this admission, the problem remained.

This controversy was surprising precisely because such good relations had prevailed between the bureau and the states. The rapid shift toward defense roads in 1941 caused some problems, but no real complaints surfaced. In fact, Maine's highway commissioner praised the BPR and district engineer A. G. Bruce, according to a state highway engineer who told Bruce in 1943: "Considering the money that has been expended and the amount of road work accomplished, he marvelled at the little disturbance and misunderstanding that has occurred in this period. He wants you to know how much he appreciates your having made this possible as well as your advice, counsel, and kindness."[64] Moreover, the decision *not* to appropriate federal-aid funds for 1944 and 1945 had actually concerned some engineers who feared it might be hard to revive the program after the war. This fear produced an outpouring of support for federal aid, including a National Highway Users Conference pamphlet entitled *Save the Federal Aid Highway Principle*.[65] Robert Moses added a warning that it would be "unwise to abandon or revolutionize a relationship between the federal government and the states which has been a fairly good model for such relationships and has been fairly free from the bureaucratic defects which have characterized so many other federal regional programs."[66]

Yet the controversy dragged on through 1943, delaying hearings with the House Roads Committee because Gentry wanted the AASHO to present its usual united front to Congress. But by October 1943, he wrote with some despair to Hadden, "I think we are getting in deep water and that our legislation is becoming harder to advocate and defend."[67] In the end, hearings began only in February 1944, and the association remained divided over its proposed legislation for the first time since 1918–21. The real effects of the Depression-era changes in road building thus appeared in highway policy determination during in 1943–44. Rather than relying on the traditional rationale that apolitical expertise shall establish highway priorities, some members of the enlarged highway community were increasingly willing to use politics to advance their own narrow interests.

The Federal-Aid Highway Act of 1944

During the "golden age" of highway building after 1920, the release of a document like *Interregional Highways* would have easily propelled the AASHO bill through Congress. To be sure, the reception of the report matched that given *Toll Roads and Free Roads* in 1939. *Better Roads* suggested that "the interregional report may come to represent a landmark in the highway history of the federal government—something comparable to the staking-out of the original interconnected federal-aid system,"[68] while Detroit traffic engineer Lloyd B. Reid summarized the opinion of most highway officials when he wrote, "I sincerely hope that this plan may serve as a guide to the Congress, Federal, State, and Local officials, in the development of our highway system."[69] Yet the bill based on the report went to Congress under a political cloud of internal AASHO squabbling.

The 1944 hearings on the bill were the most comprehensive discussion of roads in this country to that date. Between February 29 and May 3, 1944, the House Roads Committee heard from 110 witnesses whose testimony filled 1,100 pages.[70] Almost immediately, the impact of the split within the AASHO became evident. As usual, AASHO president Samuel Hadden presented the bill and the executive committee followed with orderly presentations about its main features. But events quickly departed from the norm, for instead of the usual procession of state engineers voicing agreement, the Roads Committee heard northeastern officials attack the old formula as discriminatory against cities while engineers from several rural states opposed any aid to cities. Most state highway engineers avoided these absolute positions and urged Congress not to resolve the problem by using rigid formulas to divide funds between rural and urban highways. All agreed on only one point—the need to continue the cooperative program with the BPR. Hadden attempted to lessen the impression of confusion in the technical ranks, emphasizing that most state highway engineers approved of the bill's general features. He also proposed several amendments in an attempt to unite the states.

But the damage had been done, and the testimony of business and trade groups and state and local government officials proved even more divided. Agricultural groups fought the interregional system, while mayors argued that only urban highways mattered. The general counsel for the Pennsylvania Turnpike favored toll roads, while truckers opposed them. Charles Upham presented the ARBA's plan, which called for a $1.5 billion annual appropriation, 50 percent more than the AASHO's bill. There were frequent favorable references to the National Interregional Highway Committee's report, but no one suggested giving priority to the most traveled roads as each argued that their roads should be funded first, a pattern that had been set by the eastern highway officials.

Significantly, the only constant amidst the diverse demands was the respect for Thomas MacDonald. Fred Brenckman of the National Grange even urged Congress to let MacDonald resolve the formula dispute, which the chief was willing to do. He proposed using two formulas—the traditional measure for rural roads and a strict population basis for urban funds, adding that "if Congress will trust us to agree with the States where the money [should go], it will provide flexibility."[71] He was asking, in effect, for a return to the time when respect for the BPR's expertise led Congress to leave the details up to him.

But this was not to happen in 1944, for the cooperative structure of the 1920s and 1930s had broken down and neither MacDonald nor Hadden could draw a creditable picture of consensus in their summaries. Without strong direction from the AASHO, the traditional spokesperson for federal highway legislation, individual congressmen became more involved in shaping highway policy than they had been in two decades. Importantly, several of the committee had strong reservations about the New Deal's expansion of government, and attacked the attitude that the federal government should solve local financial problems. Others, including staunch supporters of highways such as Republican Jesse Wolcott and Democrat William Whittington, were irritated by the huge demands from those outside the existing federal-aid program when the national debt was so high. Republican James Mott exploded after Charles Sells, New York's Superintendent of Public Works, urged a national policy that precisely matched the needs of his state. The committee thus reacted strongly to what it sensed were the selfish demands of road builders.[72]

This attitude and the AASHO's inability to present a unanimous front led chairperson J. W. Robinson to have the committee draft its own bill. They still relied heavily on the AASHO bill and *Interregional Highways*, and Robinson invited MacDonald to help. But the committee cut the AASHO's request in half, alloting $225 million to the regular system, $125 million to secondary roads, and $150 million to urban highways. These funds were apportioned on the basis originally proposed by the AASHO, one-half on population and the remainder split between area and post road mileage. For the first year, the federal/state matching ratio was set at 60 to 40, returning to equal matching for the second and third years. Federal involvement in land acquisition and the creation of the Interstate highway system survived. The cut in funds, opposed by Mott in the first minority report in the committee's history, was the price paid for the AASHO's disagreement about the allocation formula.[73]

To make matters worse, the Senate Post Office and Post Roads Committee also prepared its own bill after short hearings, adopting Senator Hayden's proposal for a $650 million appropriation, a sum that equaled the federal gas tax receipts identified in *Interregional Highways*. The bill provided $250 million for the regular federal-aid system and $200 million for the federal-aid secondary system, all distributed according to the traditional formula. The final $200 million was apportioned to cities on the basis of the ratio of the population of each city to the total population in cities with more than 5,000 people. Federal participation in right-of-way costs survived.[74]

The existence of two bills created a situation unprecedented in the history of highway policy. Also, the Senate's decision to violate protocol and not wait for the House version of the bill added another element of discord. Obviously, the absence of unity among road builders had opened the door for Congress to tinker with the bill. This time, the two committee chairpersons could not squash divergent ideas by announcing that the AASHO opposed them. The absence of clear signals from the highway industry also showed in the floor debates, which focused on the rural versus urban question. A few senators also questioned the size of this bill's appropriation when the national debt had ballooned during the war. In debates that reflected the differences in the highway community itself, some feared federal involvement in land purchases would raise land prices, others called for toll roads or a separate rural roads agency, and many wanted to cut the appropriation still more. Some observers recognized what this meant and called for an end to the bickering; an editorial in *Roads and Streets* argued, "From here on the diverse elements of the highway industry and profession had better present a more united front. A worse fate is possible than a bill which doesn't quite suit everyone."[75]

In the end, each chamber passed its own bill. The Senate provided only $450 million a year, with $200 million for regular federal-aid and $125 million each for rural and urban roads. The matching ratio was 50 percent, and no federal funds could be used for land purchases. The House approved $225 million for regular aid, $150 million for rural roads, and $125 million for urban roads. Left intact were federal participation in land acquisition and the original two-tiered allocation scheme for rural and urban roads. In the resulting conference, the House figures and the land section prevailed, although federal participation in right-of-way purchases was limited to 50 percent of the price, and no more than one-third of the total bill could be used for land. A more complex allocation system was created, with only primary roads using the traditional formula. Funds for city highways were distributed using urban population as a basis, while money for rural roads was allocated using geographic area, post road mileage, and rural population. The matching ratio was set at 50 to 50, and the Interstate system was authorized but was not to exceed 40,000 miles. On December 20, 1944, Franklin Roosevelt finally signed the bill that had left the AASHO's executive committee in March 1943.[76]

Once again, it seemed that MacDonald's effort to reunite the highway community had failed. Hal H. Hale, the AASHO's executive secretary, argued toward the end of the war that the visionary dreams of some advocates had given way to practical planning by the road builders themselves. Yet the broad plan of *Interregional Highways* and its urban focus had not been implemented. The disruptive attempt of eastern highway engineers to alter the allocation formula was successful, but they actually received less money than provided in the original AASHO bill. Still, confusion over the formula was not the only factor determining the outcome of the Federal-Aid Highway Act of 1944, for the magnitude of the appropriation insured that it would receive close scrutiny. As Senator McKeller had noted during the debate, the final product was shaped by political compromise. This outcome, however, became more obvious in 1944 than in earlier years because of the absence of harmony among the experts.

Yet in spite of the political maneuvering the bill still reflected, in muted form, MacDonald's vision of postwar highways. No one even mentioned altering the federal-aid concept. The Interstate system was created, and the concept of building expressways through the cities with federal money met no resistance. Steps had been taken to give priority to those roads facing the heaviest traffic demands. Federal contributions to land costs, a central piece of *Toll Roads and Free Roads*, were endorsed. MacDonald noted in his testimony before the House committee that the bill funding postwar road construction would shape highway policy for the next twenty years, and he was correct.

The mixed signals conveyed by the 1944 road bill demonstrated to the highway community the limitations of expertise as an effective tool for policy making, a lesson first learned during the struggle to establish the initial federal-aid system after 1918. When experts disagreed, they again learned, efforts to eliminate politics from decision making failed. Unity was crucial to MacDonald's vision of apolitical expertise, although paradoxically this unity rested on political concessions on funding and federalism. Further hampering MacDonald's efforts to rebuild the consensus, however, were the changes of the 1930s. Not only were more members of a larger highway community involved in the decision-making process, but the resources had apparently grown too large to permit road builders to determine their own destiny. All this conforms to Louis Galambos's description of triocratic government.

For a time, however, it seemed that MacDonald had rebuilt a consensus in the highway community by creating a bill that Representative William Whittington praised for covering all facets of road work for the first time. But road builders failed to learn the lesson of the political struggle of 1944, and what harmony did appear was short-lived. Ironically, new problems stemmed from the focus of MacDonald and his bureau on a narrowly technical determination of highway priorities. Specifically, the highway planning surveys on which he had pinned his hopes failed in the face of unforeseen postwar conditions. Instead of a Depression, the country entered a period of roaring growth, with a flood of new automobiles creating even worse congestion than had existed before the war. The pressure of the 1920s was back, amplified by the wartime neglect of American highways. MacDonald was proven a prophet when his 1944 statement that "everybody in the United States is waiting for the close of the war to get in a car to go some place" came true.[77] But rather than leading to the "translation of blueprints" that Hale had predicted in 1945,[78] this situation and other postwar events soon engulfed the highway community in new confusion regarding policy.

Chapter Nine

Shaping a New Consensus: From Postwar Struggles to Interstate Highways, 1945–56

> You have to stick to the theme—how did it happen? It happened because of the work of AASHO, and the work of the Bureau of Public Roads, working on congressional committees, testifying every biennium with authentic testimony coming in that supported the views they were advocating, and backed up in all instances by sound engineering. All the people we are talking about were engineers.
>
> *G. Donald Kennedy, 1980*[1]

By early 1945, highway officials eagerly awaited the latest federal-aid funds that were to open that new era in highway transportation, hopeful that the structure established by the 1944 legislation would guide the nation for many years. But their hopes were dashed when economic problems held construction below anticipated levels until 1949, even as postwar travel set new records every year. Worse, bickering over federal highway priorities resumed. This time, however, the struggle was more bitter, and it fully politicized highway policy making. The causes of dissension were not new, the first involving competition between rural roads and urban and interstate highways and the other focusing on toll roads. But what had begun as technical debates in the mid-1930s had become full-blown political battles. In 1946 these were exacerbated by the reorganization of the congressional roads committees into larger public works committees, even as the increasing sums devoted to highway construction drew more attention from congressmen. The relationship between the BPR, the AASHO, and a few congressmen that had guided highway policy for two decades was thoroughly disrupted, for the new committees were skeptical of and even hostile to the experts.

Resolution of this impasse began with the restoration of congressional respect for the technical expertise of state and federal highway engineers and was aided by the engineers' adoption of the political strategy first visible in 1944—give more money to everyone. This step, accepted by 1950, demonstrated the clearest contradiction in the BPR's rhetoric of apolitical expertise. Yet once this step was taken, the influence of federal engineers and the AASHO in policy matters increased steadily through 1954, ironically because the BPR had successfully appealed to its image of unbiased and even nonpolitical expertise. To be sure, a new "golden age" of road building did not appear in the 1950s; the highway community was too large to be controlled as Thomas MacDonald had done in the 1920s. But the Federal-Aid Highway Act of 1956 finally ended fifteen years of confusion by setting a new highway policy as well as reaffirming not only the BPR's philosophy of moving traffic but also many of its long-standing technical positions. In the end, continued acceptance of its image of apolitical expertise again made the BPR the arbiter of highway policy.[2]

Postwar Road Building, 1945–48

At first glance the road building community in 1946 seemed to be united by the first truly comprehensive highway legislation.[3] The task of restoring roads that had suffered heavy wear during the war further encouraged unity, as did a rapidly growing volume of traffic. Auto makers could not convert from wartime production quickly enough to meet demand for its new civilian models, selling a record 3,909,270 units in 1948, while registrations jumped from 31 million to 44.7 million between 1945 and 1949. Every year from 1946 until 1952 produced a record number of vehicle-miles; just the increase of 87 billion between 1947 and 1950 was equal to the total number of vehicle-miles in 1923. In its 1951 annual report the BPR announced, "We are being overwhelmed by a flood of traffic."[4]

There were other indications of greater unity among road builders. As the war ended, technical journals previewed construction plans that leaned heavily toward urban highways. Every large city was planning such roads, and an Urban Arterials Confab at the AASHO's 1947 meeting highlighted the shifting priorities. Construction began quite slowly, however, since only Detroit, New York, Los Angeles, and a few other cities had finished designing expressway systems during the war.[5] The BPR was thus able to play a major role in the development of urban projects, even though most efforts relied on local and state initiative and funding. Federal engineers provided advice and information on every aspect of limited-access expressways to the city and state engineers in Detroit, Chicago, and other cities. The head of the bureau's urban program, Joseph Barnett, oversaw needs studies in 30 major metropolitan areas and 135 cities with up to 50,000 people during 1944 and 1945, and the distribution of federal-aid funds for urban highways started in 1946. By late 1947, the BPR had funded surveys in 60 cities and preliminary engineering plans in 100 urban areas. Total federal funds distributed for construction or completed work rose steadily from $68.7 million in 1947 to $376 million in 1950. As MacDonald explained to the National Chamber of Commerce in 1947, "the time when highway needs of the cities might be regarded as of lesser concern than rural needs is past, and too long past. No finding of the statewide highway planning surveys is more clearly established than the fact that the tides of rural highway movement have either their origins or destinations predominantly in the urban areas."[6]

This assistance had a major impact on urban development, because in 1949 most cities were only beginning to build their freeway systems. One result was widespread acceptance by urban road builders of the BPR's belief that the new Interstate system would end congestion. To encourage states to give this program priority, even though it had no special funding, MacDonald asked state highway departments to submit route selections even before the war ended. By August 2, 1947, agreement had been reached on a system of 37,681 miles, and state engineers were already devoting a significant percentage of their urban and primary federal-aid funds to the new network. In 1948, $908.6 million had been made available in all categories of federal aid, and $193.9 million of this was assigned to 2,052 miles of Interstate highways.[7] These early Interstate roads were not, however, the nation-spanning expressways associated with the system after 1956. MacDonald explained in 1947 that "the new interstate system is in no sense a projected pattern of 'dream' highways."[8] Because he demanded that construction standards match anticipated traffic, many sections were planned for only two lanes. Existing primary routes were used wherever possible. In other ways, too, MacDonald's views had not changed from *Toll Roads and Free Roads*, for he still believed that the 2,319 miles of urban highway and the 2,882 miles that had been reserved in 1944 for by-passes and connecting arterials were of central importance. These 5,200 miles reached almost every major urban area, serving 182 of the 199 cities with more than 50,000 people.[9]

Just as the faith in urban expressways remained at the heart of the BPR's approach, so too did its planning philosophy of serving traffic. The BPR understood that decisions about highways shaped the growth of cities and at times seemed willing to open its planning process to broader social concerns. For a time, the bureau even sought to control both federal urban redevelopment and expressway funds. Yet while talking about social goals, the BPR defined "integrated transportation planning" to mean highways. No serious mention of urban mass transit can be found in the BPR's discussions; in 1948, in fact, it argued that expressways were the only answer to congestion.[10]

Given the turmoil that soon surrounded the highway community, events of the first postwar years may seem unimportant. Overall, road building got off to a slow start. Only 5,057 miles were added to the federal-aid system in 1946, and it was not until 1948 that completed mileage substantially exceeded the 12,000-mile annual average of the 1930s. *Business Week* proclaimed in 1948 that "Better Roads Are Years Off."[11] Nonetheless, the early postwar years were vital, because they saw the universal acceptance of the BPR's approach to urban highways. To be sure, there was little new here; a poll by the ASCE in 1946 found that 70 percent of engineers already favored using highways, traffic engineering, or police activities to solve urban congestion and that only 24 percent supported mass transit. The BPR was not alone with its vision, but its efforts insured the acceptance of this vision as gospel. Thus even though only 336 miles of federal-aid roads had been completed in cities through 1949 (at a cost of $70 million), the pattern was set for subsequent urban projects.[12]

Still, the slow rate of road building was a cause of concern, for it soon ruptured the uneasy agreement within the highway community. There were numerous reasons for delays. In the cities, the high cost of land was the primary obstacle, but even had ready solutions existed, other problems remained. Perhaps most damaging was inflation. From 1945 through 1949, American highway construction budgets jumped from $1.43 billion to $3.69 billion, yet the BPR estimated the larger sum bought fewer miles of finished highways. Both material costs and labor expenses almost doubled; steel and cement were in short supply at any price. Many highway departments delayed construction; others were outbid in efforts to attract the engineers to plan, design, and superintend ambitious postwar projects.[13] Finally, the Truman administration slowed construction in 1946, first by giving priority to housing over roads and then by reducing all public works expenditures to ease inflation. To a certain extent, road builders had created their own problems by capitalizing on the searing memory of the Depression to present highways as the main defense against unemployment. But the Depression proved a poor guide for postwar actions. MacDonald had fought any attempt to link highway funds and fiscal planning, claiming that needs were too great, but the connection had been made. By 1947, engineers were denouncing the use of roads to fine-tune the economy.[14]

The most damaging side effect of these delays was the creation of a huge surplus of federal-aid funds. State highway departments claimed only $146 million in federal funds in 1946 and $288 million in 1947, leaving more than $500 million unspent despite the intense demand for highways. This backlog led President Truman to ask for only $300 million a year in the 1948 federal-aid bill. Congress, however, chose to provide no new funds for 1949 and $450 million each for 1950 and 1951. This move caused no hardships, but the debate reopened the overtly political competition over the allocation of those funds and shattered, at least until 1956, MacDonald's frail consensus.[15]

Rural versus Urban Highways, 1948–50

That rural roads reintroduced dissension was not surprising. Although tensions had eased in 1944, the priority given to urban problems and the state administration (within BPR review) of federal funds for secondary roads still bothered local officials. MacDonald therefore launched the secondary program attuned to potential discord. To placate county officials about the state administration of money for rural road building, he formed a Board of Consultants with ten county highway engineers to examine the rules governing the selection, standards, and administration of the system. He also assigned a former county road engineer to head a separate secondary roads division in the bureau. State engineers retained their administrative authority, but were required to consult local road builders while selecting a system from the most important secondary roads. The board approved these procedures.[16]

This calm ended during the hearings on the 1948 legislation, when the administration proposed giving rural and urban roads equal sums of $75 million instead of adhering to the 30 percent to 25 percent split of 1944. But even after the 1944 allocation formula was reinstated, complaints continued about the requirements that limited funds to RFD, school bus, and other main rural routes and that allowed review of local construction standards. In 1944, the BPR had insisted upon both the route requirements and the standards clause, and MacDonald vigorously rejected any complaints about federal interference, citing the Board of Consultants' approval of BPR policies. He was infuriated at the complaints regarding standards, arguing that the rules for the rural roads program gave the state highway departments, not the bureau, control over standards, and that they were not too high. "When people say you can build secondary roads for $1500 to $2000 a mile, they just do not know what they are talking about."[17]

This rift resembled the problems of the late 1930s, but it was more disturbing. Both the source and the nature of the attack showed the greater prominance of political concerns, for congressmen led the attack, which openly challenged existing policy. Partly to blame was the 1946 congressional reorganization. As one observer explained, because the BPR had "always worked harmoniously" with both the Senate Post

Office and Post Roads Committee and the House Roads Committee, "an atmosphere of mutual confidence grew up through the years. The members of the congressional committees, particularly Senator Carl Hayden of Arizona and Representative Wilburn Cartwright of Oklahoma, became well acquainted with the needs in the highway field, and their experience, supplemented by the assistance of the [BPR] made the pasage of highway legislation relatively easy and unmarred by controversy."[18] Reorganization, however, disrupted the mutual confidence by adding new members unfamiliar with the bureau, and this increased the likelihood that political intrigue rather than expertise would influence policy formulation.

The shift was visible in the 1948 hearings and in the decision of both congressional committees to write their own legislation again; for the first time in a quarter-century the AASHO had not prepared the highway bill. The loss of influence by experts showed even more clearly in 1949, when two Senate bills proposed the creation of a Rural Local Roads Administration to assume the BPR's secondary roads responsibilities. In subsequent hearings, Senators John Stennis and Robert Kerr attacked the BPR's technical competence, with Kerr openly hostile and even badgering MacDonald, as the following exchange showed.

> Senator Kem. I would like to be permitted to say that I think the Commissioner is well qualified to testify on road conditions in the United States generally.
>
> Senator Sparkman. Yes, I think that would be conceded.
>
> Senator Kerr. I want the record to show that I do not concede it. I think the evidence is quite to the contrary with reference to local roads.[19]

Kerr also portrayed MacDonald as totally opposed to rural road work, and then, based on his experience as governor of Oklahoma, Kerr challenged MacDonald's claim that the BPR only approved and did not impose standards. This verbal assault convinced even friendly members of the committee that the BPR had an engineering bias against rural roads. Thus the committee failed to see the irony of a confrontation in which MacDonald, whose agency was charged with intruding into local affairs, used a federalist argument to attack the "fundamental unsoundness" of Kerr's plan to by-pass the states and place a federal agency in direct contact with local governments. MacDonald never had a rougher reception from Congress.

Yet this same hearing also laid the groundwork for removing the ability of rural roads supporters to disrupt highway discussions. After the angry claims of federal interference, several senators were surprised that most county highway officials commended the BPR's expertise and flexible administration of the secondary road program. A few local engineers blamed state highway departments, not the BPR, for imposing higher standards, and a delegation from the AASHO admitted as much by arguing that they had to prod local agencies into sound practices. The AASHO also challenged Kerr with its usual assertion that the states initiated standards and the BPR only approved them. They pointed out that existing standards permitted practices that any professional engineer would reject. The National Grange concluded that federal review was essential to improving county road agencies, and agreed with most road builders that a few minor adjustments would produce a sound secondary road program.[20]

The atmosphere surrounding the highway policy debates in the late 1940s, however, prevented technical expertise alone from determining their outcome. One highway engineer bitterly remarked that "in spite of the factual data available [from the BPR], it appears that federal legislation may be enacted to satisfy the pressures for financial aid to local governments rather than to meet the highway needs of the nation as a whole."[21] But the BPR used this widespread feeling to its advantage, answering critics with its traditional tool for shaping policy—engineering and statistical data. By the end of 1950, it had reduced, although not eliminated, this rift within the community.

The crucial step was the approval by Senator Dennis Chavez, Chairperson of the Senate Public Works Committee, of MacDonald's request during the 1949 hearings for authority to conduct a formal BPR study of rural roads. The chief asked the Board of Consultants to gather information from local highway officials while the BPR conducted studies of twelve counties. The board's report supported MacDonald's position in every detail, opposing the creation of a new local roads agency, favoring state highway department involvement with the counties on the federal-state model, demanding the employment of more county engineers, accepting the restriction of funds to the most important secondary roads, and suggesting that county governments assume the financial responsibility for secondary roads that they had abandoned in the 1930s.[22]

This document provided a factual basis for closing the debate on rural roads by framing the terms of the federal-aid hearings of 1950. It received widespread support from road builders, although farm groups, the National Rural Letter Carriers' Association, and a few others opposed the suggestion to use federal secondary road funds to pay county engineers. But the report's statistics cast the claims of rural roads supporters in a bad light, at least as measured by the deeply imbedded planning philosophy of the BPR. As AASHO president R. H. Baldock explained, "it would appear obvious that such funds as are available should be expended where they are most needed and where the expenditure will create the greatest benefits for the greatest number. . . . [It is no time] to even talk of Federal expenditures on low-class roads."[23] Rural roads proponents simply could not present a numerical justification for road construction, and another AASHO representative, Spencer Miller of New Jersey, derisively dismissed their "emotional arguments."[24] Once this distinction had been made, rural roads supporters began to face probing questions, including requests to explain where federal responsibilities should end. The final blow came from BPR figures indicating that between 1946 and 1948 federal funds had helped improve 16,300 miles of rural roads, with another 13,000 miles added in 1949, the greatest increase in any federal-aid system. Senator Kerr signaled the end of hostilities during his questioning of MacDonald: "I would like for the record to show that the junior Senator from Oklahoma was greatly encouraged in one of his most ardent desires of reaffirming if not reestablishing both diplomatic and friendly associations" with the BPR.[25]

Kerr's conciliatory gesture reflected a return to the traditional relationship between congressional committees and the BPR. One reason for the change was the effort of the AASHO and other road supporters to present a united front as they had in the 1920s. In late 1947, the AASHO's president argued that "unless and until all interested parties can sit down around the table and plan the future of highway development, so long will this pessimistic outlook prevail. . . . Engineers, if left alone, want to build roads where traffic demands . . . but they cannot combat lobbies that are formed for specific objectives."[26] One proposal for restoring the traditional cooperative atmosphere advocated by the AASHO was promotional campaigns by good roads groups in support of state highway departments. By 1949, eighteen state associations were helping state road builders create support for state and national highway legislation. Similarly, G. Donald Kennedy of the Automotive Safety Foundation developed technical and statistical statements of highway needs as public relations tools for highway departments.[27] The AASHO longed to return to the "good old days," when their reputation for apolitical expertise enabled it and the BPR to guide highway policy; the association looked wistfully back to the situation described by Representative J. W. Robinson, chairperson of the House Roads Committee, in an address to AASHO in 1946. "During the war, the members of this Association assumed the leadership in supplying to Congress factual information and expert testimony on which Congress based the Federal Highway Act of 1944."[28]

The 1950 federal-aid bill hinted at a return to "the good old days."

The AASHO's federal-aid proposal, for example, went to Congress only after a "day-long debate [that] produced a united front" at its 1949 convention.[29] At the hearings, both congressmen and witnesses praised MacDonald's leadership of the road program. Detroit's public works supervisor Glenn Richards even suggested that Congress create a discretionary fund of perhaps $75 million to eliminate major highway bottlenecks, which would be under the chief's control for "if anyone has good judgment and knows the problem over many, many years [it is] he and his staff."[30] The final bill was based on the AASHO's proposal and indicated that the public works committees of both houses had reached the same conclusion. MacDonald assisted Representative William Whittington, one of the few holdovers from the old House Roads Committee, with the House's revisions; his hand was especially visible in a clause requiring state highway departments to establish separate rural roads divisions. This provision mirrored a recommendation by the Board of Consultants and undercut critics of the BPR rural roads policy while reaffirming the role of state highway departments.[31]

Nonetheless, the final legislation showed that the experts were not completely successful in 1950. Congress reduced the AASHO's request from $810 million to $550 million, and ignored the BPR's plan to pay county engineers with federal aid. Also, funds for Interstate highways were not forthcoming, despite a BPR study of the network's problems, *Highway Needs of the National Defense*, issued in June 1949. This updated the 1941 report on the military requirements of roads, *Highways for the National Defense*, and found even greater deficiencies in the traffic capacity of the Interstate system. The BPR asked Congress to fund this program because the states could only spend $75 million per year on these highways and to change the equal matching ratio to 75 to 25 because of the national importance of the system. The AASHO had echoed these ideas in its 1950 federal-aid plan, requesting $210 million for Interstate highways. Both provisions were strongly supported by other groups during the 1950 hearings.[32]

But this plan was not passed in 1950. Senators from rural states objected to allocating these funds based on population, while the National Chamber of Commerce attacked the crisis atmosphere it saw behind the proposal. These complaints initially led Congress to cut the AASHO's request from $250 million to $70 million a year. That they survived at all was a measure of the rebirth of respect for the BPR and the AASHO. But the engineers could do little when President Truman asked Congress to reduce the road appropriation after the outbreak of the Korean War. Only then were Interstate funds removed from the 1950 legislation, producing a $550 million federal-aid allotment. In most other ways, the

1950 federal-aid bill endorsed the BPR's approach to highways by shifting the focus of highway construction toward the cities.[33]

More Money for Roads, 1950–53

While the 1950 bill showed renewed respect for the AASHO and the BPR, it also indicated that highway engineers had realized that they could not fashion an agreement about highway priorities with expertise alone. As in 1944, the sponsors of the 1950 law accepted a political approach—give something to everyone—rather than real planning by experts, which the BPR had embraced as the sole guide for highway policy through the 1940s. This approach was well described by AASHO vice-president Bertram Tallamy when Senator John McClellan asked him to choose between the different types of roads: "I think we have to give proper consideration to all."[34] The appeal of this philosophy was its avoidance of the hard choices, and it explained why the AASHO had recommended an $810 million appropriation in 1950. The *Engineering News-Record* also endorsed this figure, saying "that the time has come to ask for an increase in total federal aid is so obvious as hardly to need argument."[35] This strategy again demonstrated the gap between the rhetoric and the reality of the claims of apolitical expertise by the AASHO and the BPR. Yet by resorting to this technique they finally ended the disagreements over rural roads. Once it was clear that no one intended to take away the 30 percent of federal-aid designated for secondary roads, Congress turned a deaf ear to complaints about federal standards from rural proponents and criticized their demands for more funds as selfish.[36]

Ironically, once the problem of support for rural roads was solved via old-fashioned political compromise, the BPR had an easier time presenting its arguments about highway needs as the product of apolitical expertise. One result was a clear swing in highway priorities after 1950 toward urban roads. The crowded conditions on American streets also helped change congressional attitudes, for the BPR reported in 1949 that the mileage of new roads would not even provide parking space for the cars sold that year. This trend continued, with new car sales of 6.7 milion in 1950 and 5.3 million in 1951 pushing registrations to 51.9 million in 1951, 20 million more than in 1945.[37] By 1950, as Congress grew more receptive to urban problems, urban road building was finally progressing at a rate equal to available federal aid appropriations. Some 779 miles of new urban roads were opened in 1950, compared to forty-five miles in 1949. Continued acceleration of such work was promised by the completion of initial plans for federally funded urban arterials in 116 cities. Calls for integrated planning of mass transit and highways could be heard, but the favored solution to congestion continued to be more highways. In 1951, the BPR reported that every major city was working on arterial highway improvements and that 65 percent of the federal urban funds went to limited-access expressways. Another 14 percent was used for other types of multilane construction, and 12 percent added lanes to existing roads. As MacDonald explained in his 1951 address to the AASHO, the most fundamental shift in federal policy since 1939 was the "long overdue recognition of, and substantial assistance to, the critical problem of urban traffic."[38]

Yet there were limits to the financial solution of highway disputes, as the Korean War demonstrated. It had killed the proposal for special funding for Interstate highways in 1950, while construction delays followed the war-related imposition of steel rationing. By late 1951, the BPR, which processed all steel requests for highways, was being assigned about half of the steel plate but only a third of the structural shapes it needed. This slowed most bridge work in 1952, and a two-month steel strike extended the problem into 1953.[39]

Table 4

Federal Aid for Urban Highways, 1947–55
(in Millions of Dollars)

	Projects Programed		Projects Approved or under Construction		Projects Completed	
Year	$	Miles	$	Miles	$	Miles
1947	97.1	321.4	65.5	231.9	2.7	19.4
1948	135.5	389.6	165.2	402.6	23.1	119.3
1949	86.8	247.0	191.8	374.1	45.4	196.4
1950	143.8	636.5	251.6	841.6	124.5	778.9
1951	150.0	781.5	313.2	1150.1	106.4	554.1
1952	183.0	740.2	332.2	1097.8	137.1	772.0
1953	194.8	682.6	379.9	1063.2	149.2	758.5
1954						
Interstates	78.8	161.6	178.8	367.1	56.9	129.0
Other roads	125.9	566.6	231.3	823.9	116.7	637.5
1955						
Interstates	83.9	185.2	182.9	390.2	87.1	209.6
Other roads	152.4	706.7	267.5	1070.8	131.2	632.4

Source: *PRA Annual Report* (1947–49); and *BPR Annual Report* (1950–53).

But the Korean conflict also provided road builders with another argument for increasing highway appropriations by returning military and strategic justifications for highways to the front page. These arguments had survived in the Cold War atmosphere of the late 1940s, as indicated by the 1948 congressional request for a BPR study of defense roads that a year later became *Highway Needs of the National Defense*. The outbreak of hostilities in Korea opened the way for the more direct use of these rationales, which continued to stress that military and civilian needs were served by the same roads. The 1949 report highlighted the technical shortcomings of older roads, but MacDonald argued that "the most serious deficiency of our highways today, not only the interstate system, but others of greater or lesser importance, is their lack of capacity to provide for the ever-increasing number of motor vehicles in service."[40] Painting highway congestion as a military problem also added highly visible urgency to road building. These arguments, however, never altered the nature of highway policy. Rather, they provided an easily understood justification for larger highway programs.

They demonstrated once again that the "apolitical" experts would adopt political strategies in an effort to develop a new consensus, all the while claiming they were only presenting technical data. Stressing military needs as a means of receiving more funds was one indication that highway engineers had more openly begun playing political games, but a better sign was the claim by AASHO's executive secretary Hal Hale in 1950 that "our people do not understand the highway story."[41] Building on this idea, the AASHO encouraged state officials to expand promotional efforts on behalf of more roads. In 1951, they applauded several corporations that launched advertising campaigns to increase public awareness of highway problems and to build support for more federal funds. Slogans such as "America needs roads to meet tomorrow's need" (Mack Truck) and "America needs more and better highways for greater highway safety" (Firestone) appeared in the print media, while General Motors released a film entitled "Let's Get Out of the Muddle" and sponsored an essay contest on how to plan and pay for adequate highways.[42] From 1951 through 1953, the National Highway Users Conference sponsored a major promotional campaign, Project Adequate Roads, which was modeled on the old "out-of-the-mud" crusade. Greyhound president Arthur M. Hill chaired this project, which included the AASHO, the AAA, the ARBA, the American Trucking Association, individual firms, and even the American Transit Association.[43]

By 1952, with the "more-money" highway policy firmly entrenched, President Truman's second request for less federal aid met stiff congressional opposition. George Fallon, chairperson of the House Public

Works Committee, sympathized with MacDonald, who had to support the administration position, by saying that "this committee has the highest respect for you, Mr. MacDonald. I can only say that we feel sorry for you when you have to defend and support a measure we all know falls short of the minimum needs."[44] The *Engineering News-Record* added, "Committee questioning gave the commissioner ample opportunity to define the real scope of highway needs."[45] In the end, presidential pressure trimmed the total from the AASHO's proposed $810 million to $575 million, but this included $25 million for the Interstate system.

The importance of this outcome was that it showed that once the contentiousness over funds was removed, the highway community's traditional respect for the engineers of the BPR and the AASHO quickly returned. One senator was impressed, for example, by the unanimous agreement over the BPR's call for funding Interstate highways, and AASHO president Bertram Tallamy argued that "No one should underestimate the role of the AASHO in these developments." He added that "one of the things about the Association which I am sure each one of us cherishes . . . is the respect with which it is held by Congress." In 1952 it seemed that Congress again regarded highway engineers as the definitive sources of information and advice about road legislation.[46]

Although the 1952 legislation demonstrated the increasing harmony in the highway community, disagreements did remain. Project Adequate Roads, for example, was split apart in 1952 and 1953 by the issue of federal gasoline taxes.[47] Such differences existed because highway funds still had to be allocated, and more importantly, more money had to be found. Here again, the revival of the BPR's status as the respected apolitical expert was crucial, for the image of objectivity enabled its voice to be heard above the demands of others. One example of how the BPR combined the political strategy of funding all highways with its image of apolitical expertise to shape a revised highway policy was its solution to another problem that splintered the postwar highway community—the dispute over toll roads.

Toll Roads and Interstate Highways, 1950–53

"Self-liquidating" roads had long been proposed as an alternative to the federal-aid system; the issue figured prominently in the initial collapse of MacDonald's consensus in the late 1930s. But the 1944 Federal-Aid Highway Act had apparently ended the debate by adopting the BPR's position that free roads more efficiently served the nation's needs. Thus plans to place tolls on New York's proposed cross-state turnpike in early 1945 were successfully opposed by the highway department, which began construction with regular highway funds. Other state highway engineers were as adamant in their opposition to tolls. A Michigan State Highway Department press release of mid-1945 concluded, "One toll road won't wreck the system. Scattered toll roads will not hurt it much. But a toll road epidemic won't stop at one or a dozen toll roads. WE MUST STOP THIS EPIDEMIC NOW, IN THE PLACES WHERE WE CAN REACH IT."[48]

But by 1950, the success of the Merritt Parkway and the Pennsylvania Turnpike had destroyed the BPR's argument that the public would reject tolls, and toll roads came to be seen as a viable alternative to the "free" Interstate system.[49] As traffic increased after 1947, states seriously considered toll financing, with the Pennsylvania Turnpike as the model. Maine acted first, in 1945, launching a forty-seven-mile toll road paralleling US Route 1 by selling bonds backed only by future toll receipts. New York followed suit, authorizing tolls on the Saw Mill River and Hutchinson River parkways into New York City. MacDonald counterattacked by cautioning the states not to be influenced by the success of the Pennsylvania Turnpike, arguing that it was not a fair model because the PWA grants that had rescued it skewed any comparisons. He also warned that toll roads would attract enough traffic to destroy the financial integrity of nontoll primary routes, but not enough to pay off the bonds issued by the toll roads. He also criticized their higher cost due to interest charges on revenue bonds.[50]

But state legislatures, seeing an easy way to raise highway funds, disregarded the advice of highway builders and the toll road "epidemic" spread. In 1947, New Hampshire announced plans for a toll road to connect the Massachusetts line with the Maine Turnpike, Maryland authorized toll collection, and West Virginia created a turnpike commission. In 1948 and 1949, Pennsylvania extended its turnpike to Philadelphia and the Ohio border, New Jersey approved tolls on its turnpike between Philadelphia and New York, Ohio and Oklahoma established turnpike authorities, and Colorado approved a toll route from Denver to Boulder. There were countertrends, however, as Alabama, Kentucky, Michigan, Minnesota, Missouri, Nebraska, and Wisconsin rejected toll road plans. And while toll roads were winning acceptance, New Jersey hastened to add that it considered toll financing only an expedient for getting funds immediately, and deplored the interest charges created by this choice.[51] But the opening of the New Hampshire toll road, the beginning of construction on the New Jersey Turnpike, and the passing of legislation that approved toll financing for finishing New York's Thruway, all coming in 1950, began to erode the engineers' opposition to tolls. In typical fashion, engineering journals replaced dismissals of toll roads as totally impractical or economically ridiculous with grudging calls to study their ability to fight congestion. Further respectability came with the opening of the Colorado road in 1952 and the New Jersey and Oklahoma turnpikes in 1953. By then, toll road plans were 'Bustin' Out All Over,' " according to *Roads and Streets*. In October 1953, 762 miles of toll roads were open, with 1,077 more under construction in eleven different states.[52]

The real indication of changing attitudes about toll roads came in an AASHO report of 1952 that accepted toll financing as a potential source of revenue but urged keeping the ban on tolls on federal-aid roads. In the report the AASHO no longer stressed the financial weaknesses of toll roads but rather emphasized the failure to coordinate toll highways with state highway systems. This criticism, raised in several quarters in 1953, indicated that although state highway officials had decided they could live with toll roads, they saw them as merely an expedient that failed to serve all highway needs.[53] The BPR had also slowly reached this conclusion. In 1950, it had proposed using federal aid to retire bonds for toll roads on the Interstate system, and during the 1952 federal-aid hearings MacDonald's opposition to tolls was noticeably less harsh. Even so, he continued to argue that toll roads had a limited place in the nation's highway network, backing his statements with a report that concluded toll construction had reached its limit since the most heavily traveled routes were already under development.[54]

Indeed, 1953 was the high point for toll roads, although the peak of construction came in 1954, when nine states were building such facilities. Thereafter, this financing scheme lost ground, and only 1,000 of the 2,663 miles of toll roads proposed but not started by October 1953 were built. Much of this mileage was on the New York, New Jersey, and Pennsylvania systems. Several routes proposed after 1953 were added, but almost 3,000 of the 3,557 miles of toll roads open in 1963 had been planned before 1954. This was substantially less than the 12,000 miles of toll highways that had been proposed.[55]

The Federal-Aid Highway Act of 1956, which again banned tolls on federal-aid roads and provided significant funding for the Interstate network, finally killed the toll road movement, but the limits had been set by 1953, as the BPR predicted. The bureau's proposal for repaying toll bonds with federal aid was not an endorsement of tolls but a way to insure that tolls would not be permanent. The rejection of this plan during the toll road boom was largely due to the BPR's general opposition to tolls on both bridges and roads. By the early 1950s, toll financing was accepted only as a means of constructing expensive express highways when no other resources were available. The BPR's warnings that revenue bonds were not a panacea for congestion, supported by the AASHO and most road users, had been heeded.

In 1953, congressional hearings proved that the BPR could limit the acceptance of toll roads even if it could not prohibit them. A bill to give Connecticut the option of imposing tolls on its Interstate roads was introduced with the hope that other states might follow suit. At the hearings, however, most witnesses argued that free roads should be the main strength of the highway system, although tolls were not rejected completely. The BPR's testimony only pointed out the obstacles this bill would create in Connecticut's efforts to work within the federal-aid framework. The committee sympathized with the state's plea for flexibility in financing, but rejected the plan to alter existing policy.[56]

Progress on the Interstate system also diluted support for toll financing. By 1953, one-third of the primary and one-half of the urban federal-aid appropriation was being spent on these roads. In 1952, 1,232 miles were completed and 1,620 miles were under construction, using $197.6 million in federal aid. By the end of 1953, $489.4 million in federal aid had contributed to building 5,620 miles of Interstate roads. Even though 76 percent of the network was still deemed inadequate, the BPR was proving that toll financing was not essential for construction of this system. Indeed, much of the progress was made on highways that were inappropriate for tolls, such as urban expressways where toll booths would have increased traffic congestion. In 1953, the bureau observed, "It is difficult to imagine the magnitude of the congestion which would now be present in many cities if the urban arterials now completed and in use were not available."[57]

The debate over toll roads was thus largely resolved by 1953. The BPR had shifted its position after discovering that its predictions of low toll-road usage had been invalidated by the insistent public demand for uncongested roads. The bureau should have known better than to place such weight on traffic counts, for in New York Robert Moses had learned that traffic quickly used roads to their capacity during the 1930s.[58] The BPR's reliance on traffic predictions encountered problems similar to those experienced by the mathematically oriented research projects of the 1920s and 1930s, but neither Moses nor the Bureau appreciated the shortcomings of traffic counts as a guide to future construction. Even so, the technical problems about traffic usage that the bureau forecasted did prove correct, for toll highways were limited to special applications in populous areas between cities. With the notable exceptions of northern New Jersey, the New York metropolitan area, and Connecticut, toll roads did less to remedy congestion than promised. Once again, the BPR had shaped what became the accepted view by 1953. Significantly, it was at this point that the highway community seemed more united than at any time since 1935. Evidence of the formation of a new consensus on highway policy came in a House hearing on the state of American highways in 1953 and in the biennial hearings on federal aid.

The Emergence of a New Consensus, 1953–54

In mid-1953, the House Public Works Committee's Subcommittee on Roads opened hearings that were more far reaching than those in 1944. Chairperson J. Harry MacGregor listed fourteen questions about the organization, administration, and financing of the federal highway program that would guide the inquiry. These included the place of tolls in the system, the funding of Interstate roads, the role of the BPR, federal participation in highway maintenance, the size of federal gasoline taxes and their linkage to highway expenditures, the military utility of highways, the effect of trucks on roads and other matters related to construction standards, and the impact of by-passes on cities. The chief of the BPR opened the hearing, but for the first time in thirty-four years, Thomas MacDonald was not in the witness chair. After retiring in April 1953, he was replaced by the chairperson of the Delaware Highway Commission, Francis V. du Pont. In his first appearance before Congress, du Pont gingerly sidestepped most questions, explaining that BPR studies on tolls and financing were underway. Du Pont began, however, by commenting, "I doubt that there has been a time in the long life of the Bureau of Public Roads when there have been so many and divergent opinions voiced as to the role of the Federal government in our highway program."[59]

Although the usual postwar diversity of views on highways was heard at the hearing, the striking fact was the emerging agreement on several key issues. Almost no one challenged the importance of the Interstate system, and discord on the traditionally divisive subject of rural roads was muted. Even Keith Seigmuller, spokesperson of the National Association of County Officials and a constant critic of the federal secondary program, observed,

> Over the nine years since the Federal-aid program began we have made considerable adjustments with the Federal Bureau. We have learned considerable about the Bureau, and the Bureau learned considerable about us. We feel . . . the relationship is in pretty good shape. We are satisfied with it. We may have minor complaints from time to time, but that is purely a matter of administration.[60]

A narrowing of opinions was even shown on the question of paying for highways, about which enormous differences existed. Toll financing found few supporters, while fuel taxes seemed to be the key for providing more federal aid. Initially, several witnesses, including truckers and the nation's governors, called for an end to the federal gasoline tax, arguing that gasoline levies ought to be a state matter. But this proposal faced harsh opposition from the committee, which asked how road construction funds should be raised if not from gas taxes. Congressmen were noticeably unsympathetic to the American Trucking Association's argument that it was a "federal responsibility to provide roads," without directly taxing fuel and vehicles. The belief that few states would raise gas taxes if the federal levy were repealed doomed this proposal.[61] But the challenge to federal gasoline taxes brought into relief their potential for supporting federal highways.

In addition to these financial developments, the hearing showed three other areas of almost complete agreement. Every witness concurred that a larger federal highway program was needed. Moreover, most agreed that the Interstate system should become the focus of federal attention because of its transcontinental role and its ability to ease urban congestion. This argument returned to the logic of the 1921 highway bill and accepted the BPR's planning philosophy—federal aid should build the roads of first importance as measured by traffic. Finally, there was almost complete agreement about maintaining the BPR as the guardian of the federal-aid program. Praise for the bureau came from both witnesses and congressmen, with Representative and former governor of New Mexico John Dempsey defending the BPR the few times criticisms were voiced.[62]

The continuing recovery of the BPR's reputation after the harsh denunciations only four years before was interesting in itself, but the praise also helped to explain the emergence of these points of agreement. Every statistic about highway needs, construction progress, or potential costs that surfaced at the hearing came from the BPR. Raleigh V. Gamble, Milwaukee's director of expressways, observed, "I am opposed to any curtailment of the activities of the Bureau of Public Roads. . . . I do not know how we would get along without the engineering and research material that is coming out, and the cooperation of the Bureau."[63] This attitude largely accounted for the acceptance of the Interstate system, the deemphasis of rural roads, and the end of the fascination with tolls. Not one of these points represented a new position for the BPR, although the bureau was not the only source of these ideas, nor their only supporter. The origin of any particular point usually cannot be traced to one individual or organization. But the BPR was relentless in its promotion of the positions it backed, and its information-gathering ability built the cases that convinced the highway community to accept them.

The manner in which the BPR helped build a new consensus was demonstrated by its role in making fuel taxes a basic element in any program to build more roads. The BPR's crucial contribution in this matter was the construction of the connection between these revenues and roads. From the first appearance of state gasoline taxes in Oregon in 1919, they had been justified by the argument that they would raise funds for road construction, a position that received early support from the BPR.[64] But almost as quickly, state legislatures were attracted by the rapidly growing amounts of money these imposts raised, and the BPR actively tried to stop the diversion of highway-user taxes to nonhighway purposes. By the time that the National Highway Users Conference was organized by General Motors head Albert Sloan in 1932 to combat such diversion, federal and state highway officials had been fighting the nonhighway use of this revenue for several years. The decision, approved in the Hayden-Cartwright Act of 1934, to penalize states for diversion by withholding federal-aid funds, aided their efforts. The favored procedure was a state constitutional amendment that earmarked all fuel taxes and occasionally automobile registration and other fees for road construction. Kansas and Missouri passed the first of these amendments in 1928; by 1940, eleven states had followed, and by 1949, ten more. California formalized this plan by creating the first highway trust fund as the repository for gas taxes in 1947. The steady increase in total gasoline tax revenues from $773 million in 1945 to $1.6 billion in 1950 and a similar jump in combined motor vehicle taxes from $1.2 billion to $2.6 billion added a strong incentive to this effort. In sum, by the 1930s state tax policies and penalties against diversion urged by highway engineers had convinced the public that taxes related to motor vehicles should go only to highway work. Steady lobbying by the automobile industry only reinforced this position.[65]

It was more difficult to transfer this policy to federal highway-user taxes, a move that budgetary officials stoutly resisted. The first one-cent federal fuel tax was enacted as a general revenue measure in 1932 and increased to $.015 per gallon during World War II. The steadiness of this source of revenue explained its popularity in Congress as in the states, and calls for its repeal during the federal-aid hearings of the early 1950s went unheeded. By that time, many people informally linked the level of federal highway expenditures to federal fuel tax receipts, and efforts to formally connect them had steadily gained strength. The National Interregional Highway Committee had first noticed such a correlation in 1940, but opposition in the Treasury and the Budget Bureau forced the committee to abandon its attempt to tie construction expenditures to revenues. In 1948, this logic reappeared in a federal-aid proposal that argued it was unfair to punish states for diversion when the federal government had failed to use all of its $800 million in gas tax revenue for roads. In its 1950 and 1952 legislative proposals, the AASHO used gasoline tax receipts as a benchmark in its decision to request $810 million in federal aid.[66]

But the state highway officials and the BPR deliberately avoided *directly* linking tax revenues and federal aid, for they remembered that state highway departments had been crippled when gas tax collections plummeted after rationing was imposed in World War II. Instead, they repeated the Interregional Committee's argument that gas tax receipts were a perfect *indicator* of highway funding needs. After 1950, the AASHO based all of its federal-aid requests on this indirect linkage, and in 1953 the ARBA also endorsed this method. At the congressional hearings that year, both the ARBA and the AASHO claimed that the federal government had collected almost $900 million in gasoline taxes and another $700 million in excise taxes related to vehicles, but had spent only $575 million on highways. Given the widely accepted belief that state highway-user taxes should be used for roads, many members of the House committee also agreed, almost without thinking, to some form of linkage between revenue and federal aid. There were, of course, groups that intensely opposed this concept. Rural representatives felt it was unfair to use taxes on fuel used in farm tractors for roads, while truckers wanted to eliminate the tax altogether. But the linkage was a tidy method of dealing with the rising tide of traffic, since it increased funds without appearing to increase taxes. Not only did critics find these arguments hard to refute, but they also had no serious alternative to present.[67]

To convince Congress to base highway expenditures on user taxes, the BPR typically relied on unassailable statistics and its cooperative ties to the AASHO. The same approach built the other aspects of the new consensus that were visible in the 1953 hearings. The BPR was thus largely, but by no means completely, responsible for the progress toward agreement on the basic issues. The positive tone of 1953 set the stage for the 1954 Federal-Aid Highway Act, which adjusted highway policy in several significant ways. As soon as the marathon hearings of 1953 had adjourned, proposals for the 1954 bill surfaced. All called for more federal aid, and the AASHO submitted one of the first recommendations, this time for $900 million, split into $292.5 million for primary roads, $195 for secondary roads, $162.5 million for urban road, and $250 million for Interstate highways, on a 75 to 25 federal/state matching ratio. The *Engineering News-Record* was cautiously hopeful about this plan, guessing that the request might be too high but adding that "Congressmen promise some surprises when the bill comes up."[68] By February 1954, five other federal-aid bills, calling for $800 million to $850 million based on federal gas tax collections, had been introduced in the House. Each also provided Interstate funds, on a 60 to 40 ratio, and the AASHO's executive secretary noted in April 1954 that the Interstate system was now recognized as "the nerve center" of the economy in both peace and war.[69]

Both the House and Senate held hearings on these bills in February 1954. Because they covered the same ideas presented in 1953, they lasted only three days in the House and six days in the Senate, the latter delayed by rural roads supporters who felt threatened by the emphasis on the Interstate network. But most witnesses believed that the development of urban highways was the primary goal and that the Interstate system was the best means of achieving it. Moreover, they kept advocating the gas tax as the way to finance a larger highway program. Rural opponents of this policy faced pointed demands to suggest an alternative. Even truckers renounced their previously vociferous opposition to federal gas taxes and endorsed the AASHO's plan.[70]

Two issues, however, indicated that the new harmony was not complete. The first was opposition to the state highway officials' argument that population offered the best basis for allocating Interstate funds where they were most needed. In the ten years since the Interregional Committee had adopted this position, it had been endorsed by the BPR, various congressmen, and road supporters; it had been thoroughly discussed during the 1944 hearings and supposedly disposed of in that year's legislation. But senators from rural states still denounced any plan that tampered with the traditional allocation formula of population, area, and road mileage, even when engineers from these states offered no objection. The second issue—utility companies' requests for reimbursements for moving gas, water, and electrical lines during highway construction—was new. Most road builders felt that these newest members of the highway community had little merit, but with policy decisions still centered in Congress, their demands were not easily dismissed.[71]

Nonetheless, agreement about the nature of highway problems and their solutions seemed to be appearing. The hearings, as well as the subsequent deliberations by each house, were relatively brief and amicable. The only commotion was caused by rural senators who opposed giving special attention to Interstate highways, but they were defeated on the Senate floor. The two bills had only minor variations: The Senate version provided $910 million in federal aid, $110 million more than the House, and ignored the House's connection between Interstate funding and the continuation of a $.02 federal gas tax. The conference committee resolved these discrepancies, with the only problem caused by the backstage maneuvering of officials of the administration. Clearly, the divisiveness of the late 1940s was gone, and these bills passed Congress "almost without debate."[72]

The Federal-Aid Highway Act of 1954 provided $875 million in federal aid, divided as follows:

Primary system	$315 million
Secondary system	$210 million
Urban system	$175 million
Interstate system	$175 million

Half of the Interstate funds were allocated using the traditional formula of road mileage, geographic area, and population, and half using only population, with the federal government providing 60 percent of the money. Moreover, the availability of these funds was not contingent on continuing the federal gas tax, thus avoiding a direct link between tax revenues and appropriations. All of these steps meant that the Interstate system was indeed accepted as the focus of the nation's highway policy. Because the states also could allocate both their primary and urban funds to the Interstate system, these roads were finally receiving top priority. This bill, then, came closer to accepting the AASHO's recommendations than any other postwar road legislation. As the *Engineering News-Record* noted, "Congress Fulfills Highway Chiefs' Hopes."[73] Moreover, the bill also commissioned BPR studies on two highway policy issues that had not been fully resolved—the status and prospects of toll roads and utility relocation costs caused by highway construction. The BPR was also asked to codify existing highway laws and draft proposals to simplify the array of legislative requirements that had grown up since the original 1916 bill.

Thus the 1954 legislation was a major step for the highway program. It generally adopted the policy positions of highway engineers. Obviously, the BPR had regained the respect of Congress as the primary source of information on highways, and legislators were again letting the experts use these statistics and other data to answer the difficult questions confronting highway policy. The 1954 federal-aid bill, in combination with events that began immediately after it was signed by President Eisenhower, marked a turning point in American highway policy. The way was clear for building the highway network that the BPR had first advocated almost twenty years before.

The National System of Interstate and Defense Highways, 1954–56

Between 1954 and 1956, the final steps in developing the Interstate highway system that now grids the American landscape were taken. These events are the most widely known chapter in American highway history, thanks to a number of studies that either document the passage of the Federal-Aid Highway Act of 1956, which culminated the process, or use this legislation as the beginning point in an analysis of the impact of modern highways on American society, especially its cities.[74] Therefore, a detailed treatment of the 1956 road bill need not be repeated here. Yet certain aspects must be considered because at first glance it will seem that in the two-year effort to produce the 1956 legislation experts were again rejected as the arbiters of policy and highway policy making was completely politicized for a second time. Appearances can be deceiving, however. It was clear that influence such as Thomas MacDonald had wielded during the 1920s and 1930s was not held by one man in the mid-1950s. After MacDonald retired in 1953, the *Engineering News-Record* predicted, "One thing seems sure: [New bureau chief Francis V.] du Pont is not going to have the free hand in running BPR that MacDonald did."[75] Even so, du Pont and the BPR played a role in shaping a new highway consensus not unlike MacDonald's during 1921.

In many respects, it is incorrect to see 1956 as the pivotal year in highway development, since the basic outlines of a new consensus were clear by 1954. Nonetheless, several crucial details were added in the next biennial bill. The chronology of events leading to these changes begins just months after the 1954 legislation was passed, with a speech by Vice-President Nixon to the National Governor's Conference in July 1954.[76] Speaking for President Eisenhower, Nixon outlined a dramatic $50 billion program to solve highway congestion, a move that for the first time, shifted the initiative on highway policy to the White House and away from the BPR and Congress, thus apparently completing the transformation of roads into a partisan political issue. Intensifying this perception was the inability of White House advisers to agree on specific details, an impasse that led Eisenhower to appoint the Advisory Committee on a National Highway Program, headed by General Lucius Clay. But even while Clay's committee held hearings, quarreling in the administration continued. A final indication that road policy had become a creature of politics was du Pont's resignation as BPR commissioner to become a special assistant to the secretary of commerce, a post from which he intended to push the president's plan through Congress. Du Pont's change of title seemed a clear signal that the control of highway policy had passed from the hands of the engineers to the politicians.

This political maneuvering in Washington was equaled by the activities of the National Governors' Conference, which, like others concerned with roads, was determined to take advantage of the momentum that the president's involvement had provided. The governors announced support for the administration's highway policy and began preparing a plan that renounced their earlier efforts to repeal the federal gas tax, emphasized the importance of the Interstate system, and created a national road financing authority to issue revenue bonds as a supplement for existing funds. Clay's committee heard these proposals, and its report closely resembled the governors'. This was not a coincidence, because du Pont had worked with both groups. Thus the Clay report recommended the creation of a Federal Highway Corporation to finance a ten-year road building program costing $101 billion. About $25 billion was assigned to Interstate roads, with the federal government providing 90 percent of the funds. The corporation, to be composed of the secretaries of the Treasury and commerce and three presidential appointees, had authority to issue $20 billion in bonds and to settle any disputes between the BPR and the states involving Interstate construction and location. Federal aid was to be frozen at existing levels for ten years.[77]

The Clay plan, however, failed to end the highway controversy; in fact, it launched the most overtly political policy struggle in the history of the federal-aid road program. Disagreement emerged along party lines, as Democrats claimed the bond proposal created a "hidden debt." Democratic Senator Harry Byrd, an influential member of the Finance Committee, condemned Clay's plan as "pork-barrel" politics while Republican Senator Albert Gore responded with a bill to provide more than $1 billion in federal aid annually. Even the AASHO disliked giving the corporation veto power over the BPR. At congressional hearings, du Pont, Commerce Secretary Sinclair Weeks, and Clay faced a "barrage of critical questions."[78]

Not surprisingly, the House and Senate passed the most widely divergent highway legislation since 1921. An amended version of the Gore bill that simply increased federal aid emerged from the Democratic Senate. In the Republican House the committee initially accepted the Clay plan, but then redrew the bill completely after hearing sixty-six witnesses from April 18 through June 1, 1955. The resulting compromise, assembled largely by the Chairperson of the House Public Works Committee, George Fallon, still contained pieces of the Clay plan but eliminated the Federal Highway Corporation and bond financing provisions. Even within the committees, however, the disagreements were deep, and for the first time since 1916, both released minority reports. Several senators denounced the Gore bill, claiming it scattered "billions of politically guided dollars over the country for the next five years as though they were shot from a blunderbuss." House committee members produced four reports that attacked specific features of Fallon's plan, especially its financial arrangements. Thus divided, the House rejected both the original Clay plan and Fallon's compromise bill. The impasse was worsened by House irritation at the Senate for initiating highway legislation in violation of the protocol that appropriation bills originate in the House. Before Congress adjourned in August, every road bill, especially Clay's was dead.[79]

Both highway engineers and the press decried this outcome, stressing that it was the result of partisan politics. An editorial in the *New York Herald Tribune* complained, "One suspects that the proposal's really fatal flaw was that it was too good, that the opposition couldn't bear to help the President solve the problem so cleanly and get the credit for it." An editorial in *Life* magazine entitled "Dirty Work on Roads" added that the failure of the Eighty-Fourth Congress "to pass a new highway bill was as scandalous a show of political incompetence as our democracy has seen in years."[80] Highway analyst Duane L. Cronk noted that the deadlock left everyone "stunned," for "the motoring public, primed with high hopes for new highways, was dismayed. Newspaper and magazine editors fumed in exasperation. Congress shrugged its shoulders and adjourned without a hint of what it would do about the national traffic jam in 1956."[81] Yet most observers also expressed optimism about the future. The *Engineering News-Record* believed that "the Eisenhower Administration has done a masterful job on getting across to the nation the demand for a first-class highway system. There is no longer any serious argument, even in Congress, about the need." Moreover, it praised Fallon's bill as a basis for compromise in the next session.[82] The BPR made the same point in its annual report for 1955:

> For years, the Bureau and numerous highway agencies have pointed to the growing highway deficiencies and urged that highway needs be measured, the cost of meeting them determined, and funds provided to do the job within a reasonable period of years. Some additional money was . . . no real solution to the problem. Everyone had agreed that the highway conditions were bad—that something should be done—but there has been no compelling demand from the public that the whole broad problem be faced and measures adopted that would provide a complete solution.

But now, the report continued, the problem had begun to be tackled, and the future looked better because differences had been confined to financial matters.[83]

The optimists proved correct. Fallon prepared both a highway measure and a revenue bill for the Ways and Means Committee to avoid any procedural difficulties about House prerogatives in financial matters, while the president demanded progress on highways in his 1956 State of the Union Address. Eisenhower later endorsed Fallon's plan. By February, one observer said that despite opposition from some groups about three issues—the link between gasoline taxes and federal aid, the application of minimum wage provisions to federal-aid highway contracts, and the reimbursement of utilities for disruptions caused by road construction—"the consensus among many Congressmen is 'Let's get the program and a reasonable plan of financing authorized so the states know where they stand. We can settle the details at leisure next year.' "[84] As a result, the bill sailed through Congress without the acrimony and turmoil of 1955. The House Ways and Means Committee approved higher federal excise taxes on tires and vehicles and a $.01 increase in fuel taxes, but more importantly, it created a highway trust fund for all highway-user taxes, thus completing the linkage of highway spending with highway-user revenues in a way that did not tie each year's expenditures to revenue collections. Fallon shepherded a federal-aid bill through the House that provided for partial utility reimbursement, higher usage fees for heavy trucks, a federal cost-sharing ratio of 90 percent for Interstate highways, and larger appropriations to all other federal-aid programs. The Senate amended the Gore bill to bring it into line with the Fallon's plan. A conference committee had little to iron out, and the president signed the bill on June 29, 1956.

The final plan provided $25 billion for Interstate roads, on a 90 to 10 matching basis. For 1957, 1958, and 1959, half of the funds were distributed on the traditional formula of mileage, area, and population and half on population alone. Thereafter, funds were to be allocated according to the needs of the system, which was expanded by 1,000 miles and renamed the National System of Interstate and Defense Highways. It was to be completed in 1969, with the taxes lapsing three years later. The bill also reaffirmed advance acquisition of land with federal funds and increased regular federal-aid appropriations as well. Instead of the $700 million in federal-aid originally provided in the 1954 bill, $825 million was designated for 1957, $850 million for 1958, and $875 million for 1959.

Highway observers praised this achievement; one commentator suggested that "the unsavory political aura that had hung over the bill since it was first introduced by Congressman Fallon in January cleared away amazingly fast as 'good roads' advocates on both sides of the aisle came to their feet in praise of the measure all through the two days of debate.[85] Congress congratulated itself for an effort marked by moderation and compromise. But Mark Rose has shown that the reality was simpler: The bill provided something for everyone without forcing any radical shifts in principle.[86] This was the perfect political solution, as the highway engineers had recognized earlier, but it also suggested that the long-time goal of having experts plan a rational highway program based on genuine needs had been subordinated to political calculations. To be sure, all earlier federal-aid highway programs contained such compromises, but not to the extent of the 1956 legislation. As if to cement the apparent submergence of highway policy into politics, Congress accepted a plan from the administration to establish a federal highway administrator, appointed by the president, as head of the BPR. The commerce secretary argued that the bureau's commissioner should handle the immense day-to-day task of overseeing federal-aid construction, while the president's appointee should tackle larger administrative problems and policy questions. Moreover, this administrator was not required to have a technical background; political considerations assumed much more importance in this appointment. The bill passed easily.[87]

Politics or Expertise?

The partisan struggle over the 1956 Federal-Aid Highway Act should not obscure the role played by technical expertise in shaping this new policy. Political maneuvering certainly affected the timing of the bill's passage, and the dual financial strategies of more money for all programs and a 90 percent federal contribution was an essential aspect of its acceptability. Almost lost in the political battle, however, were the bill's contents, which matched the ideas of highway engineers at the BPR and the AASHO in every detail. One factor that disguised their role was Francis V. du Pont's enforcement of an interpretation of the concept of apolitical expertise that seemed to differ from his predecessor's. MacDonald had always claimed to be guided by the general principle that highway matters should be left to experts. In practice, however, this meant that he used his personal ties to Congress and the BPR's distinguished reputation as the engineering leader of American highways to dominate the political arena. For MacDonald, the rhetoric of expertise worked, even when he accepted such political concessions as the federal-aid allocation formula, which gave every state money. He justified these deviations from the ideal as being in the interest of a smoothly running administrative system, recognizing the practical impossibility of producing a plan that distributed funds strictly on the basis of needs.

Du Pont, however, assumed that engineers were nonpolitical rather than apolitical; they should stay out of politics completely and not attempt to replace politicians as policy makers. This attitude explained his resignation as commissioner of the BPR in December 1954. To du Pont, highway policy had become a strictly political problem that only Congress could resolve, so when President Eisenhower asked him to act as liaison to Congress and general adviser on highway legislation, he moved to a political post in the Commerce Department. Du Pont made his thinking clear in a 1953 address to the AASHO's annual convention:

> As you well know, Public Roads is purely an administrative agency. It has no policy making role or legislative responsibility. These are the joint responsibility of the administration and the Congress. . . Regardless of what may be said at this and subsequent meetings during this convention, the Congress and only the Congress, with Presidential approval or over the veto, can tell the future Federal highway program.[88]

Du Pont placed special stress on this point, drawing an apparently sharp distinction between his and MacDonald's administration of the BPR. At another point in that 1953 address, he refused even to predict what Congress might do about highways, and in 1954 du Pont told the state officials to concentrate on building the Interstate system with existing funds and not to campaign for more money. Implicitly, he suggested they ignore politics.[89]

In spite of these apparent differences, du Pont's position was in reality a modern restatement of MacDonald's Progressive ideal that apolitical expertise should shape policy. Obviously matters had changed since the 1920s and 1930s, when the BPR had largely convinced the Congress to leave highway matters to the experts. Du Pont lacked MacDonald's lofty Progressive vision of the proper role of experts and, like most engineers trained after 1920, he considered experts not a replacement for, but as subservient to policy makers. Engineers stood above the political fray, ready to implement the policy decisions made by others. This view seems to contradict MacDonald's completely, but actually there was a fundamental connection between the two positions, for American engineers before and after retained a self-image rooted in MacDonald's philosophy—an expert was unbiased and objective. Thus du Pont, like MacDonald, believed that experts should provide Congress with unbiased information and objective advice.

Not surprisingly, MacDonald and du Pont adopted different strategies for performing this task. Du Pont lacked MacDonald's enormous personal reputation and the contacts that came with thirty-four years in the same position. Therefore his circumspect dealings with Congress contrasted sharply with MacDonald's willingness to dispute views he considered unwise or dangerous. When the Senate Roads Committee asked du Pont to prepare language to remedy a problem in one 1953 bill, he demurred, saying he had to wait for the administration to define its program.[90] Although MacDonald observed similar boundaries, he often provided such services. Du Pont, on the other hand, presented the BPR as a neutral source of statistical information and engineering advice that lawmakers could fashion into policy. This strictly technical, nonpolitical guise, was perhaps a response to MacDonald's role as both a technical adviser and a principal in the political dispute over rural roads. Yet by brandishing a more narrowly defined image of apolitical expertise and apparently disengaging the BPR from politics, du Pont actually strengthened the ability of federal engineers to affect highway policy.

This outcome was once again attributable to a gap between the rhetoric and the reality of expertise and politics. In this case, du Pont, despite his public statements about nonpolitical engineers, never stood aside and left the formulation of highway policy to Congress. As commissioner of the BPR, he was a leading participant in the 1954 White House discussions that led Dwight Eisenhower to enter the fray over highways. His primary policy aim never varied, for both as commissioner and as special assistant to the commerce secretary, he argued that the Interstate system was important enough to justify even 100 percent federal funding. He continued MacDonald's opposition to tolls, fending off the argument of Under Secretary of Commerce for Transportation Robert Murray that tolls should pay for the Interstate network. As always, du Pont's ability to provide statistical data, estimates, and the supporting opinions of state highway engineers strengthened his position. State highway officials, including New York's Bertram Tallamy, were twice called upon by the White House for information on highway needs during the 1954 discussions that shaped Eisenhower's highway plan, and they always echoed the BPR.[91] In short, the nonpolitical engineer Francis du Pont was following in MacDonald's footsteps.

The studies that Congress had directed the BPR to prepare in 1954 gave the bureau additional leverage in policy questions. The first was a thirty-year projection of highway needs and finances, based on a cooperative study with the state highway departments. The BPR had placed existing deficiencies at $101 billion, with the Interstate system accounting for $23.2 billion.[92] The other study was a report on toll roads. One engineering journal had predicted that with du Pont's arrival, the "uncompromising opposition to toll roads on the part of the BPR may be a thing of the past." But the only serious alteration in the BPR's attitude was the end to "uncompromising opposition."[93] The bureau believed more miles of toll roads were economically viable than *Toll Roads and Free Roads* had proposed in 1939; the new total was 6,900 miles, of which 2,621 were built or under construction. But the BPR's study also reiterated that parallel free roads remained essential, for toll roads could serve only part of the traffic. Thus the bureau continued to oppose using federal-aid funds for toll roads, although it urged including existing toll roads in the Interstate system. As if to verify the BPR's position, in 1955 the West Virginia Turnpike failed to make its bond payments, the revenues on the Pennsylvania Turnpike declined for the first time, and truckers threatened to boycott the Ohio Turnpike in a protest over tolls. One observer echoed the BPR's position of 1939 by concluding that "the advantages of thousand-mile distances without traffic lights or intersections have apparently been much over-rated."[94]

These BPR reports provided the basic information on which the framers' of the legislative proposals of 1955 and 1956 depended, beginning with the Clay committee. The BPR explained in 1955 that "throughout the period of intense public study and debate, the Bureau was constantly called upon for basic factual information necessary for intelligent study of the problem. As a result of an earlier directive from Congress, the Bureau was able to supply the President's Advisory Committee with data as to the extent of needs on the various highway systems.[95] As another account said, the Clay committee relied on "statistical ammunition being compiled by the Bureau of Public Roads."[96] Clay adopted without challenge such basic information as the BPR's figure on the federal aid needed for highways—$101 billion. Moreover, the plan Clay outlined for the Interstate network almost certainly reflected du Pont's influence. This same data, which was the product of a joint BPR-AASHO project, also guided the National Governors' Conference report, for the governors turned to the state highway officials "for advice as to what type and size of road-building program they should endorse."[97]

The respect given the BPR's data was such that even though the financial schemes proposed by Clay and the governors were roundly attacked in Congress, the information on which they relied went unchallenged. Thus both roads committees also relied on the BPR for data. According to the bureau, "Four separate reports on highway matters were presented to Congress and considered by committees. When committees began to consider specific legislation, the Bureau was called upon to prepare statistical reports showing the division of proposed funds among the States according to various methods of apportionment, analysis of bond financing and retirement proposals, and the yield of possible sources of revenue over a period of years."[98] In other words, BPR data shaped the 1956 bill.

The BPR provided more than statistics, for it was not simply the collection agency of du Pont's description, gathering data that others turned into policy statements. Instead, the reports clearly interpreted the material, outlining the BPR's position on the issue at hand. The report on toll roads, for example, defended the BPR's suggestions that construction of toll roads with federal aid be prohibited and that existing toll roads be included on the Interstate system. The report on highway needs clearly showed the BPR's emphasis on the Interstate network. The only report that avoided policy statements was the survey of utility relocation costs, which were shown to be extensive on Interstate construction in the cities. The bureau took no stand on this issue, but because of the report, Congress permitted the use of federal aid for as much as half of these expenses. In the end, both the data in the BPR reports and the positions they supported were embodied in the new law.[99]

Other elements of the 1956 bill also embraced positions that the BPR had promoted over the years. The highway trust fund, for example, extended the BPR-AASHO position on the diversion of highway-user fees in a way that preserved an indirect link between such revenue and federal aid. The bill permitted federal funds to be used for advance acquisition of rights of way, an idea the BPR had suggested in *Toll Roads and Free Roads* in 1939. The federal-aid administrative arrangement—state construction under BPR review—was also retained; no other administrative plan received consideration. A study by President Eisenhower's Commission on Intergovernmental Relations even endorsed the BPR as a model for efforts to limit federal intervention in the states. The importance of the existing administrative style was brought out in the debate about creating a federal highway administrator, as one observer noted that "a number of Congressmen acceded to the move on the condition" that no change would occur in "the historic working relationship between the states and the bureau.[100]

Clearly, Congress had found renewed faith in both the BPR's administration of federal aid and its policy positions. During a 1955 hearing on the fourth BPR report to Congress, a draft bill consolidating the thirty-six highway bills passed since 1916, Congressman George Dondero of Michigan stated, "I have always had such implicit faith in the Bureau of Public Roads that I dare not hazard a question lest I might exhibit my ignorance of what you are trying to do."[101] It was not surprising that in 1956 Congress asked the BPR to study the one remaining problem—how to reimburse the states for toll roads that were to be included in the Interstate system.[102]

The BPR thus influenced most major features of the Federal-Aid Highway Act of 1956, including the routes chosen, the extent of the system, the determination of highway needs using traffic statistics, the focus on urban congestion, the rejection of toll financing, and the indirect connection of highway-use taxes and federal expenditures. Many of these issues had been discussed first in *Toll Roads and Free Roads*, although the 1939 study was not a true blueprint for the system because MacDonald's vision had never included cross-country superhighways. Du Pont had a different perspective that reflected the changed conditions of the postwar period. Traffic projections had always fallen far short of actual traffic use, and he was determined that the system would not be obsolete when completed in 1969. This outlook explained the bill's only significant departure from the BPR's initial plan for interregional highways—its provision requiring uniformly high standards for Interstate roads, regardless of traffic projections. However, the most basic feature of the 1939 plan—its urban orientation—was finally assured by the 1956 legislation.

Federal highway engineers had at last shaped a new consensus for the highway community, as G. Donald Kennedy explained in the quotation at the opening of this chapter: "How did it happen? . . . All the people we are talking about were engineers." [103] The key was financing, for like the 1921 highway legislation, everyone was given a fixed share of the available resources. Moreover, the cooperative system still made everyone *feel* they had a stake in the entire structure. To be sure, politics was more important in highway policy than it had been before the war, for such elements of the consensus as utility relocation reimbursements and the funding procedure revealed the compromises required to build the agreement. Similarly, rural road work continued despite calls by experts outside the BPR for focusing federal involvement on the roads of national importance.[104] No concentration of funds took place, yet this "failure" put all highway problems under the federal-aid umbrella and largely accounted for the first real consensus since 1935. As they had always done, to greater or lesser degrees, the leading engineers at the Bureau of Public Roads resorted to politics, but without disturbing their reputation as objective experts. This reputation was more important than the political strategy in accounting for their success, for with it they resolved the issues left open by the expedient of throwing money at highway problems, such as how and where to spend the funds. As historian Richard O. Davies explained, they had again achieved true unity:

> Virtually no one challenged the [1956 highway] system when it was established. . . . Even in the heyday of Republican fiscal conservatism few dared challenge the enormous costs of the program, despite the fact that it would be the biggest and most expensive public works project in history; the much-maligned expenditures of Franklin Roosevelt's entire New Deal paled in comparison. No less an authority than the archconservative Secretary of the Treasury George M. Humphrey gave it his fiscal blessing: "America lives on wheels, and we have to provide the highways to keep America living on wheels and keep the kind and form of life we want." [105]

With these words, the most impressive highway boom in American history was launched.

Conclusion

Highway Engineers as Policy Makers

American highway development from the 1890s through 1956 followed a path with many twists and turns, yet basic continuity can be found in three main features. The first was the steady shift from a rural to an urban focus; the second was the ability of a group of experts in the federal government's highway agency to guide both the general goals and the details of highway policy. The third constant was nearly unanimous public support for road building. This last feature undergirded the entire process and formed the background against which policy was framed and roads were built.

The most constant element of American highway policy was the continuously changing justification for roads. The agitation for better roads that began with demands by bicyclists for routes into the countryside grew into a Progressive reform movement to improve the quality of rural life through easier access to markets and towns. Roads were given both economic and social importance and labeled an agent of democracy bringing farmers the rights, privileges, and luxuries urban Americans took for granted. After 1910, this reforming impulse swung from demands for action by local and state governments to calls for federal assistance, which was finally provided in 1916. Within five years, however, the Progressive Era's moral concerns were overshadowed by a new rationale for highway work—the need to serve the automobile. The Federal Highway Act of 1921 centered federal activity on the construction of a national network of primary roads. The initial rural emphasis was not totally abandoned, for the new system stopped at city lines. Thus while the federal-aid system began to connect cities, it retained its original goal of linking rural America to these cities. Equally important, federal aid after 1920 produced professional highway departments in most states that adopted the BPR's newer and narrower technical definition of road building. Indeed, the rapid acceptance of the automobile gave this work the air of a popular crusade, with the only consistent complaints being raised by railroads, and there were easily dismissed as selfish.[1]

Construction of the primary system united the road community until the Depression turned road construction into the country's largest public works project. Both urban and secondary highways were included in the New Deal's unemployment programs, a decision that legitimized demands to add these highways to the federal-aid system. The growing traffic congestion in the cities also supported this extension of federal funds. After 1936, the BPR worked to shift attention to urban highways, but progress was tortured and slow. A BPR master plan in 1939 was ignored, in part because the approaching war turned highway work into preparedness programs. In 1944 legislation did recognize city needs, but outright political resistance to funding urban highways insured that support for rural roads was not disturbed. Only after this opposition was mollified and deflected did urban road building receive unchallenged first priority. The culmination of this process was the Interestate highway system, which funded urban expressways as part of a network of trunkline highways.

The second constant in this history was the role played by experts, primarily federal highway engineers, in guiding both road construction and overall policies. For sixty years, four directors built and maintained the capability, skills, and information base that gave the BPR undisputed technical leadership. From the beginning, this agency was a model of the Progressive belief that unbiased, technically trained specialists were the ideal shapers of society. Thus engineer Logan Page, portraying himself as removed by definition from politics, promoted the state and federal legislation that created highway systems and funded construction from 1905 through 1918. The depth of his involvement in politics and even his tactics seemed to contradict his claim of objectivity, but his desire to base the crucial decisions about roads on technical data and the absence of partisan political goals conformed to the Progressive style. This same Progressive ideal has since surrounded American engineers and shaped their self-definition. The BPR exemplified the survival of this image, as Thomas MacDonald consistently presented the BPR as a model of unbiased expertise. The acceptance of this image, which relied on the technical superiority evident in the bureau's research and standardization programs, enabled his priorities to become the national policy: to use federal aid to build a system of primary roads as a joint project of the state highway departments and the BPR. The strength of this consensus showed in both its nearly unanimous acceptance from 1921 through the mid-1930s and the widespread belief that state highway departments were equal partners in this effort despite obvious indications the BPR shaped technical and most other decisions.

The breakdown of MacDonald's consensus in the mid-1930s seemed to reflect a breakdown in the ability of expertise to determine highway policy. The problems began with the disintegration of unity in the highway community because of technical disputes over toll roads, transcontinental superhighways, and the relative highway needs of cities versus rural areas. The BPR attempted but failed to resolve these conflicts with its superior expertise. Similarly, efforts to base a new policy on highway planning data were unable to reunite the highway community, and controversies between experts spilled into the political arena.

The BPR's inability to rebuild a consensus after 1936 should not be seen, however, as a repudiation of the ideology of experts. First, road builders were confronting constantly changing conditions. In rapid succession, the Depression, World War II, the postwar inflationary boom, and finally rationing during the Korean conflict imposed severe strains on the highway community. In only sixteen years, highway priorities shifted from building a primary system to providing jobs to conducting a very limited defense road program to building roads as fast as possible after several years of inattention amidst rapid inflation and materials shortages. These changes encouraged the reemergence of overt political struggles over highways. Second, changes in the highway community itself, which expanded in size steadily after 1932, hampered the efforts of experts to shape policy. First, urban and secondary roads appeared in the federal-aid picture in 1936 and 1938. After 1940, urban planners, social thinkers, chambers of commerce, economists pursuing Keynesian theories of government spending, bureaucrats in departments such as State, Agriculture, and Interior, and advocates of other special interests such as utilities and labor all demanded a voice in shaping policy. In June 1955, President Eisenhower's Commission on Intergovernmental Relations described the situation perfectly:

> There is now some tendency to overlook the original purposes of the Federal-aid program and to regard them as merely a means of providing financial assistance to the States for highway construction. Actually as the total program has grown by superimposition of one program upon another, it has become difficult to discern its underlying objectives. It is now a program of broad scope that materially assists in highway improvement throughout the country, but which never reaches any definite goal.[2]

Yet in spite of the renewed competion within the highway community, the policy that evolved between 1950 and 1956 was inspired by technical experts in the AASHO and the BPR. By employing a political strategy—giving more money to all—that removed the most contentious issue, they were free to decide the technical issues and even such potentially divisive matters as the allocation of the larger sums. The BPR had first suggested most features of the new highway policy in its 1939 master plan, although some had been modified by 1956. Essential to the bureau's success was the highway engineers' appeal to their image of apolitical experts.

The striking aspect of this outcome is its congruence with John G. Gunnel's discussion of technocracy. Technocracy usually conjures up an image of technology or technological systems threatening the existence of democracy, especially as experts have pursued technological rationality. Gunnel, however, suggested that the real problem was the inability of democratic systems to resolve the difficult problems associated with complex, large-scale technologies. When political institutions become deadlocked on these issues, he argued, the proposals of experts are seen as promising solutions. Their technical expertise, which is assumed to be politically neutral, grants them and their solutions legitimacy.[3]

Gunnel's analysis explains the resolution of both periods of turmoil in American highway policy: the 1918–21 era and the protracted struggle of 1938–56. In both instances, competition among advocates of various types of roads forced the issue into Congress, where it became tangled in politics. This in itself is significant, for the engineers who helped shape the initial structure of federal highway policy always considered the congressional roads committees to be members of the highway community. But once the system was functioning and the major questions had been answered, engineers expected the politicians, in the best Progressive manner, to step back and leave the "technical" decisions about highway location, standards, and priorities to the cooperative structure of federal aid. The emergence of competing technical priorities during periods of unusual conditions—World War I, the Depression, and World War II—twice destroyed this system. Yet once the disputes found their way back to Congress, that body also proved unable to resolve them. Congressmen soon tired of trying to appease everyone, as was evident in the lament of Senator John McClellan during the federal-aid hearings in 1950. "You can well appreciate the problem of this committee and of the Congress, when one witness comes before us and urges that rural roads to get the farmer out of the mud is the most critical of our problems, and the next witness says we have to do something about the cities, and we have to handle this congestion in the urban areas, and then you present to us the argument that the interstate roads should have priority. I wonder how we are going to decide it."[4]

Eventually, Congress answered McClellan by turning to the recognized expert, the Bureau of Public Roads and its sidekick, the American Association of State Highway Officials. After 1918, Thomas MacDonald assembled a consensus that guided road builders from 1921 through the 1930s. He succeeded in building a sense of shared interests among road builders and their supporters and then maintained this agreement through cooperative structures that lessened the likelihood of controversy. By preserving harmony within the field, MacDonald hoped to keep highway decisions away from politicians and permit experts, who were presumably immune to corruption, to find the most efficient technical answers to problems. His success rested on the BPR's reputation for unbiased expertise, at least until 1936.

The highway policy controversy that raged from the late 1930s through 1956 was somewhat different in that the issues had become more complex and the highway community had increased in size. This situation accounted for the length of the dispute. Yet in the end, the groundwork for shifting highway policy from a rural to an urban orientation was laid by MacDonald. From the late 1930s, engineers claimed that the rational, impartial study of current and projected traffic conditions held the solution to highway problems. MacDonald summarized this approach in his last address to the AASHO in December 1952. "It is important before deciding upon the guides to follow in determining our recommendations for legislation or new administrative policies, to review the factual and research data from the studies of the State Highway Departments and the Bureau of Public Roads which will provide a sound foundation for public understanding of our critical highway situation and clear some of the foggy ideas now prevalent.[5]

By the mid-1950s, many people found the BPR's narrowly technical view of highway policy an appealing solution to the seemingly endless political debate over roads, especially because they could already witness the results of relying on the experts at the state level. In 1942, the BPR had devised a system for determining how well a highway served traffic by using a point scale that weighed structural condition, traffic capability, and safety. According to *Business Week*, this was "a scientific, statistical measure of highway conditions and needs." By ranking the scores for each road, state highway departments could identify those most in need of attention without causing a furor. Twenty-two states had adopted this approach to determining highway priorities by 1952 because it was "a way to take the road building program out of the pork barrel." In Arizona, this sufficiency formula transformed the annual highway department announcement of its construction program from "hectic affairs" with crowds stretched into the hallway near the hearing room into "routine, orderly sessions attended by only a handful of people,"[6] As Gunnel suggested, the system preferred such order to chaos. Legislators complained in 1956 that "the highways bill has become a gigantic grab bag. We're through drafting a road program; now we're debating 'pork'—pure and simple."[7] An *Engineering News-Record* editorial that had earlier predicted "normal politics will not produce the abnormal action that is needed" sounded correct.[8] Congress turned to the BPR, which had a plan, the crucial features of which were not new but which purported to be a neutral technical answer. Highway policy, then, offers an instructive insight into the way American society has chosen to solve the problems created by technological change.

This presentation of the role of expertise, however, fails to convey the complete story of the BPR's ability to use its image of objectivity to gain dominance in policy circles. A significant gap existed between their rhetoric and their actions, for clearly political activities were intimately involved in their guidance of policy choices. The BPR engineers were always active in the political arena, although they claimed they were different from their fellow participants because of their technical training. MacDonald especially meshed the cooperative aspects of both the Progressive and associative political ideologies with the image of expertise to shape policy. This combination was visible in both 1921 and the 1950s, for the key to maintaining consensus proved to be systems that prevented biennial squabbles over money. Although presented in technical terms, this was certainly a political answer, a compromise of technical efficiency to provide stability. In 1916 and 1921, the formula of population, land area, and post-road mileage was installed for allocating federal aid, and it insured every state a share of money without opening a divisive debate every two years. The establishment of the highway trust fund in 1956, which earmarked taxes on gasoline, tires, and diesel fuel for road building, ended the dissension over highway priorities by ending a period of limited resources. Moreoever, the money in the trust fund was still divided using variations on the original federal-aid formula. Both plans attempted to limit political maneuvering over highways by using apparently rational financial structures that actually represented sophisticated political compromises. When combined with another political stabilizer, the cooperative system of sharing power with the states, or federalism, the system proved politically harmonious.

This analysis forces a consideration of the political context of technological developments, especially in a democracy. Often this becomes only a precondition for discussing innovation, but the case of highways clearly shows how political activity impinges on technology. Indeed, what emerges is an understanding of the sensitivity of technological undertakings to political and social changes, for shifting economic conditions proved especially capable of buffeting the best plans of road builders. Even more unpredictable was the impact of wars, for both major conflicts upset not only road construction but also the entire structure of the highway community. The conditions under which these experts operated were thus delicately balanced. Even a master juggler such as Thomas MacDonald was still at the mercy of events beyond his control, regardless of the momentum of his technology and the astounding public support it enjoyed.

Thus while experts dominated highway policy, they always operated within narrow constraints. They were not technocrats in the usual sense, for their influence stemmed from the failure of democratic institutions to make reasoned choices about technology. Even in 1956, there existed alternatives to the BPR's ideas for highway construction, but the image of the BPR's superiority eliminated all contenders. Yet since 1956, the results of the BPR's narrowly technical style have become apparent. The "planless planning" that characterized the accommodation of the automobile and that failed to consider comprehensive transportation systems suggests the lack of vision of the experts to whom highway policy was entrusted. Yet while the experts can be blamed for being wroing, they did not assume their policy-making roles because of surreptitious plotting. Indeed, their approach to highway construction found significant support in American society. Hence in 1921 and 1956, Congress removed itself from the highway policy arena by returning control to the experts, who acted as if the action of democracy must be limited to establish a working technological system. As if to prove the engineers were correct, both contentitious highway disputes gave way to long-lived road-building booms.

Yet 1956 was the year of the BPR's last starring performance in the policy arena. Road builders moved ahead on the Interstate system in a fashion that resembled the building of the original primary system during the 1920s. The results, measured by technical standards, were impressive, as the 41,000-mile system spread across the landscape with little attention to topography or climate. The ease of interregional travel was the obvious benefit of these roads, but the completion of sections of Interstate expressways through cities was also applauded, accompanied as they were by the first studies predicting an increase in the value of adjoining property. More generally, with the end of the rancorous political discord that had become a hallmark of postwar highway policy, Congress returned to the position it had occupied in the 1920s. Hearings became reviews of progress and discussions of the AASHO's recommendations for the next biennial federal-aid bill, not challenges to the plans and approaches outlined by the BPR.

But below the surface, the BPR's traditional position as arbiter of highway policy was slipping away in the 1960s. Among the reasons for this change was the first serious public criticism of highway policy, particularly of the urban segments of the Interstate system, which displaced more people than any previous highway construction and caused vehement opposition in those neighborhoods split by concrete and steel corridors. Building for high-speed travel, a goal that consumed huge areas of land, was the culprit, and it eroded the support of a significant number of Americans, although by no means a majority, for road building. Protests focused on the failure to contemplate alternatives to highways and on the exclusion of the public from the planning process. At times, opponents discovered that the federal-aid system, with its ambiguities regarding ultimate responsibility, permitted both state and federal engineers to claim that the decisions regarding routes and standards had been made by the other. As road work continued despite these protests, opposition increased with this sense of powerlessness. The BPR belatedly implemented a planning process in the mid-1960s that considered broader social factors, and environmental impact statements were required after 1969. But public support had already been damaged, as the system that had limited the action of democracy to produce a working technology had altered many people's opinions about highway policy. One scholar examining roads in Syracuse, New York, condemned the 1956 legislation as the most counterproductive federal program for the cities.[9]

Yet this erosion in the nearly unanimous support for highway construction did not by itself explain the BPR's loss of influence. Another factor was its increasing bureaucratization, symbolized by the installation of a political appointee as head of the bureau in 1956 and the creation of a Department of Transportation in 1968. The latter move granted the BPR, later known as the Federal Highway Administration, a higher political profile. Ironically, it also lowered its stature in the policy realm, partly because of the increasing size of the highway community, one of the factors that had made the building of a postwar consensus so difficult. Despite the calls for a focused federal program, the Federal Highway Administration continued to oversee all facets of road building. When highways were placed under the control of the Transportation Department, new competitive pressures emerged as highways competed with mass transit for funds and priority.

The most important factor in the fall of the BPR from its position as arbiter of highway policy, however, was the decline in its technical stature. This slow process operated over the entire postwar era, and was evident only by comparison with the bureau's commanding technical reputation before World War II. One indication was the AASHO's emergence as a truly equal research partner, symbolized by the 1954 decision to stop appointing BPR engineers as secretaries of AASHO committees. The BPR's encouragement of the development of fully staffed research facilities in every state department had made this possible, and the results appeared in the association's ability to conduct major research projects during the 1950s that laid the basis for Interstate construction standards.[10]

As the BPR's reputation declined, the position of research arbiter was gradually assumed by the Highway Research Board. The changing balance of technical expertise was indicated by the HRB's supervision of several major postwar investigations and the formation of the Research Correlation Service after World War II. Under this program, each state highway department gave the HRB a portion of its federal-aid funds earmarked for research to permit more rapid dissemination of research results. With this guaranteed income, the HRB finally became independent of the BPR. Every year from 1920 through 1945, the bureau had furnished more than half of the board's budget but in the postwar period 60 percent of the HRB's funds came from state highway departments and only 14 percent from the BPR. The board's reliance on BPR engineers as researchers declined as well, as the HRB began to maintain its own staff. Moreover, fewer BPR engineers served as members and chairpersons of HRB research groups. In 1944, 30 percent of the board's committee chairs and 17 percent of its committee assignments had been held by BPR engineers, and these numbers had not changed since 1920. But in 1954,

Table 5
BPR Support of the HRB, 1945–70

Year	BPR Contribution	Total HRB Budget
1945	$ 20,000	$ 39,974
1946	20,000	107,863
1948	20,000	134,722
1951	30,000	183,837
1956	36,000	251,800
1958	45,000	308,519
1960	56,250	420,030
1961	61,875	454,562
1962	86,875	759,147
1965	156,375	900,126
1968	322,625	1,376,000
1970	322,625	1,675,000

Source: National Academy of Sciences, National Research Council, HRB, *Ideas and Actions: A History of the Highway Research Board, 1920–1970* (Washington, D.C., 1971), p. 136.

the numbers were 26 percent and 11 percent, respectively, and in 1964, 17 percent and 8 percent, respectively. By 1968, the BPR held only 12 percent of the committee chairs and 5 percent of the total committee posts.[11]

The growing importance of the Highway Research Board and the maturing of the research efforts in state highway departments were symptoms, not causes, of the declining stature of the BPR as a research institution. This is not to say that the BPR's research programs collapsed, for it still conducted the largest and most diversified research effort in the country. But in addition to seeing its near-monopoly of research end after the war (again, a change it encouraged), the BPR watched the departure of the individuals who had built that research reputation. By 1950, the engineers who had begun their careers in the BPR during the Progressive period were reaching mandatory retirement age. In 1950, L. I. Hewes, the BPR's west coast deputy commissioner since 1916, died, and in 1954, E. W. James, who handled all of the BPR's overseas programs and had drawn up the first federal-aid rules, retired. H. S. Fairbank, MacDonald's most trusted assistant, left about the same time. But with MacDonald's retirement in 1953, the accolades rained down. He was called "Mr. Highways" in one account and the man who "revised our road maps" in another; the *Engineering News-Record* said that "he originated, developed, and operated a post of great public trust with integrity and according to principles that produced a vast road system with remarkable efficiency and economy. The 'Chief' retires, but his precepts and examples will continue to influence roadbuilding and administration for many years to come."[12]

The technical ability of these men built the reputation of the BPR; they were the reason that the bureau could report in 1954 that "states continued to consult specialists of the Bureau in the Division headquarters offices on the design of large and complex projects such as expressways and major bridges, but they did this, in general, because they desire the consultation rather than to meet a requirement. In some instances the advice of the Bureau was sought on plans for non-Federal-aid projects."[13] One of Francis du Pont's first announcements requested this "old guard" to step aside. Although he did not precipitously change the administrative structure, he also commissioned an outside management audit and then slowly rearranged the various divisions. Part of this was in keeping with the economy-minded goals of the Eisenhower administration, for du Pont reduced overhead expenses in 1953 by $400,000, mostly by eliminating 128 permanent positions. One of the areas affected was research, for du Pont had directed the auditors to examine the program and cut projects that had little promise of producing immediate results. These cuts combined with retirements meant that some of its most respected engineers left the BPR at about the same time.[14]

Probably more important than personnel turnover in undercutting the bureau's reputation for expertise was the change in the mission of the BPR, especially after the passage of the 1956 legislation. The crucial shift was precipated by the allocation of an unprecedented amount of money for highways. Increases for regular federal-aid systems would have caused significant administrative problems as the BPR attempted to maintain its traditional standards of inspection and its emphasis on integrity. But the addition of an entirely new network of 41,000 miles of expressways built to the highest standards raised new challenges. Inspecting the plans and designs of these highways became the primary item on the BPR's agenda. To meet these demands, the BPR further decentralized its approval procedures, giving more authority to its division offices and expanding the district offices in each state. But the task also called for more engineers. In 1956, 2,200 of the BPR's 3,100 employees were in the nine division and fifty-two district offices. Of these, 1,100 were engineers, but the BPR hoped to hire 900 more and was permitted to appoint nineteen new supergrade administrators when the post of Federal Highway Administrator was created. Their assignment was not to conduct research but to inspect and approve state specifications, plans, and estimates.[15]

This administrative task, although always central to the BPR, now became even more important as the Bureau annually authorized more than $2.5 billion in highway funds by 1958. By way of contrast, Thomas MacDonald had overseen the distribution of $6.6 billion in federal-aid highway funds from 1919 to 1953.[16] The mere size of the highway appropriations for which it was responsible after 1956 all but demanded major changes in how the bureau defined its mission. But a greater threat to the BPR's position and reputation was posed not by the greater administrative load but by the change in the ratio of state to federal funds in the Interstate program. Connecticut highway engineer William Cox had predicted in 1943 that a federal contribution above 50 percent would alter the basic relationship between the federal and state highway engineers, and Thomas MacDonald later worried about the effects of a 90 to 10 split.[17] Both proved correct, for as the federal share of such funding rose, a cornerstone of federalism—cooperation—disappeared. In 1957, the BPR announced it was hiring appraisers to insure that federal involvement in land sales did not produce outrageous prices, and it also formed a project examination division to spot check the procedures used by both the state and federal engineers. Inevitably, federal inspectors felt more responsible for seeing that projects funded almost completely with federal aid were designed and built correctly. Even when inspections continued to be conducted in the cooperative style established under MacDonald, the large federal presence in the state highway departments disrupted the smooth relationships of the prewar era.

The BPR, in short, had become a huge bureacracy in which the balance of state-federal authority was tipped by the power of the purse. Ironically, this came when the states were much more capable of assuming a more equal position in the system. No longer could the BPR intrude in state highway departments without provoking complaints, as it had in the 1920s. Appearances now mattered, but as an enormous bureaucracy the BPR was simply less willing to consider state sensibilities. For that reason, the cadre of federal engineers who had held the respect of the entire highway profession was missed even more after 1956, because the BPR had few people of similar stature to replace them. Because the best engineers were attracted to higher-paying posts in the industrial and trade groups, most of the new engineers hired by the BPR were recent college graduates. The balance of expertise had shifted away from the BPR, and it has not recovered. In the 1980s, according to one civil engineering professor, a job with the Federal Highway Administration is an entry-level position for highway engineers. Many stay only long enough to get some experience and then move on to state transportation departments or industrial or trade associations.[18]

In the wake of these changes, the harmonious relations between the BPR and the state highway departments were often strained and at times acrimonious. A major irritant was the assignment of inexperienced federal engineers to examine the work of state engineers with years of experience, and the close, constant attention soon rankled. By 1963, W. O. Wright, the president of the western branch of the AASHO entitled an address "Jonah and the Whale," suggesting that the smooth relationship of the past was gone."

> I do not wish to imply that there are extraordinary areas of disagreement between the Bureau of Public Roads and the Highway Departments. Nor do I wish to infer that we are now on as sound a footing as we have been in the past. . . . I speak up in protest, however, against a trend—a system—which is in the developing state and which bids fair to wreck the excellent relationship of long standing between the Bureau and the states.[19]

In fact, Wright's fears were founded. An employee of the Iowa Department of Transportation said in 1980 that the relationship between the state's highway officials and federal representatives was frequently adversarial, especially on large projects. The Federal Highway Administration maintained an office with about twenty people in Ames that was responsible for inspecting state plans for about half the state. It was, he suggested, "A mini-Department of Transportation, except they don't build any roads."[20]

After 1956, Iowa was not the only source of complaints about federal interference that had not been heard during the 1920s or the Depression, when federal funds provided a larger percentage of highway funds. Federal officials blamed this discontent on the rearrangement of highway agencies and their responsibilities at both the state and federal levels. According to the Federal Highway Administration, the formation of the U.S. Department of Transportation and the concomitant creation of state transportation departments responsible for all transport systems, not just roads, created more complex patterns in state-federal relationships, and this had "an expecially trying effect on the Bureau's longtime partner, the American Association of State Highway Officials, which eventually absorbed or was absorbed by the State DOT officials as the name of the organization was changed to the American Association of State Highway and Transportation Officials. The cordial relations between the Bureau, the States, and the Congress of earlier, simpler, times was lost, perhaps never to be regained."[21]

Yet the changes in transportation agencies were not the only source of the problem, for a more important reason was the fundamental shift in the BPR's position as technical leader of the highway industry. By the time the Department of Transportation was created, the BPR's technical reputation had already been diminished by circumstances that had made it more of a bureaucratic than a technical and research organization. As the BPR's technical dominance within the AASHO faded, the basis of state-federal cooperation was also threatened. Earlier, respect for the BPR's technical expertise had always been translated into its influence in all highway matters. But when state highway engineers came to see the BPR as a bureaucratic hurdle rather than a research partner, the essential mechanism for the BPR's indirect control of highway policy was lost. Not every change in the BPR's status in the highway community can be attributed to this factor, but it is significant that as technical respect for the BPR eroded, the bureau relied more on the power of the federal-aid purse and less on cooperation. With this change came the resentment that explained the "Jonah and the Whale" analogy of W. O. Wright in 1963. The bureau's relationship with state highway engineers had been so vital to the ability of both Logan Page and Thomas MacDonald to act as arbiters of highway policy. It is in the demise of that relationship, built on technical expertise, that it is possible to see just how important expertise had been to the BPR in the shaping of highway policy from 1893 through 1956.

Abbreviations, Notes, Note on Sources, and Index

Abbreviations

The following abbreviations are used in the text and the notes:

AAA: American Automobile Association

AAHI: American Association for Highway Improvement

AASHO: American Association of State Highways Officials

America's Highways: U.S., Department of Transportation, Federal Highway Administration, *America's Highways, 1776–1976: A History of the Federal-Aid Program* (Washington, D.C., 1977)

ARBA: American Road Builders' Association

ARBA Proceedings: *Proceedings of the Annual Meeting of the American Road Builders' Association*

ASCE: American Society of Civil Engineers

ASCE Trans.: *Transactions of the American Society of Civil Engineers*

ASTM: American Society for Testing Materials

BPR: Bureau of Public Roads

BPR Records: Records of the Bureau of Public Roads, Record Group 30, National Archives, Washington National Records Center, Suitland, Md.

BPR Annual Report: U.S., Department of Agriculture, Bureau of Public Roads, "Report of the Chief of the Bureau of Public Roads," in U.S., Department of Agriculture, *Annual Report of the Secretary of Agriculture* (Washington, D.C., 1918–38); U.S., Federal Works Administration, Public Roads Administration, *Work of the Public Road Administration* (Washington, D.C., 1939–49); and U.S., Department of Commerce, Bureau of Public Roads, *Annual Report* (Washington, D.C., 1950–68)

Cartwright Papers: Papers of Wilburn C. Cartwright, Carl Albert Congressional Research and Study Center, University of Oklahoma, Norman

CCF, BPR Records: Classified Central File, 1912–50, Records of the Bureau of Public Roads, Record Group 30, National Archives, Washington National Records Center, Suitland, Md.

Chapin Papers: Papers of Roy D. Chapin, Michigan Historical Collection, Bentley Historical Library, University of Michigan, Ann Arbor

Crosby Lectures: W. W. Crosby Lectures, Special Collections, Folger Library, University of Maine, Orono

CWA: Civil Works Administration

EN: *Engineering News*

ENR: *Engineering News-Record*

ER: *Engineering Record*

FDR Library: Franklin D. Roosevelt Library, Hyde Park, New York

FERA: Federal Emergency Relief Agency

GC, BPR Records: General Correspondence, 1893–1916, Records of the Bureau of Public Roads, Record Group 30, National Archives, Washington, D.C. National Records Center, Suitland, Md.

HIA: Highway Industries Association

Historical Statistics: U.S., Bureau of the Census, *Historical Statistics of the United States: From Colonial Times to 1957* (Washington, D.C., 1960)

HRB: Highway Research Board

Joy Papers: Papers of Henry B. Joy, Michigan Historical Collection, Bentley Historical Library, University of Michigan, Ann Arbor

Kennedy Interview: Author's Interview with G. Donald Kennedy, former Michigan State Highway Engineer and President of the Portland Cement Association, August 18, 1980, Wilmette, Ill.

Kennedy Papers: Papers of G. Donald Kennedy, Michigan Historical Collection, Bentley Historical Library, University of Michigan, Ann Arbor

LAW: League of American Wheelmen

MacDonald Papers: Papers of Thomas H. MacDonald (uncatalogued), University Archives, Texas A & M University, College Station

MacDonald Reports: Reports of Thomas H. MacDonald, Bentley Historical Library, University of Michigan, Ann Arbor

McKellar Papers: Papers of Kenneth D. McKellar, Memphis Public Library

MSHD: Michigan State Highway Department

MSHD Records: Records of the Michigan State Highway Department, Michigan State Archives, Lansing

NGRA: National Good Roads Association

NIRA: National Industrial Recovery Act

NLGR: National League for Good Roads

North Carolina Geological Survey Records: Records of the North Carolina Geological Survey, Southern History Collection, University of North Carolina at Chapel Hill

OPR: Office of Public Roads

OPR Annual Report: U.S., Department of Agriculture, Office of Public Roads, "Report of the Director of the Office of Public Roads," in U.S., Department of Agriculture, *Annual Report of the Secretary of Agriculture* (Washington, D.C., 1905–17).

OPRI: Office of Public Road Inquiry

OPRI Annual Report: U.S., Department of Agriculture, Office of Public Road Inquiry, "Report of the Director of the Office of Public Road Inquiry," U.S., Department of Agriculture, in *Annual Report of the Secretary of Agriculture* (Washington, D.C., 1899–1904)

ORI: Office of Road Inquiry

ORI Annual Report: U.S., Department of Agriculture, Office of Road Inquiry, "Report of the Special Agent and Engineer for Road Inquiry," in U.S., Department of Agriculture, *Annual Report of the Secretary of Agriculture* (Washington, D.C., 1893–98)

PRA: Public Roads Administration

PRA Annual Report: U.S., Federal Works Administration, Public Roads Administration, *Works of the Public Roads Administration* (Washington, D.C., 1939–49)

Pratt Papers: Papers of Joseph Hyde Pratt, Southern History Collection, University of North Carolina at Chapel Hill
PWA: Public Works Administration
RFC: Reconstruction Finance Corporation
RFD: Rural Free Delivery
Spoon Papers: Papers of William Luther Spoon, Southern History Collection, University of North Carolina at Chapel Hill
Tugwell Papers: Papers of Rexford G. Tugwell, Franklin D. Roosevelt Library, Hyde Park, New York
USDA: United States Department of Agriculture
Van Wagoner Papers: Papers of Murray D. Van Wagoner, Michigan Historical Collection, Bentley Historical Library, University of Michigan, Ann Arbor
Waldon Papers: Papers of Sidney D. Waldon, Burton Historical Collections, Detroit Public Library
WPA: Works Progress Administration
Yearbook of Agriculture: U.S., Department of Agriculture, *Yearbook of Agriculture* (Washington, D.C. [published annually]).
Ziegler Papers: Papers of Charles M. Ziegler, Michigan Historical Collection, Bentley Historical Library, University of Michigan, Ann Arbor

Notes

Introduction

1. Two studies that generally link highway development to cars are John B. Rae, *The Road and the Car in American Life* (Cambridge, Mass., 1971); and James B. Flink, *The Car Culture* (Cambridge, Mass., 1975). The broader political process that produced American highways is shown in Mark Rose, *Interstate: Express Highway Politics, 1914–1956* (Lawrence, Kans., 1979). On urban developments, see Mark S. Foster, *From Streetcar to Superhighway: American City Planners and Urban Transit, 1900–1940* (Philadelphia, 1981); and Paul Barrett, *The Automobile and Urban Transit: The Formation of Public Policy in Chicago, 1900–1930* (Philadelphia, 1983).

2. A few of the important sources include Thorstein Veblen, *Engineers and the Price System* (New York, 1921); Albert Speer, *Inside the Third Reich*, trans. Richard and Clara Winston (New York, 1970), pp. 344–45, 520; Jacques Ellul, *The Technological Society*, trans. John Wilkinson (New York, 1964); Langdon Winner, *Automonous Technology: Technics Out of Control as a Theme in Political Thought* (Cambridge, Mass., 1977); David F. Noble, *America by Design: Science, Technology, and the Rise of Corporate Capitalism* (New York, 1977); Samuel Florman, *The Existential Pleasures of Engineering* (New York, 1975); and John Kenneth Galbraith, *The New Industrial State* (New York, 1968).

3. Robert Wiebe, *The Search for Order, 1877–1920* (New York, 1967).

4. See ibid.; Samuel P. Hays, *The Response to Industrialism, 1885–1914* (Chicago, 1957); Alfred D. Chandler, Jr., *The Visible Hand: The Managerial Revolution in American Business* (Cambridge, Mass., 1977); and Daniel Nelson, *Managers and Workers: Origins of the New Factory System in the United States, 1880–1920* (Madison, Wisc., 1975).

5. See Martin V. Melosi, *Garbage in the Cities: Refuse, Reform, and the Environment, 1880–1980* (College Station, Tex., 1981); Stanley K. Schultz and Clay McShane, "To Engineer the Metropolis: Sewers, Sanitation, and City-Planning in Late Nineteenth Century America," *Journal of American History* 65 (September 1978): 389–411; Joel A. Tarr, ed., "The City and Technology," *Journal of Urban History* 5 (May 1979); and James E. O'Connell, *Chicago's Quest for Pure Water*, Essays in Public Works History, No. 1 (Chicago, 1977). On social scientists, see also Donald T. Critchlow, *The Brookings Institution, 1916–1952: Expertise and the Public Interest in a Democratic Society* (DeKalb, Ill., 1985), pp. 17–40.

6. The following draws heavily on John C. Burnham, "Essay," in John D. Buenker, John C. Burnham, and Robert M. Crunden, eds., *Progressivism* (Cambridge, Mass., 1977), pp.3–29.

7. Gabriel Kolko, *The Triumph of Conservatism: A Reinterpretation of American History, 1900–1916* (New York, 1963), showed that not all reformers were altruistic.

8. Critchlow, *Brookings Institution*, discusses the antimajoritarian attitudes of some reformers. On the emergence of codes of ethics, see Edwin T. Layton, Jr., *The Revolt of the Engineers: Social Responsibility and the American Engineering Profession* (Cleveland, 1971), pp. 53–178.

9. On city managers, see Richard J. Stillman, *The Rise of the City Manager: A Public Profession in Local Government* (Albuquerque, 1974); Wiebe, *Search for Order*; Raymond Callahan, *Education and the Cult of Efficiency: A Study of the Social Forces That Have Shaped the Administration of the Public Schools* (Chicago, 1962); and Samuel P. Hays, *Conservation and the Gospel of Efficiency: The Progressive Conservation Movement, 1840–1920* (Cambridge, Mass., 1959).

10. T. F. Hickerson, "Relation of the Engineer to Successful Road Location, Construction, and Maintenance," Paper read at the Road Institute, University of South Carolina, March 4, 1920, in Engineering Series, Folder 19, Papers of T. F. Hickerson, Southern History Collection, University of North Carolina at Chapel Hill.

11. See Ellis Hawley, ed., *Herbert Hoover as Secretary of Commerce: Studies in New Era Thought and Practice* (Iowa City, 1981), pp. 78–79. On the attitude of engineers during the 1920s, see Layton, *Revolt of the Engineers*, pp. 210–48.

12. John G. Gunnel, "The Technocratic Image and the Theory of Technocracy," *Technology and Culture* 23 (July 1982): 392–416.

13. Critchlow, *Brookings Institution*, p. 8. In the last fifteen years, a vast literature on Hoover's New Era has emerged that stresses his Progressive origins and his emphasis on voluntary cooperation. A good overview can be found in Hawley, ed., *Hoover as Secretary of Commerce*; and in Ellis Hawley, "Herbert Hoover, the Commerce Secretariat, and the Vision of an 'Associative State,'" *Journal of American History* 61 (June 1974): 116–40.

14. On FDR, see Robert Leuchtenberg, *FDR and the New Deal* (New York, 1963); on Joseph Eastman, see William R. Childs, *Trucking and the Public Interest: The Emergence of Federal Regulation, 1914–1940* (Knoxville, 1985), pp. 120–24; on technocracy, see Henry Elsner, *The Technocrats: Prophets of Automation* (Syracuse, 1967).

15. See Lowell Tozer, "A Century of Progress, 1893–1933: Technology's Triumph over Man," in Hennig Cohen, ed., *The American Culture: Approaches to the Study of the United States* (Boston, 1963), pp. 206–9.

16. See Daniel Kevles, *The Physicists: A History of a Scientific Community in Modern America*, (New York, 1971), pp. 365–68.

17. Alvin M. Weinberg, "Social Institutions and Nuclear Energy," *Science* 177 (July 7, 1972): 27–34.

18. Ferguson's thoughts on engineers as experts can be found in Eugene S. Ferguson, "Enthusiasm and Objectivity in Technological Development," Paper read at the American Association for the Advancement of Science Symposium "Technology: Nuts and Bolts or Social Process," Washington, D.C., December 29, 1970; and idem, "The Nature of Technological Invention: Play and the Intellect," Paper read at the Symposium on Technology and the Evolution of Human Culture, American Anthropological Association, Washington, D.C.,

1980. The phrase is from Daniel F. Greenberg, "Don't Ask The Barber Whether You Need a Haircut," *Saturday Review* 55 (November 25, 1972): 55–59.

19. "The Science Court Experiment: An Interim Report," *Science* 193 (August 20, 1976): 653–56.

Part I

1. On early roads, see George Rogers Taylor, *The Transportation Revolution, 1815–1860* (New York, 1951), pp. 39–51; Thomas H. MacDonald, "The History and Development of Road Building in the United States," *ASCE Trans.* 92 (1928): 1181–95; *America's Highways,* pp. 3–37; and W. Turrentine Jackson, *Wagon Roads West: A Study of Federal Road Surveys and Construction in the Trans-Mississippi West, 1846–1869* (Berkeley, 1952).

Chapter One

1. *ORI Annual Report* (1893), p. 585.

2. See Phillip Mason, "The League of American Wheelmen and the Good Roads Movement" (Ph.D. diss., University of Michigan, 1957), pp. 83–149; Michael Berger, *The Devil Wagon in God's Country: The Automobile and Social Change in Rural America, 1893–1929* (Hamden, Conn., 1979), pp. 13–35; and Hugh Myron Hoyt, Jr., "The Good Roads Movement in Oregon, 1900–1920" (Ph.D. diss., Univeristy of Oregon, 1966), pp 161–206. Plank roads used wooden boards set in the roadbed to provide a smooth surface. See Taylor, *Transportation Revolution*, pp. 29–31.

3. On the LAW's role, see Mason, "American Wheelmen;" Charles L. Dearing, *American Highway Policy* (Washington, D.C., 1942), pp. 225–26; and Stone Correspondence, 1890–92, Files 2 and 28, GC, BPR Records.

4. Charles F. Johnson to Eugene D. White, March 22, 1897, File 80, GC, BPR Records.

5. Unless otherwise cited, information on the ORI is from *ORI Annual Report*; GC, BPR Records; and W. Stull Holt, *The Bureau of Public Roads: Its History, Activities and Organization* (Baltimore, 1923).

6. See A. Hunter Dupree, *Science in the Federal Government: A History of Policies and Activities to 1940* (Cambridge, Mass., 1957), pp. 157–83; T. Swann Harding, *Two Blades of Grass: A History of Scientific Development in the U.S. Department of Agriculture* (Norman, Okla., 1947); and Schultz and McShane, "To Engineer the Metropolis," pp. 389–411.

7. *ORI Annual Report* (1893), p. 596.

8. The only treatise on road building published before 1890 was William M. Gillespie, *A Manual of the Principles of Road Making*, 8th ed. (New York, 1854 1st ed. ca. 1847). Later books included Clemens Herschel and Edward P. North, *Road Making and Maintenance* (New York, 1894); and Nathaniel P. Shaler, *American Highways* (New York, 1896). One can gain access to ORI publications

through OPR, *Publications of the Office of Public Roads*, Circular No. 88 (Washington, D.C., 1908).

9. Martin Dodge, "Steel-Track Wagon Roads," in *Yearbook of Agriculture, 1898* (1899), pp. 291–96; and Dodge's Correspondence, 1895–98, Files 96 and 270, GC, BPR Records.

10. Arthur W. Dean, "Massachusetts Highways," *Proceedings of the Boston Society of Civil Engineers* 16 (December 1929): 493–510; and Charles E. Rosenberg, "Science, Technology, and Economic Growth: The Case of the Agricultural Experiment Station Scientist, 1875–1914," in George H. Daniels, ed., *Nineteenth-Century American Science: A Reappraisal* (Evanston, 1972), pp. 181–209.

11. *OPRI Annual Report* (1904), p. 432; see also Roy Stone, "Object-Lesson Roads," in *Yearbook of Agriculture, 1898* (1899); pp. 373–82; and *America's Highways*, pp. 45–48, 64–66, 72–75.

12. On Stone's efforts, see *ORI Annual Report*; and the various ORI bulletins and circulars.

13. On these cooperative ties, see Correspondence, 1895–98, Files 3, 103, and 134, GC, BPR Records; and Mason, "American Wheelmen," pp. 173–75. Examples of LAW efforts include Albert A. Pope, "Good Roads for Wheelmen and the People," *Independent* 48 (February 1896): 174; and Otto Dorner, "Good Roads and State Aid," *Forum* 26 (February 1899): 668–72. The ORI's position can be seen in Roy Stone, "Best Roads for Farms and Farming Districts," in *Yearbook of Agriculture, 1894* (1895), pp. 501–4; and M. O. Eldridge, "Construction of Good Country Roads," in *Yearbook of Agriculture, 1898* (1899) pp. 317–24.

14. See Wayne E. Fuller, *RFD: The Changing Face of Rural America* (Bloomington, Ind., 1964), pp. 17–48.

15. Quotation from *ORI Annual Report* (1895), p. 195; see also Stone Correspondence, File 28, GC, BPR Records; ORI, *Addresses on Road Improvements*, Circular No. 14 (Washington, D.C., 1894); and Holt, *Bureau of Public Roads*, p. 7.

16. Sterling Elliot to Martin Dodge, February 19, 1898, File 103, GC, BPR Records.

17. See Files 12, 28, 90, and 134, GC, BPR Records; ORI, *Must Farmers Pay for Good Roads*, Circular No. 31 (Washington, D.C., 1898); and idem, *State Aid to Road Building in Minnesota*, Circular No. 32 (Washington, D.C., 1898). Mason ("American Wheelmen," p. 173) commented that "on a number of occasions, [the LAW] served as a mouthpiece for the director of the OPRI."

18. *ORI Annual Report* (1898), p. 160. For examples of press coverage, see Roswell P. Flowers, "Good Roads for Farmers," *Independent* 48 (February 1896): 48; and "Good Roads in Relation to Farm Values," *Scientific American* 76 (March 1897): 162.

19. USDA, *Annual Report* (1895), p. 51.

20. Mason, "American Wheelmen," p. 149.

21. See "Biographical Sketch of Hon. Martin Dodge," File 3, GC, BPR

Records; *America's Highways*, p. 191; Dodge, "Road Building in Ohio," in ORI, *Historical and Technical Papers on Road Building in the United States*, Bulletin No. 17 (Washington, D.C., 1895), pp. 16–23; and idem, "The Need for Good Roads," *North American Review* 161 (July 1895): 125–28.

22. *OPRI Annual Report* (1900), p. 288. The following information is from Correspondence, 1900–1905, Files 28 and 90, GC, BPR Records; Letters Sent, E. G. Harrison, 1900, BPR Records; and *OPRI Annual Report* (1900–1904).

23. M.O. Eldridge to A. R. Shattuck, November 5, 1904, File 90, GC, BPR Records.

24. OPRI, *Social, Commercial, and Economic Phases of the Road Subject*, by W. H. Moore, Circular No. 34 (Washington, D.C., 1900).

25. See Pope, *Wagon Roads as Feeders to Railways* (Boston, 1892); "The Railroads and Highway Improvement," *Railroad Gazette* 23 (February 13, 1891): 114–15; ORI, *Information Regarding Road Materials and Transportation Rates in Certain States West of the Mississippi*, Bulletin No. 5 (Washington, D.C., 1894); idem, *Information Regarding Roads, Road Materials, and Freight Rates in Certain States North of the Ohio River*, Bulletin No. 6 (Washington, D.C., 1894); idem, *Information Regarding Roads and Road-Making Materials in the Eastern and Southern States*, Bulletin No. 7 (Washington, D.C., 1894); Dearing, *American Highway Policy*, pp. 226–28, 260–62; and Roy V. Scott, *Railroad Development Programs in the Twentieth Century* (Ames, Iowa, 1985), pp. 33–39.

26. *OPRI Annual Report* (1901–05); "Report of the Work of the 'Good Roads Train' over the Illinois Central Railway System," 1901, File 137, GC, BPR Records; and Files 56 and 68, GC, BPR Records; "The Good Roads Train of the Southern," *Railroad Gazette* 33 (November 8, 1901); 769–70; Earl Mayo, "The Good Roads Train," *World's Work* 2 (July 7, 1901): 956–60; and Dodge, "Government Participation in Object-Lesson Road Work," *Yearbook of Agriculture, 1901* (1902), pp. 409–14.

27. W. H. Potts to Dodge, May 21, 1901, File 28, GC, BPR Records. Information about the six bulletins and other OPRI publications can be found in OPRI, *Publications*, Circular No. 36.

28. James W. Abbott to Logan W. Page, November 11, 1905, File 104, GC, BPR Records; see also other complaints and the secretary's responses in Files 2, 56, 68, and 104, GC, BPR Records.

29. Harrison to Moore, December 22, 1900, Letters Sent, E. G. Harrison, BPR Records; see also Files 2, 3, 29, and 104, GC, BPR Records; and Holt, *Bureau of Public Roads*, p. 9.

30. See *America's Highways*, pp. 215–16; W. P. Brownlow, "National Aid to Road Improvement," *Cosmopolitan* 34 (January 1903): 355–58; Files 2 and 4, GC, BPR Records; and W. P. Brownlow to W. L. Spoon, January 9 and 31, 1903, Folder 80, Spoon Papers.

31. Eldridge to Dodge, October 24, 1903, File 4, GC, BPR Records; see also File 153, GC, BPR Records; OPRI, *Proceedings of the National Good Roads Convention, Held at St. Louis, Mo., April 27 to 29, 1903*, Bulletin No. 26 (Washington, D.C., 1903); and U.S., Congress, Senate, Agriculture and Forestry

Committee, *Hearings on S. 3477: To Establish in the Department of Agriculture a Bureau of Public Highways . . .* , 58th Cong., 2d sess., 1904, S. Doc. 204.

32. Dodge to A. R. Shattuck, December 1, 1904, File 90, GC, BPR Records; see also Files 2, 56, and 153, GC, BPR Records; and J. W. Wilson to Moore, April 11 and 13, 1904, Letterbook 86, Letters Sent, Secretary of Agriculture, Records of the Secretary of Agriculture, Record Group 16, National Archives, Washington, D.C.

33. *OPRI Annual Report* (1900–1905; Logan W. Page, "Object-Lesson Roads," in *Yearbook of Agriculture, 1906* (1907), pp. 137-50; and OPRI, *The Construction of Sand-Clay and Burnt-Clay Roads*, Bulletin No. 27 (Washington, D.C., 1906).

34. See Correspondence: Series A, January 1901–June 1911, Spoon Papers; *OPRI Annual Report* (1900–1904); and articles by the agents in *Yearbook of Agriculture* (1900–1905).

35. USDA, *Annual Report* (Washington, D.C., 1900), p. lv; see also *OPRI Annual Report* (1900–1905); and Files 5, 27, and 59, GC, BPR Records.

36. See *OPRI Annual Report* (1901), p. 243; *BPR Annual Report* (1919), pp. 424–26; *America's Highways*, pp. 64, 191–92; and Ira O. Baker, *A Treatise on Roads and Pavements* (New York, 1907), pp. 185–89.

37. File 59, GC, BPR Records, contains eight boxes of such correspondence for 1896–1912; File 10 has two more boxes, and in Letters Sent, Division of Laboratory Tests and Investigations, 1901–5, are six press copybooks. See also "The Organization and Equipment of the Office of Public Roads," *ER* 54 (August 25, 1906): 205–07.

38. John Meigs, Jr., to Dodge, February 18, 1905, File 7, GC, BPR Records.

39. See Dean, "Massachusetts Highways," pp. 493–510; M. C. Krarup, "From Coast to Coast in an Automobile," *World's Work* 8 (May 1904): 4740–54; and *America's Highways*, pp. 60–62.

Chapter Two

1. Dabney, "Letter to the Editor," *EN* 76 (November 23, 1916): 1001.

2. See OPRI, *List of National, State, and Local Road Associations and Kindred Organizations in the United States*, Circular No. 36 (Washington, D.C., 1902); idem, *Public Road Mileage, Revenues and Expenditures in the United States in 1904*, Bulletin No. 32 (Washington, D.C., 1907); and OPRRE, *Public-Road Mileage, Revenues, and Expenditures in the Unied States, 1914—A Summary*, Department Bulletin No. 390 (Washington, D.C., 1917). On state improvements, see MacDonald "History of Road Building," pp. 1181–1205.

3. Interestingly, James B. Flink, *America Adopts the Automobile, 1895–1910* (Cambridge, Mass., 1970), pp. 202–13, does not even mention the ORI or OPR in his brief discussion of roads.

4. This summary draws heavily on Burnham, "Essay"; and Daniel T. Rogers, "In Search of Progressivism," *Reviews in American History* 10 (December 1982): 113–32.

5. USDA, *Annual Report* (Washington, D.C., 1905), pp. cxvi, cxxii, 429.

6. Page to E. E. Britton, February 16, 1918, File 3.7, CCF, BPR Records; see also Files 3, 4 103, and 104, for 1905–10, GC, BPR Records.

7. Cushman to D. W. King, January 24, 1907, File 69; and Page to E. W. Scripps, February 17, 1909, File 369, GC, BPR Records.

8. See David P. Thelen, *Robert M. La Follette and the Insurgent Spirit* (Boston, 1976); and Melvin I. Urofsky, *Louis D. Brandeis and the Progressive Tradition* (Boston, 1981).

9. See John E. Wiley, "History of the A.R.B.A.," *American Road Builder* 54 (March-April 1977): 17–72; and Clay McShane, "Transforming the Use of Urban Space: A Look at the Revolution in Street Pavements," *Journal of Urban History* 5 (May 1979): 289–307.

10. Throughout this chapter, the main source for OPR activities is *OPR Annual Report* (1905–16); see also A. N. Johnson, "Aims and Purposes of the Office of Public Roads, United States Department of Agriculture," *ER* 52 (September 16, 1905): 326; and Holt, *Bureau of Public Roads*.

11. Charles T. Harrison to Page, March 27, 1908, File 270, GC, BPR Records; on the road drag, see File 69, GC, BPR Records; and USDA, *The Use of the Split-Log Drag on Earth Roads*, Farmers' Bulletin No. 321 (Washington, D.C., 1908).

12. Fuller, *RFD*, pp. 190–98; File 67, GC, BPR Records, contains the OPR inspection reports.

13. USDA, *Mileage and Cost of Public Roads in the United States in 1909*, Bulletin No. 41 (Washington, D.C., 1912); see also Reports Concerning Economic Effects of Road Improvement, 1910–18, BPR Records; OPR, *Public Road Mileage and Revenues in the United States, 1914*, Department Bulletin No. 390 (Washington, D.C., 1917); and "Help from Uncle Sam," *Colliers* 52 (October 10, 1913): 21–22.

14. "Object Lesson in Road Building," *Scientific American* 106 (March 16, 1912): 232; "The Organization and Equipment of the Office of Public Roads, United States Department of Agriculture," *ER* 54 (August 25, 1906): 205–7; "The United States Office of Public Roads," *ER* 62 (December 17, 1910): 709–11; and USDA, *Object-Lesson and Experimental Roads*, Department Bulletin No. 53 (Washington, D.C., 1913).

15. See Page, "Effects of Motors on Macadam Roads," *ER* 58 (September 26, 1908): 53; idem, "The Motor Car and the Road: Destructive Effect of High Speed," *Scientific American* 102 (January 15, 1910): 46–47; and "Road Construction and Maintenance: Use of Bituminous Materials and Mixing Methods," *ASCE Trans.* 75 (1912): 572–648.

16. On standardization, see Bruce Sinclair, *Centennial History of the American Society of Mechanical Engineers, 1880–1980* (Toronto, 1980), pp. 50, 144–57; Noble, *America by Design*, pp. 69–83; and George V. Thompson, "Intercompany Technical Standardization in the Early American Automobile Industry," *Journal of Economic History* 14 (Winter 1954): 1–20.

17. More detailed treatment of OPR standardization efforts can be found in

Bruce E. Seely, "Engineers and Government-Business Cooperation: Highway Standards and the Bureau of Public Roads, 1900–1940," *Business History Review* 58 (Spring 1984): 53–60. See also numerous OPR Bulletins and Circulars and USDA Department Bulletins released between 1911 and 1918.

18. See "Report of the ARBA Committee on Standards," December 11, North Carolina Geological Survey Records; "American Society of Civil Engineers Annual Meeting," *EN* 73 (January 28, 1915): 186–87; W. W. Crosby, "Standardizing Road Building Terminology," Folder 4, Crosby Lectures; and "Committee on Lincoln Highway Ideal Section Named," *ENR* 85 (September 2, 1920): 576.

19. "Object Lesson in Road Building," p. 232. On OPR influence, see A. E. Johnson, "Committees of AASHO: The Real Strength" in AASHO, *AASHO—The First Fifty Years, 1914–1964* (Washington, D.C., 1965), p. 104; and numerous articles on state highway standards in engineering periodicals after 1913.

20. Files 5, 27, 29, and 330, GC, BPR Records; and File 470, CCF, BPR Records.

21. See, for example, the discussion of the OPR's cooperation with the ASCE on dust prevention from 1907–15 in "Bituminous Materials for Roads: Special Committee of the ASCE," *EN* 71 (January 29, 1914): 265–26; and File 335, GC, BPR Records.

22. OPR-ASTM ties can be traced in Letters Sent Regarding the Society for Testing Materials and File 470, CCF, BPR Records; ASTM, *Fifty Year Index of ASTM Technical Papers and Reports, 1898–1950* (Philadelphia, 1952); and *ASTM Yearbook* (Philadelphia, 1905–15).

23. On the Road Congresses see Files 322 and 458, GC, BPR Records; and File 2, CCF, BPR Records. Appreciations of the OPR can be found in "Help from Uncle Sam," pp. 21–22; and H. Willsie, "Two Million Miles of Road," *Harper's Weekly* 58 (February 28, 1914): 18–20.

24. Harrison to Page, March 17, 1908, and Page form letter, March 4, 1908, both in File 270, GC, BPR Records. See also Page, "Problems of Road Administration: A Plea for Centralized Control of Highway Construction and Maintenance," *Engineering Magazine* 38 (November 1909): 265–67.

25. See *OPR Annual Report* (1906–11); and Holt, *Bureau of Public Roads*, pp. 10–13.

26. See USDA, *Annual Report* (Washington, D.C., 1912), p. 210; File 369, GC, BPR Records; and Files 430 and 440 for each state, CCF, BPR Records.

27. "After Investigation, What?" *EN* 71 (February 19, 1914): 424. On early highway departments, see Mason, "American Wheelmen," pp. 214–15.

28. Holt, *Bureau of Public Roads*, p. 12.

29. Page to George R. Chatburn, January 21, 1905, File 28, GC, BPR Records.

30. See Correspondence Regarding the North Carolina Good Roads Association, Pratt Papers; Harry Wilson McKown, "Roads and Reform: The Good Roads Movement in North Carolina, 1885–1921" (M.A. thesis, University of North Carolina at Chapel Hill, 1972); Robert Craaco, "History of the South

Dakota Highway Department, 1919–1941" (M.A. thesis, Northern State College, 1969); Iowa State Highway Commission, *Annual Report* (Ames, 1905–16); Hoyt, "Good Roads Movement in Oregon;" "New York State Road Organization," *EN* 70 (November 13, 1913): 982–84; and "Politics vs. Business in New York State Road Work," *EN* 71 (March 26, 1914): 686.

31. "Letter to the Editor," *EN* 71 (April 2, 1914): 745; see also "A Proper Plan for Promoting Publicity," *EN* 71 (January 22, 1914): 210–12.

32. USDA, *Annual Report* (Washington, D.C., 1910), p. 153.

33. Quotation from Page to Marston, March 1, 1911, File 28, GC, BPR Records; also see address by Page, December 4, 1906, File 326, GC, BPR Records.

34. A. S. Cushman to William M. Hager, February 7, 1908, File 90, GC, BPR Records; see also L. W. Page, "Competitive Examination Should Govern Highway Appointments," *ER* 73 (January 29, 1916): 139.

35. See Files 28 and 90, GC, BPR Records; and Page, "State Aid in Roadbuilding," *ER* 62 (December 24, 1910): 741–43.

36. See File 3.9, CCF, BPR Records.

37. On the AAHI, see Files 28, 528, and 530, GC, BPR Records; "The A.A.H.I.," *Municipal Engineering* 41 (December 1911): 484–85; American Highway Association, *The Official Good Roads Yearbook of the United States* (Washington, D.C., 1912–17); and John M. Goodell, "The American Highway Association and the Mission It Fulfilled," *ENR* 78 (November 8, 1917): 870–71.

38. See Hays, *Conservation and the Gospel of Efficiency*, pp. 199–260; and Dupree, *Science in the Federal Government*, pp. 157–81.

39. The Lincoln Highway Association, *The Lincoln Highway: The Story of a Crusade That Made Transportation History* (New York, 1935), p. 121; and Dearing, *American Highway Policy* pp. 29–45. For examples of the articles, see "The Ideal Highway Commission," *ER* 63 (February 25, 1911): 206; A. N. Johnson, "State Control of Road Construction," *ER* 70 (November 11, 1914): 568–69; idem, "State Road Laws Should Centralize Authority," *ER* 72 (September 18, 1915): 351; and "Essentials of Proper Laws for Highway Work," *Good Roads*, n.s., 10 (October 2, 1915): 193–94.

40. MSHD, *Biennial Report of the Michigan State Highway Commissioner* (Lansing, 1905/6—1915/16); and Frank F. Rogers, *History of the Michigan State Highway Department, 1905–1933* (Lansing, 1933), pp. 14–26.

41. This information is drawn from "Report Concerning State Management of Highways," 1915–18, BPR Records.

42. See W. P. White, "Good Roads for People," *Outing Magazine* 51 (November 1907): 221–28; "Roadmaking As a Reform Measure," *Survey* 26 (April 22, 1911): 157–59; and Mrs. Frank de Garmo, "The Humanity of Highways," *Journal of Proceedings and Addresses of the National Education Association of the United States* 51 (Washington, D.C.; 1912): 301–7.

43. USDA, *Annual Report* (Washington, D.C., 1905), p. cxvi.

44. Page to A. R. Mann, May 20, 1909, File 446, GC, BPR Records.

45. The gospel of good roads was evident in most OPR publications on rural

roads released by the Agriculture Department. For example, see "Importance of Good Roads to New Settlers, Irrigated Lands," in *Yearbook of Agriculture, 1912* (1913), pp. 490–94; and USDA, Office of Experiment Stations, *Syllabus of Illustrated Lectures on Roads and Road Building*, Farmers' Institute Lecture No. 7 (Washington, D.C., 1907).

46. See File 411, GC, BPR Records; OPR, *Descriptive Catalog of the Road Model Exhibit* (Washington, D.C., 1909); and "Teaching Road-Making by Models," *World's Work* 27 (November 1913): 116–18.

47. On educational trains, see Scott, *Railroad Development Programs*, pp. 25–39; on Good Roads Trains, see Files 503 and 528, GC, BPR Records; and R. Franklin, "Teaching Good Roads by Special Train," *Technical World* 17 (June 1912): 448–51.

48. Scripps to Page, February 9, 1909, File 369, GC, BPR Records.

49. See Peter Hugill, "Good Roads and the Automobile in the United States, 1880–1929," *Geographical Review* 72 (July 1982): 332–40; A. A. Pope, "Good Roads and Automobiles," *Munsey's* 29 (May 1903): 1167–70; and material for 1910 in Box 1, Waldon Papers.

50. A. S. Cushman to Robert M. McWade, January 25, 1909, File 4, GC, BPR Records. This claim was repeated frequently by the OPR.

51. See Holt, *Bureau of Public Roads*, p. 14; and Mason, "American Wheelmen," pp. 200–204.

52. See File 528, GC, BPR Records; and File 3.6, CCF, BPR Records; see also "The A.A.H.I.," pp. 484–85.

53. See Lincoln Highway Association, *Lincoln Highway*, pp. 9–18; Scrapbook No. 29, Lincoln Highway, Joy Papers; Henry B. Joy, "Transcontinental Trails," *Scribner's Monthly* 55 (February 1914): 160–72; Records of the Yellowstone Highway Association, 1914–39, South Dakota Historical Resources Center, Pierré, and Charles Henry Davis, *Arguments for and against National Highways versus Federal Aid* (Washington, D.C., 1914).

54. American Highway Association, *Good Roads Yearbook* (1913), pp. 14–19; and "Federal Legislation in Aid of Road Improvements," *EN* 68 (September 26, 1912): 592.

55. W. W. Crosby, "National Aid," Folder 1911, Crosby Lectures. On legislation, see n. 54 above; *EN* 67 (April 18, 1912): 750; and Page to J. W. Parmley, May 5, 1912, File 2, GC, BPR Records.

56. See George C. Diehl, "For Good Roads," *Harper's Weekly* 56 (April 6, 1912): 12; and Roy D. Chapin, "The Automobile Manufacturer in Relation to Good Roads," pp. 3–4, Miscellaneous Correspondence File, 1914, Chapin Papers. On the AAA as a lobbying group for Detroit, see David J. St. Clair, "Entrepreneurship and the American Automobile Industry" (PH.D. diss., University of Utah, 1979), pp. 203–5; "Report of the Good Roads Committee to the Directors of the NACC," December 1, 1915, Correspondence File, 1915, Chapin Papers; and Flink, *America Adopts the Automobile*, pp. 145–63.

57. W. W. Crosby, manuscript for an article in *Good Roads*, October 2, 1913, Folder 4, Crosby Lectures.

58. "Federal Aid for Highway Improvements," *EN* 66 (December 28, 1911): 776.

59. See Files 2 and 528, GC, BPR Records; "Federal Appropriations in Aid of Good Roads," *EN* 67 (April 28, 1912): 750; and Austin F. MacDonald, *Federal Aid: A Study of the American Subsidy System* (New York, 1928).

60. Bankhead to Page, March 2, 1910; see also Correspondence, 1912, File 2, GC, BPR Records; and U.S., Congress, Senate, *Experiments on Certain Highways* . . . , 61st Cong., 2d sess., 1910. S. Rept. 438.

61. See Diehl, "For Good Roads," p. 12; and Chapin, "The Automobile Manufacturer in Relation to Good Roads."

62. U.S., Congress, House, *Report of the Joint Congressional Committee on Federal-Aid in the Construction of Post Roads*, 63d Cong., 3d sess., (1915), H. Doc. 1510, pp. 30–1, 232–35, 244–54; and American Highway Association, *Good Roads Yearbook* (1913), p. 14.

63. Compare U.S., Congress, House, . . . [on] *Post Roads Report*, pp. 19, 24, with Page's ideas in USDA, *Annual Report* (Washington, D.C., 1914) pp. 30–32.

64. "Federal Appropriations for State Road Construction," *EN* 71 (February 26, 1914): 472; see also "Federal Aid for Road Construction," *EN* 72 (August 6, 1914): 313. On the legislative developments, see American Highway Association, *Good Roads Yearbook* (1915), pp. 255–70; D. W. Shackleford, "Roads for All the People," *Colliers* 52 (March 7, 1914): 19–20; J. Bourne, Jr., "National Aid to Good Roads," *North American Review* 198 (May 1913): 320–31; and File 740.1, CCF, BPR Records.

65. *EN* 67 (May 6, 1912): 938.

66. "Federal Appropriations for State Road Construction," p. 472; see also Joseph A. Allen and Whidden Graham to Sidney D. Waldon, February 11, 1910, Box 1, Waldon Papers.

67. Batchelder to Chapin, April 12, 1916, Correspondence File, Chapin Papers. On the AASHO, see A. E. Johnson, "History of the Origin, Development and Operation of the American Association of State Highway Officials," in AASHO, *AASHO: The First Fifty Years*, pp. 49–53; Mississippi Valley Conference of State Highway Departments, "Historical Highlights, 1909–1974," mimeographed (Iowa Department of Transportation Library, Ames); and Correspondence, File 740.1.1, CCF, BPR Records; The following legislative history is largely drawn from these sources.

68. The course of the bill can be followed in U.S., Congress, House, *Debates on Rural Post Roads, H.R. 7617*, 64th Cong., 1st sess., 1916; idem, *Congressional Record*, 64th Cong., 1st sess., vol. 53 (1916); and the various House, Senate, and Conference Committee reports on H.R. 7617 released during 1916. See also "Road Congress Advocates Continuity of Policy," *EN* 72 (September 18, 1915): 367. On Page's efforts, see Files 481, 740.1.1, and 740.1.2, CCF, BPR Records.

69. For a summary of the bill, see Holt, *Bureau of Public Roads*, pp. 15–18.

70. Although the OPR had constructed 457 miles of road in 28 counties in 17 states with the 1912 funds, this was accomplished only with constant skirmishing

over rules. See U.S., Congress, House, *Joint Report on the Progress of Post Road Improvement*, 63d Cong., 1st sess., 1915, H. Doc. 204.

71. On the OPR's support of the gospel of good roads, see "The Importance of Good Roads to Farmers," *Yearbook of Agriculture, 1914* (1915), pp. 43–46; Page, "How Small Communities May Have Good Roads," *Scientific American* 112 (January 2, 1915): 1–5, 38; idem, *Roads, Paths and Bridges* (New York, 1913); and Dearing, *American Highway Policy*, pp. 255–62. On rural engineering, see *OPR Annual Report* (1915–17); *BPR Annual Report*, (1918–21); a wide array of BPR publications through the early 1920s on such topics as the use of tractors on farms; and Holt, *Bureau of Public Roads*, pp. 38–40, 51–55.

72. "Henry Ford, the Road Promoter," *ENR* 79 (September 16, 1917): 434.

73. See Page, "The Motor Car and the Road," 46–47; idem, "Our Highways and the Motor Car," *Collier's* 48 (January 10, 1912): supp. 34; idem, "Automobiles and Good Roads," *Scientific American* 109 (September 16, 1913): and 178; *ER* 61 (February 26, 1910): 231; See also D. F. Houston, "Better Roads Mean Better Schools," *Delineator* 85 (September 1914): 15; 'Road That Takes Children to School," *Outlook* 107 (July 18, 1914): 637–38; T. M. R. Von Keler, "Farmer and the Motor Car," *Collier's* 54 (January 9, 1915): supp. 34; and "Building Roads and Morals," *Literary Digest* 48 (March 14, 1914): 576–78.

74. John M. Goodell, "An Apostle for Good Roads, Logan Waller Page," *The American Review of Reviews* 59 (March 1919): 302–4. A number of other magazines ran articles that portrayed Page as the perfect Progressive expert.

75. For a discussion of the contradictions of Progressivism, see Rogers, "In Search of Progressivism."

76. Page to J. J. Murray, April 3, 1916, File 740.1.1, CCF, BPR Records.

Chapter Three

1. Mehren, "A Suggested National Highway Policy and Plan," *ENR* 81 (December 19, 1918): 1113.

2. *America's Highways*, p. 108.

3. See *OPR Annual Report* (1916); and *America's Highways*, p. 117.

4. *Historical Statistics*, pp. 458–59; and *OPR Annual Report* (1917), p. 360.

5. Transcript of an interview with Page, August 1916, File 740.1.1, CCF, BPR Records.

6. "Rules for Applying Federal-Aid Road Improvement Act," *EN* 76 (September 15, 1916): 521; see also "Progress for Federal Aid for State Highway Work," *EN* 76 (November 30, 1916): 1054; and Files 481 General and 481 Rules and Regulations for 1916, CCF, BPR Records.

7. On standards, see File 470, CCF, BPR Records; BPR, *Standard and Tentative Methods of Sampling and Testing Highway Materials*, Department Bulletin No. 949 (Washington, D.C., 1921); *OPR Annual Report* (1917), p. 360; and *BPR Annual Report* (1918), p. 384, (1920), p. 519.

8. W. A. Crossland to J. T. Bullen, May 16, 1917, File 440 Special Advice, CCF, BPR Records.

9. J. T. Voshell, Memorandum, July 30, 1920, File 481 General, CCF, BPR Records. On Michigan, see File 481 PS & E—Michigan, CCF, BPR Records.

10. See *BPR Annual Report* (1920), p. 500; and "Good Progress Made in Operation of Federal-Aid Road Act," *ENR* 79 (December 20, 1917): 1148–50.

11. See Conference in Secretary's Office, November 10, 1916, File 481 General, CCF, BPR Records; and *America's Highways*, p. 101.

12. Quotations from MacDonald, *Federal Aid*, p. 90; and Conference in the Secretary's Office, p. 11, File 481 General, BPR Records; also see T. W. Gregory to Secretary of Agriculture, March 2, 1917; and Memorandum to the Secretary, July 19, 1918, both in File 481 General, CCF, BPR Records. Rae, *The Road and the Car*, pp. 36–37, noted the ambiguity.

13. See *BPR Annual Report* (1920), pp. 395–96; Files 481 Rules and Regulations and 481 General, and Legislative Drafts, File 740.1.1, CCF, BPR Records.

14. On the government's highway material priorities, see numerous articles in *ENR* 78–80 (1917–18); and *OPR Annual Report* (1917–18).

15. Again, articles in vols. 78–80 of *ENR* (1917–18) detail the use of trucks during the war; see also Robert F. Karelovitz, *This Was Trucking* (Seattle, 1966), pp. 49–67.

16. "A New Era in Highway Transportation," *ENR* 81 (November 28, 1918): 967.

17. Prevost Hubbard, "Efficiency of Bituminous Surfaces and Pavements under Motor-Truck Traffic," *Public Roads* 1 (February 1919): 25; also see *BPR Annual Report* (1918), pp. 373–75; and "Our Highways and the Burden They Must Carry," *Public Roads* 1 (June 1918): 4–29.

18. See *BPR Annual Report* (1920), pp. 494, 500; Andrew P. Anderson, "Highway Expenditures in the Year of 1918," *Good Roads*, n.s., 16 (October 18, 1919): 171–74, 176; and Mississippi Valley Conference, "Historical Highlights," p. x-35.

19. "Post-War Highway Drive Begins," *ENR* 81 (November 14, 1918): 878–79.

20. Robert Cuff, *The War Industries Board: Business-Government Cooperation during World War I* (Baltimore, 1973), discusses the end of the Progressive Period.

21. "Military Strategy and the Good Roads Propaganda," *ENR* 78 (June 7, 1917): 514–15.

22. Mehren, "Wanted: A Connected National Highway System," *ENR* 81 (October 3, 1918): 608; see also P. S. Bond, "Roads, a National Not a Local Question," *ENR* 79 (December 27, 1917): 1217; U.S., Congress, Senate, *S. 4993: A Bill to Provide for the Taking over, Improvement, Relocation, Construction, and Maintenance of a System of National Highways and State Highways*, 65th Cong., 2d sess., 1918.

23. These events can be followed in numerous articles in vols. 80–81 of *ENR* (1918), especially E. J. Mehren, "A Suggested National Policy and Plan," 81 (December 19, 1918): 1112–17; and in File 481 General, CCF, BPR Records; Correspondence File for 1918, Chapin Papers; *BPR Annual Report* (1919), pp.

34–36; and Mississippi Valley Conference, "Historical Highlights," pp. x-37–x-38, x-48–x-52.

24. See *OPR Annual Report* (1920), pp. 395–96; File 481 General, CCF, BPR Records; and "Federal-Aid Highway Appropriation Goes through," *ENR* 82 (March 20, 1919): 592.

25. The campaign for a national commission can be traced in the correspondence of Roy Chapin, head of the National Automobile Chamber of Commerce Highway Committee, in Corespondence File for 1919–21, Chapin Papers; and "Pyke Johnson," *American Highways* 48 (April 1969): 16. See also *BPR Annual Report* (1919), p. 424; and *America's Highways*, pp. 216–17.

26. See "Large Committee Will Promote Federal Highway Commission," *ENR* 82 (February 6, 1919): 303–4; "The Townsend Road Law," *Good Roads*, n.s., 16 (July 2, 1919): 7; M. O. Eldridge, "The Need of a Nation," *Good Roads*, n.s., 16 (October 1, 1919): 163, 166; J. M. Richie, "Motor Transportation by the Army," *Good Roads* n.s., 16 (July 23, 1919): 33–35; and Correspondence File for 1919, Chapin Papers.

27. This information is drawn from a series of articles in *ENR* 84 (1920); and Correspondence File for 1920, Chapin Papers.

28. Coleman to MacDonald, January 31, 1919, MacDonald Papers. On MacDonald, see Iowa State Highway Commission, *Annual Report* (1906–19); George S. May, "The Good Roads Movement in Iowa, *The Palimpest* 46 (February 1965): 65–128; "To Direct Federal-Aid Road Work," *ENR* 82 (April 3, 1919): 647; William E. Lind, "Thomas H. MacDonald: A Study of the Career of an Engineer and His Influence on Public Roads in the United States, 1919–1953" (M.A. thesis, The American University, 1965); and *America's Highways*, pp. 102–5.

29. See MacDonald to D. F. Houston, January 9 and March 20, 1919, MacDonald Papers.

30. *BPR Annual Report* (1919), pp. 397–98.

31. MacDonald, "Classification and Uses of Highways and the Influence of Federal-Aid Acts," *ENR* 83 (December 11–18, 1919): 985–88; and USDA, *Annual Report* (Washington, D.C., 1920), p. 45. On the classification itself, see "Highway Classification undertaken by Bureau of Public Roads," *ENR* 84 (March 25, 1920): 635–37. The map's name came from MacDonald's decision to involve the Army in the mapping process, and General John Pershing reviewed the final map when it was presented to Congress.

32. See *BPR Annual Report* (1920), p. 494; and transcript of a conference with state highway officials, March 1919, File 481 General, CCF, BPR Records.

33. See "Criticism of Bureau of Public Roads Unfounded," *ENR* 82 (March 27, 1919): 597; *BPR Annual Report* (1920), pp. 493–96; MacDonald, "Four Years of Road Building under the Federal-Aid Act," *Public Roads* 3 (June 1920): 3–14; and idem, "Federal Aid Accomplishments," *Public Roads* 3 (December 1920): 11–15.

34. MacDonald to All Engineers, May 25, 1919, File 481 General, CCF, BPR Records.

35. *BPR Annual Report* (1919), p. 424, see also *BPR Annual Report* (1920), pp. 44–45; "Committee Appointed to Cooperate with Bureau of Public Roads," *ENR* 82 (May 22, 1919): 1034; and *ENR* 84 (June 3, 1920): 1128.

36. Sherwood K. Booth, "The History of Federal-Aid Legislation," in AASHO, *AASHO: The First Fifty Years*, pp 152–55; *America's Highways*, pp. 178, 543; and MacDonald, *Federal Aid*, pp. 114–15.

37. See File 481 General, especially L. I. Hewes to MacDonald and reply, March 5 and 29, 1920, CCF, BPR Records; on Arkansas, see "Federal Inquiry on Roads," *ENR* 86 (April 7, 1921): 577–78; "Aid to Arkansas Good Roads Halted by Federal Government," *Commercial and Financial Chronicle* 112 (May 21, 1921): 2142–43.

38. Coleman, "Relations of the States with the Bureau of Public Roads," *Public Roads* 2 (January-February 1920): 3–9.

39. See the twelve article series "Trends in Highway Development—A Survey," *ENR* 85 (November 11, 1920)–86 (January 27, 1921); C. S. Hill, "Developments for 1919 in the Use of Machinery for Highway Construction," *ENR* 84 (January 1, 1920): 8–9; and regular reports on road expenditures in *ENR* 84–86 (1920–21).

40. See *BPR Annual Report* (1919–20); and *Historical Statistics*, p. 462.

41. See "Summary of Highway Construction Conditions in Various States," *Engineering and Contracting* 53 (January 7, 1920): 12–14; "Trends in Highway Development," "Highway Construction in 1919 and Its Lessons for 1920," *ENR* 84 (January 1, 1920): 5; "State Highway Construction in 1920 and 1921," *Engineering and Contracting* 55 (February 2, 1921): 100–106; and S. E. Bradt, "Difficulties Experienced by the States in the Matter of Rail Transportation," *Good Roads*, n.s., 21 (March 2, 1921): 131–132.

42. See *BPR Annual Report* (1920); *ENR* 85 (1920); and "State Highway Construction in 1920 and 1921," pp. 100–106.

43. See File 481 General for 1918–21, CCF, BPR Records; see also *BPR Annual Report* (1920–21).

44. See "Report Constructively Criticizes California's Highway Policy," *ENR* 86 (February 3, 1921): 207–08; "California Highway Commission Investigated by Federal Bureau," *ENR* 86 (March 17, 1921): 423–25; BPR, *Report of a Study of the California Highway System* (Washington, D.C., 1922); and File 440 California, CCF, BPR Records.

45. See "Secretary Houston Discusses Federal Road Commission Bill," *Public Roads* 2 (June 1919): 3–5; White to MacDonald, August 5, 1919, File 481 General, CCF, BPR Records; *ENR* 83 (October 30–November 6, 1919): 836; MacDonald to E. G. Harrison, November 19, 1919, MacDonald Papers; and Correspondence File for 1919–20, Chapin Papers.

46. See Correspondence File for 1919-20, Chapin Papers; Mississippi Valley Conference, "Historical Highlights," pp. x-38–x-41; and File 481 General, 1920, CCF, BPR Records.

47. Mackall to MacDonald, June 7, 1920, File 481 General, CCF, BPR Records.

48. See several articles on AASHO's efforts to develop a highway bill in *ENR* 85 (1920); *BPR Annual Report* (1920), pp. 508–59; and P. St. J. Wilson to W. A. Crossland, March 28, 1921, File 481 General, CCF, BPR Records.

49. The development of contacts between MacDonald, Johnson, and Chapin can be followed in this file. Chapin to MacDonald, July 16, 1920, Correspondence File for 1919–20, Chapin Papers.

50. "Federal Highway Appropriations," *ENR* 86 (January 13, 1921): 50–51.

51. On the bill's passage and the lobbying efforts on its behalf, see Correspondence File for 1921, Chapin Papers; Files 481 General and 740.1.1, CCF, BPR Records; and *ENR* 86–87 (1921).

52. Quoted in "Phipps Highway Bill Goes to Senate Committee," *ENR* 87 (July 7, 1921): 35–38.

53. See the extensive correspondence for 1920 and 1921 in Files 481 General and 740.1.1, CCF, BPR Records; and in the Townsend Bill File, 1919–21, MacDonald Papers.

54. These changes can be followed in the coverage of *ENR* 86 (1921); and in Files 481 General and 740.1.1, CCF, BPR Records.

55. *ENR* 87 (August 25, 1921): 337–38.

56. See a series of pieces by MacDonald, including "Rural Transportation," May 18, 1921, and "The Problem to be Studied," 1921, both in MacDonald Reports; "Farm to Market Roads," paper presented to the American Farm Bureau Federation, July 12, 1921, File 481 General, CCF, BPR Records; and *BPR Annual Report* (1921).

57. Chapin to W. M. Kinney, November 9, 1921, Correspondence File; Chapin Papers; see also other correspondence in the Chapin Papers; and W. D. Uhler, "Summary of AASHO Poll," October 10, 1921, MacDonald Papers.

58. Eldridge to MacDonald, January 7, 1922, and reply, January 10, 1922, File 481 General, CCF, BPR Records; also H. S. Fairbank to MacDonald, January 5, 1924, File 740.1.1, CCF, BPR Records; and Chapin Papers, Correspondence File for 1921–29.

59. Hirst to MacDonald, September 21, 1921, File 740.1.1, CCF, BPR Records.

Part II

1. U.S., Congress, *Report . . . [on] on Post Roads*, p. 24.

2. Samuel P. Hays, "Political Choice in Regulatory Administration," in Thomas K. McCraw, ed., *Regulation in Perspective: Historical Essays* (Cambridge, Mass., 1981), p. 127.

3. An editorial, "The Age of Good Roads," *Good Roads* 71 (December 1928): 647, conveyed the same idea.

4. Among the works examing the 1920s in light of the associative ideas are Hawley, "Hoover," pp. 116–40; idem, ed., *Hoover as Secretary of Commerce*; J. Joseph Huthmacher and Warren I. Susman, eds., *Herbert Hoover and the*

Crisis of American Capitalism (Cambridge, Mass., 1973); Jerry Israel, ed., *Building the Organizational State: Essays on Associational Activity in Modern America* (New York, 1972); and Louis Galambos, "The Emerging Organizational Synthesis in Modern American History," *Business History Review* 44 (Summer 1970): 279–90.

5. The clearest statement of this view may be in Louis Galambos, *America at Middle Age: A New History of the United States in the 20th Century* New York, 1983).

Chapter Four

1. Booth, "Federal-Aid Legislation," p. 152.

2. MacDonald, "The Need for Highway Research," paper presented to the Highway Research Council, November 11, 1920, New York, File 001.11, CCF, BPR Records. A large literature examines the enthusiastic reception given the automobile, including Flink, *America Adopts the Automobile*; idem, *Car Culture*; Rae, *The Road and the Car*, pp. 40–59, 87–169; Warren Belasco, *Americans on the Road: From Autocamp to Motel, 1910–1945* (Cambridge, Mass., 1979).

3. *Historical Statistics*, p. 461.

4. MacDonald, "Highway Administration in the United States," *Good Roads* 68 (November 1925): 277.

5. *Historical Statistics*, pp. 458–61; John C. Burnham, "The Gasoline Tax and the Automobile Revolution," *The Mississippi Valley Historical Review* 48 (December 1961): 435–59.

6. One can follow state developments during the 1920s in *ENR*, *Good Roads*, *Engineering and Contracting*, and the annual reports of state highway departments.

7. On the mechanization of highway building, see Reynold Wik, "Benjamin Holt and the Invention of the Track-Type Tractor," *Technology and Culture* 20 (January 1979): 90–107; Robert LeTourneau, *Mover of Men and Mountains* (New York, 1960); Robert Harelson Selby, "Earthmovers in World War II: R. G. LeTourneau and His Machines" (Ph.D. diss., Case Western Reserve University, 1970); J. A. Allhands, *Tools of the Earthmover* (Huntsville, Tex., 1951); and *ENR* 84–124 (1920–40).

8. See *BPR Annual Report* (1922–30); *Historical Statistics*, p. 458; and MacDonald, "Earnings of Improved Highways Far Exceed Their Cost," *Automotive Industries* 49 (December 20, 1923): 1237–39; see also "Pennsylvania Adds Local Roads to State System," *ENR* 107 (December 31, 1931): 1028–29; and "Centralized Road Control in North Carolina," *ENR* 110 (January 19, 1933): 82–84.

9. See *BPR Annual Report* (1925); *Historical Statistics*, p. 459; L. I. Hewes, "Federal Roads of the West," *Professional Engineer* 14 (July 1929): 18–19; and E. McGaffey, "What Federal Aid Means to the West," *Sunset* 56 (March 1926): 36–37.

10. MacDonald to Secretary of Agriculture Henry Wallace, March 6, 1922, File 740.1.1, CCF, BPR Records.

11. MacDonald, "Resume of Federal Aid Work and Future Policies," *Engineering and Contracting* 57 (January 4, 1922): 13–16; see also idem, "Federal and State Policies in the Construction of the Federal Aid Highway System," *Engineering and Contracting* 61 (January 1924): 26–33; and *BPR Annual Report* (1922–25).

12. See *BPR Annual Report* (1922), pp. 462–63; and P.St.J. Wilson to L. R. Pests, May 15, 1922, File 740.1.1, CCF, BPR Records.

13. MacDonald to District Engineers, December 2, 1921, CCF, BPR Records; see also File 481 General for 1921–22, CCF, BPR Records; and *America's Highways*, p. 108.

14. See transcript of San Francisco Conference, July 25, 1922, File 481 General, CCF, BPR Records; and *BPR Annual Report* (1922–24).

15. Johnson to Chapin, January 4, 1922, Correspondence File for 1922, Chapin Papers.

16. Quoted in R. J. Windrow, "The Business of Handling Federal Aid from the States' Standpoint," and "Comments," *American Highways* 2 (June 1923): 7–10;

17. MacDonald to Wilson, September 7, 1922, File 481 General, CCF, BPR Records.

18. MacDonald to L. A. Gillett, September 19, 1922, MacDonald Papers.

19. MacDonald, *Federal Aid*, p. 97; see also File 481 General for 1921–40, CCF, BPR Records.

20. MacDonald, *Federal Aid*, p. 97; see also MacDonald to L. I. Hewes, March 29, 1920, and MacDonald to District 4, May 12, 1921, File 481 General, CCF, BPR Records.

21. R. A. Gomez, *Intergovernmental Relations in Highways* (Minneapolis, 1950), p. 114.

22. Quotation from MacDonald to Gillett, September 19, 1922, MacDonald Papers; see also Files 481 Regulations and 481 Right of Way, CCF, BPR Records.

23. *America's Highways*, p. 207; on complaints, see discussion by W. H. Neel in Windrow, "Handling Federal Aid," p. 10.

24. Johnson, "Committees of AASHO," p. 82. Between 1914 and 1964, the BPR provided 50 of 69 AASHO secretaries, 12 of its committee chairmen, and 5 members of its Executive Committee; see ibid., pp. 70–126; and "Chairmen of AASHO Committees," in AASHO, *AASHO: The First Fifty Years*, pp. 343–51.

25. On signs, see Mississippi Valley Conference, "Historical Highlights," pp. xi-1–xi-36; A. E. Johnson, "A Story of Signing," in AASHO, *AASHO: The First Fifty Years*, pp. 129–46; File 014 General, CCF, BPR Records; and issues of *American Highways* and *ENR* from 1925 to 1930. An example of the pressures to include certain roads is in D. E. Colp to the Glacier-to-Gulf Highway Association, June 12, 1925, D. E. Colp Papers, Eugene C. Barker Texas History Center, University of Texas in Austin.

26. Johnson, "History of the AASHO," p. 56; see also correspondence between MacDonald and Markham in File 481 General, CCF, BPR Records, and in the MacDonald Papers.

27. On toll bridges, see *ENR*, *American Highways*. and *Good Roads* during the 1920s; and File 740.1.1, CCF, BPR Records.

28. Gomez, *Intergovernmental Relations*, p. 58.

29. See Rogers, *History of the MSHD*; MSHD, *Biennial Report of the State Highway Commissioner* (Lansing, 1905/6–1945/46); *BPR Annual Report* (1921, 1930); Kennedy interview; and Kenneth Earl Peters, "The Good Roads Movement and the Michigan State Highway Department, 1905–1917" (Ph.D diss., University of Michigan, 1972).

30. See MSHD, "Financing Michigan Highways, Roads and Streets, 1910–1946," Near Print Records, 1936–49, RG 62-20, MSHD Records.

31. This and the following information from Files 481 Correspondence—FAS Michigan and 481 PS & E—Michigan, CCF, BPR Records.

32. Material drawn from the Van Wagoner, Kennedy, and Ziegler papers and Kennedy interview.

33. Voshell to H. K. Bishop, June 5, 1924, File 481—D-7, CCF, BPR Records.

34. R. R. Havens to Voshell, August 6, 1945, File 481 PS & E—Michigan, CCF, BPR Records.

35. See MacDonald, *Federal Aid*, pp. 109–10; A. R. Losh to MacDonald, January 17, 1925, File 740.1.1, CCF, BPR Records; James W. Wadsworth, "Let's Stop This 'Fifty-Fifty' Business," *Nation's Business* 14 (March 1926): 23–24; and Benjamin A. Arneson, "Federal Aid to the States," *American Political Science Review* 16 (August 1922): 443–54.

36. Quoted in MacDonald, *Federal Aid*, pp. 114–15.

37. Ibid., pp. 109–11, 114–15.

38. Quoted in Mississippi Valley Conference, "Historical Highlights," pp. x-105–x-106.

39. On MacDonald, see Kennedy interview; *America's Highways*, pp. 176–79; on Arkansas, see "Arkansas Clears the Way to Sanity in Road Improvement," *ENR* 96 (May 6, 1926): 724–26.

40. See *BPR Annual Report* (1922–1940); articles in *Public Roads* 2–21 (1920–40); and *America's Highways*, pp. 486–513.

41. Kennedy interview.

42. Yenne to MacDonald, June 9, 1925, File 440 for Illinois, CCF, BPR Records; see also File 440 for each state, CCF, BPR Records.

43. Kennedy interview; see also Thomas H. MacDonald, "The General Road Situation in the United States," *Engineering and Contracting* 59 (May 1923): 977–83.

44. Baynard Kendrick, *Florida Trails to Turnpikes, 1914–1964* (Gainsville, 1964) p. 16.

45. John Shaver, "Road Progress in the South Atlantic and Gulf States," *ENR* 95 (December 3, 1925): 917. On state developments, see idem, "Road Progress," *ENR* 95–96 (December 3, 1925–January 4, 1926; "Highway Progress and Problems in the Mid-South," *ENR* 91 (July 26–August 16, 1923); Thomas H. MacDonald, "Federal Influence and Authority," *ENR* 99 (September 1, 1927): 342–43; and Henry G. Shirley, "The Development of Highways in the Southern States," *Good Roads* 68 (May 1925): 127–28.

46. See MacDonald, "Federal Influence and Authority"; File 440 for each state, CCF, BPR Records; and Dearing, *American Highway Policy*, p. 86.

47. For the traditional interpretation of the New Deal, see Richard Hofstadter, *The Age of Reform: From Bryan to F.D.R.* (New York, 1955) and Arthur M. Schlesinger, Jr., *The Coming of the New Deal* (Boston, 1959). For the newer viewpoint, see Martin A. Fausold and George T. Mazuzan, eds., *The Hoover Presidency: A Reappraisal* (Albany, 1974); Hawley, ed., *Hoover as Secretary of Commerce*; and idem, "Hoover."

48. *Historical Statistics*, p. 458.

49. See *BPR Annual Report* (1932–34); and "Three Types of Highway Building Favored by President's Commission for Employment," *American City* 43 (December 1930): 158.

50. Quoted in "Accelerating Road Construction," *ENR* 103 (December 4, 1930): 874–75. On roads as public works, see Edward A. Williams, *Federal Aid for Relief* (New York, 1939), pp. 20–57; and J. Joseph Huthmacher, *Senator Robert F. Wagner and the Rise of Urban Liberalism* (New York, 1968), pp. 59–100.

51. See Huthmacher, *Wagner*, pp. 95–100; James Stuart Olson, *Herbert Hoover and the Reconstruction Finance Corporation* (Ames, Iowa, 1977), pp. 62–75; and *BPR Annual Report* (1933), pp. 1–3.

52. *BPR Annual Report* (1932), pp. 2–7; (1933), pp. 1–3.

53. See MacDonald's correspondence for 1932 in File 481 General, CCF, BPR Records; "Hard Facts," *ENR* 109 (August 11, 1932): 174; and "Wages and Salaries in Road Building," *American Highways* 11 (April 1932): 1–4.

54. On FDR and public works, see Elliott Rosen, *Hoover, Roosevelt, and the Brain Trust: From Depression to New Deal* (New York, 1977), pp. 144–45; and Robert Sherwood, *Roosevelt and Hopkins: An Intimate History*, rev. ed. (New York, 1950), pp. 44–59. See also Files 481 and 740.1.1 for early 1933, CCF, BPR Records; letters to Franklin Roosevelt in Official File No. 129 (Roads and Highways, 1933), FDR Library; and *BPR Annual Report* (1934–35).

55. Ickes, *Back to Work: The Story of the PWA* (New York, 1935), p. 82, see also p. 24.

56. See Files 481 General and 740.1.1, and "Highway Projects Approved during the Week Ending November 11, 1933," File 001.13, CCF, BPR Records; see also "Special Provision for Highway Projects," June 29, 1933, File 42, Box 7, Cartwright Papers.

57. MacDonald, "Informal Memorandum re: Proposed Public Works Legislation," June 8, 1933, File 42, Box 7, Cartwright Papers; see also Files 481 General and 740.1.1, CCF, BPR Records; and "Progress on PWA Highways to January 20, 1934," *American City* 49 (February 1934): 5.

58. MacDonald to Frederick Hoitt, October 16, 1933, File 481 General, CCF, BPR Records.

59. Ickes, *Back to Work*, p. 82; figures from U.S., Federal Works Administration, *Final Report of the WPA Program* (Washington, D.C., 1946), pp. 27, 35, 122, 131; MacDonald to Senator Kenneth McKellar, September 30, 1936, File

481 General, CCF, BPR Records; and John Charnow, *Work Relief Experience in the United States,* Pamphlet No. 8 (Washington, D.C., 1943), pp. 82–84.

60. See Michigan Good Roads Federation, *Highway Needs in Michigan* (1948), pp. 44, 48–53; and Records Relating to the Michigan Good Roads Federation and the Highway Study Committee, RG 58-5, MSHD Records.

61. "In This Issue," *ENR* 112 (January 18, 1934): 4.

62. "Emergency Roadbuilding," *ENR* 112 (January 18, 1934): 70.

63. E. W. James, "How National Planning Can Help in the Development of Major Highway System," *Roads and Streets* 77 (April 1934): 154–56.

64. Ickes, *Back to Work* pp. 81–83; see also MacDonald to C. B. Baldwin, August 11, 1934, File 481 General, CCF, BPR Records; Federal Works Administration, *Final Report of the WPA*, pp. 2, 3–6; Harry Hopkins, *Spending to Save: The Complete Story of Relief* (New York, 1936); and William W. Bremer, "'Along the American Way': The New Deal's Work Relief Programs for the Unemployed," *Journal of American History* 62 (December 1975): 636–52.

65. See File 481 General for 1934–36, CCF, BPR Records; "The Works Program," Release 2-99, Division of Applications and Information, Department of Agriculture, File 3, Box 8, Cartwright Papers; and *BPR Annual Report* (1935–36).

66. See Files 740.1.1 for 1935–39 and 481 Correspondence—FAS Michigan for 1938, CCF, BPR Records; *American Highways* 8–18 (1928–39); and *ENR* 118–19, 122–23 (1937, 1939).

67. See Official File No. 129 (Roads and Highways, 1935), FDR Library; "Use of Relief," *ENR* 86 (July 4, 1935): 26; and BPR correspondence to District 14, March 12, 1936, File 481 General, CCF, BPR Records.

68. Arthur Hyde to George S. Graham, August 1, 1930, File 740.1.1, CCF, BPR Records. Hyde was the BPR's solicitor, and was repeating a sentiment found in numerous other letters in this file.

69. MacDonald's views can be seen in U.S., Congress, House, Roads Committee, *Hearings on H.R. 8096: Roads*, 73d Cong., 2d sess., February 23, 1934, pp. 7–8.

70. Quoted in U.S., Congress, House, Roads Committee, *Hearings on H.R. 11687: Roads*, 74th Cong., 2d sess., 1936, pp. 39–40. MacDonald's influence in this decision can be traced in File 740.1.1, CCF, BPR Records; U.S., Congress, House, *Hearing on H.R. 8096*; U.S., Congress, Post Office and Post Roads Committee *Hearing on H. R. 8781; Emergency Construction of Public Roads*, 73d Cong., 2d sess., May 14, 1934; and *BPR Annual Report* (1934).

71. Arthur W. MacMahon et al., *The Administration of Federal Work Relief* (Chicago, 1941), p. 62; see also pp. 57, 62–63, 322–23. Later, WPA funds were restricted to non-federal-aid system roads and the federal-aid rules lapsed.

72. Comment of Senator George McGill, quoted in U.S., Congress, House, *Hearings on H.R. 8781*, p. 11. See also U.S., Congress, Senate, Post Office and Post Roads Committee, *Hearings on H.R. 8096: Roads*, 73d Cong., 2d sess., 1934.

73. The papers of Cartwright, McKellar, and Hayden contain distressingly

little information about their respective committees, but letters do show the close personal and professional ties built by MacDonald. See also File 740.1.1, CCF, BPR Records; and author's correspondence with Archivist Susie Sato, Arizona State University, Tempe, September 1983.

74. Tugwell, *Roosevelt's Revolution: The First Year—A Personal Perspective* (New York, 1977), pp. 152–53; see also Kennedy interview; and Official File No. 129 (Roads and Highways, 1935, 1939), FDR Library.

75. See "The U.S. Highway System," *Fortune* 23 (June 1941): 92–99, 106, 108; see also Rae, *The Road and the Car*, pp. 73–83.

76. "Trends in Highway Departments," *ENR* 96 (January 1, 1926): 5.

77. W. R. Nelson, "Adequate Country Roads," *Road Builders' News* (August 1939): 12.

78. Cuff, *War Industries Board*, p. 265.

79. Paul Conkin, *Tomorrow a New World: The New Deal Community Program* (Ithaca, 1959), pp. 305–25.

80. Hawley, ed., *Hoover as Secretary of Commerce*, p. 79.

81. MacDonald to District Engineers, May 23, 1919, File 481 General, CCF, BPR Records.

82. MacDonald, "Resume of Federal Aid Work," p. 14.

Chapter Five

1. Howe, "Industrial Research for Sharp Answers to Definite Questions," *ENR* 81 (July 4, 1918): 46.

2. "Extracting Stories from Stone: Roadbuilding Reduced to a Science," *Scientific American* 118 (March 2, 1918): 196.

3. *BPR Annual Report* (1922), pp. 461–62.

4. See ibid. (1920–40); *Public Roads* 2–21 (1921–40); and File 810 General, CCF, BPR Records. See also National Academy of Sciences, National Research Council, HRB, *Ideas and Actions: A History of the Highway Research Board, 1920–1970* (Washington, D.C., c. 1970); AASHO, "Report of the Committee on Highway Research Activities," paper presented to the 14th Annual Meeting of the AASHO, Chicago, November 12–15, 1928; ibid., paper presented to the 18th Annual Meeting of the AASHO, Milwaukee, November 14–19, 1933, both in Iowa State University Library, Ames; and U.S., National Resource Committee, *Research—A National Resource*, vol. 2, *Relation of the Federal Government to Research* (Washington, D.C., 1938), pp. 72–73.

5. "Publications of the Bureau of Public Roads," *Public Roads* 13 (May 1932): 53–54, is a complete listing of the BPR's research publications.

6. MacDonald, "The Bureau of Public Roads," *Civil Engineering* 1 (March 1931): 494; see also H. S. Fairbank quoted in *America's Highways*, p. 180.

7. See Burnham, "The Gasoline Tax;" and *America's Highways*, pp. 243–46. For the bureau's involvement in gasoline taxes and antidiversion efforts, see Files 740.1.1 and 481 General, CCF, BPR Records; *BPR Annual Report* (1931–40); and summaries of gas tax figures printed at frequent intervals in *Public Roads*, especially in the 1930s.

8. See "State Highway Testing Laboratories," *Public Works* 57 (August 1926): 254–56.

9. See HRB, *Ideas and Actions*, p. 11; and AASHO, "Report of the Committee on Highway Research Activities," p. 3.

10. In addition to numerous books by BPR engineers, *Industrial Arts Index, 1922–23*, vol. 11 (New York, 1923), listed 28 articles by BPR engineers under the heading of "Highway Engineering," and later volumes showed a similar volume of publications.

11. MSHD, *Thirteenth Biennial Report* (1929–30), p. 83. See also F. H. Eno, *A Decade of Progress in Highway Research*, The Ohio State University Engineering Experiment Station Circular No. 20 (1930). Requests for help can be found in Files 407 General, 481 for each state, 481 Correspondence, 810 General, 810.1 Sand and Gravel, and 810.2 Bridge Impact, CCF, BPR Records.

12. See "Road Research Station," *ENR* 119 (September 23, 1937): 527.

13. See *OPR Annual Report* (1905–16); Prevost Hubbard, "Test and Research Investigations of the Bureau of Public Roads," *Public Roads* 2 (July 1919): 28; and Dupree, *Science in the Federal Government*, pp. 157–83.

14. The development of research has attracted several leading historians of technology in the last decade, including Leonard S. Reich, *The Making of American Industrial Research: Science and Business at GE and Bell, 1876–1926* (New York, 1985); George Wise, *Willis R. Whitney, General Electric, and the Origins of U.S. Industrial Research* (New York, 1985); and Noble, *America by Design*, pp. 110–28.

15. Willis R. Whitney, "Research as a National Duty," *Science*, n.s., 43 (May 5, 1916): 629. On the National Research Council, see Daniel J. Kevles, "George Ellery Hale, the First World War, and the Advancement of Science in America," in Nathan Reingold, ed., *Science in America since 1820* (New York, 1976), pp. 278–88.

16. On the impact of the war on science, see Dupree, *Science in the Federal Government*, pp. 323–27; Ronald S. Tobey, *The American Ideology of National Science, 1919–1930* (Pittsburgh, 1971); and Kevles, *The Physicists*, pp. 102–54.

17. See "Objectives of Highways Research," paper presented to the HRB, November 9, 1923, MacDonald Reports.

18. This project can be traced in articles in the BPR's research journal, *Public Roads* 1–4 (1918–21); and *BPR Annual Report* (1919–21). I have explored this issue in greater detail in "The Scientific Mystique in Engineering: Highway Research at the Bureau of Public Roads, 1918–1940," *Technology and Culture* 25 (October 1984): 798–831.

19. A. T. Goldbeck, "The Present Status of Impact Tests on Roadway Surfaces," *Public Roads* 2 (October–November 1919): 19.

20. Idem, "Highway Researches and What the Results Indicate," *Good Roads* 62 (April 9, 1922): 238.

21. Idem, "Investigations of Road Subgrades," *Society of Automotive Engineers Journal* 8 (April 1921): 346.

22. *BPR Annual Report* (1926), p. 33.

23. See articles in *Public Roads* 8–12 (1927–31); National Research Council, HRB, *Proceedings of the Annual Meeting of the Highway Research Board*, vols. 15–25 (1935–45); and Files 407 General and 407 for each state for 1925–40, CCF, BPR Records.

24. See E. B. Smith, "The How and Why of Motor Truck Impact," *Public Roads* 3 (July 1920): 16–18; and MacDonald to District Engineers, April 20, 1920, File 407 General, CCF, BPR Records. On the Bates Road, see Clifford Older, "Illinois Begins Traffic Endurance Test," *ENR* 87 (August 18, 1921): 274–76; idem and H. F. Clemmer, "Preliminary Report on the Bates Road Experiment," *Public Roads* 4 (September 1921): 3–13; Illinois Department of Public Works and Buildings, Division of Highways, *Bates Experimental Road; or Highway Research in Illinois*, Bulletin No. 21 (Springfield, 1924); and Files 407 General and 407 for Illinois, CCF, BPR Records.

25. See "Progress Made in Highway Engineering," *ENR* 90 (January 18, 1923): 55; "Concrete Pavement Sections with Thickened Edges," *Engineering and Contracting* 60 (July 4, 1923): 20–21; and "Practical Lessons from the Bates Road Tests," *ENR* 90 (January 18, 1923): 57–61.

26. See P. Leonard Capper and W. Fisher Cassie, *The Mechanics of Engineering Soils* (New York, 1949), p. 26; Charles Terzaghi, *From Theory to Practice in Soil Mechanics* (New York, 1960), p. 60; and "Progress in Highway Research Reported at Baltimore Meeting," *ENR* 127 (December 18, 1941): 887–88.

27. Mylrea to BPR, December 8, 1938, File 890 Tests—Bridge Impact, CCF, BPR Records.

28. Charles Terzaghi, "Concrete Roads—A Problem in Foundation Engineering," in Boston Society of Civil Engineers, ed., *Contributions to Soil Mechanics, 1925–1940* (Boston, 1940), p. 61.

29. A large body of literature traces the changing relationship between science and engineering, including two special issues of *Technology and Culture*: "Science and Engineering," 2 (Fall 1961): 305–99; and "The Interaction of Science and Technology in the Industrial Age," 17 (October 1976): 621–742. Other issues of this journal also contain articles on this subject. On changes in engineering after 1918, see James Kip Finch, *The Story of Engineering* (Garden City, N.Y., 1960), pp. 387–88.

30. Norris, "Research as Applied to Engineering," *Civil Engineering* 5 (May 1935): 408. See Alfred D. Flinn, "Engineering Foundation: Division of Engineering of National Research Council—Their Origin, Work, Plans, and Needs," *ASCE Trans.* 86 (1923): 1286–91; and idem, "Research Advances Civil Engineering, *Civil Engineering* 1 (October 1930): 15.

31. On the Yadkin River Bridge Test, see Correspondence, 1925–29, File 810.2 for North Carolina, CCF, BPR Records; and Clyde T. Morris, "Yadkin River Bridge Tests," *Proceedings of the ASCE* 55 (March 1929): 608–17.

32. *BPR Annual Report* (1924), p. 9.

33. See U.S., National Resource Committee, *Research*, vol. 2, pp. 72–73.

34. Crum, "Highway Research," *ENR* 114 (January 17, 1935): 75. Noted highway engineer W. W. Crosby had called for this approach as early as 1910,

arguing that "it may eventually, if not soon, be possible to reduce a large number of the problems of highway engineering to a mathematical solution, particularly if the Science underlying the Art is properly developed." See Crosby, "The Present Status of Laboratory Tests and the Use of Materials in Road Construction," paper presented to the American Academy for the Advancement of Science, December 1912, Crosby Lectures.

35. W. K. Hatt, "The Coordination of Highway Research," *American Highways* 1 (January 1922): 29. On trade association research, see National Industrial Conference Board, *Trade Associations: Their Economic Significance and Legal Status* (New York, 1925); H. E. Howe, "Scientific Research as an Association Activity," *Bulletin of the American Ceramic Society* 7 (June 1924): 209–21; "How These Industries Conduct Their Research," *Brick and Clay Record* 65 (September 2, 1924): 307–12; and "Research among Trade Associations," *Special Libraries* 16 (April 1925): 117–19.

36. *BPR Annual Report* (1922), p. 490. See also File 810.2 General, CCF, BPR Records; and *BPR Annual Report* (1920–40).

37. See "Activity in Highway Research," *Public Roads* 6 (1925): 130; AASHO, "Report of the Committee on Highway Research Activities" (1928, 1933); and *BPR Annual Report* (1921–40). For an example of the BPR's active role in university investigations, see Reinforced Concrete Slab Investigations File for 1936, Engineering Experiment Station Subject File, 1909–61, Series 11/2/1, University Archives, University of Illinois at Urbana-Champaign.

38. See T. Warren Allen to P.St.J. Wilson, April 3, 1920, File 481 General, and other correspondence from 1920–40 in File 407, CCF, BPR Records; F. H. Jackson and C. E. Proudley, "Observations on the Operations of Material Testing Laboratories," *Public Roads* 5 (June 1924): 14–17; and J. W. Powers, "Materials Testing Laboratory of Arizona Highway Department," *Roads and Streets* 68 (October 1928): 499–500.

39. Hirst, "Memorandum," *Public Roads* 2 (August–September 1919): 37; see also T. R. Agg, "Comprehensive Investigation in Highway Engineering Needed," ibid., pp. 35–37.

40. HRB, *Ideas and Actions*, p. 11.

41. Kevles, "George Ellery Hale," pp. 282–88; Dupree, *Science in the Federal Government*, p. 323; and Charles Whiting Baker, "Organizing America's Research Work on a National Scale," *ENR* 80 (June 6, 1918): 1079–80.

42. HRB, *Ideas and Actions*, p. 12; see also Program of the Highway Research Conference, November 11, 1920, Correspondence File, Chapin Papers; and Minutes of the Highway Research Conference, November 11, 1920, File 001.11, CCF, BPR Records.

43. Hatt, "Moot Questions in Highway Transportation," paper presented to the Faculty of Economics, Princeton University, December 1, 1922, File 001.11, CCF, BPR Records. Hatt gave 28 lectures in 1922 and 48 in 1923, while *Industrial Arts Index*, vol. 11, p. 697, listed 19 articles for 1922–23.

44. See HRB, *Ideas and Actions*, pp. 32–55; "Report of the Director," in HRB, *Proceedings of the Annual Meeting*, vols. 6–20 (1926–40); annual summa-

ries of HRB meetings in *ENR* each November or December from 1925–40; and File 001.11, CCF, BPR Records.

45. On state research activities, see numerous articles in *Public Works*, *Public Roads*, *Municipal and County Engineering*, and *Roads and Streets* during the 1920s and 1930s; and AASHO, "Report of the Committee on Highway Research Activities" (1928, 1933).

46. HRB, *Proceedings of the Annual Meeting*, vols. 5–20 (1925–40); *BPR Annual Report* (1922–26); and the bulletins of the Engineering Experiment Stations at Purdue, Illinois, and Iowa State universities, 1921–40.

47. See "Activity in Highway Research," p. 130; AASHO, "Report of the Committee on Highway Research Activities," (1928, 1933); and HRB, *Ideas and Actions*, pp. 44, 191.

48. Figures from HRB, *Ideas and Actions*, pp. 11, 44; AASHO, "Report of the Committee on Highway Research Activities," (1928, 1933); and *BPR Annual Report* (1922–38).

49. R. W. Crum, "Report of the Director of the Advisory Board for Highway Research," (November 23, 1922), p. 8, in File 001.11, CCF, BPR Records.

50. See File 001.11, CCF, BPR Records; and HRB, *Ideas and Actions*, pp. 25, 47–49, 134–37.

51. See "Some Recent Developments in Highway Research," *ENR* 93 (December 11, 1924): 958.

52. See *BPR Annual Report* (1925–40); and HRB, *Proceedings of the Annual Meeting*, vols. 5–20 (1925–40).

53. Hatt to Marston, October 20, 1921, File 001.11, CCF, BPR Records; see also other correspondence in this file; and HRB, *Ideas and Actions*, pp. 138–39, 208.

54. Brousseau to MacDonald, January 25, 1922, File 001.11, CCF, BPR Records.

55. MacDonald to State Highway Departments, November 29, 1929, File 001.11, CCF, BPR Records. For another view of status concerns by engineers regarding science, see Stuart W. Leslie, "Thomas Midgley and the Politics of Industrial Research," *Business History Review* 54 (Winter 1980): 480–503.

56. Flinn, "Highway Research: Its Relation to Highway Bonds and Banks," February 21, 1921, File 001.11, CCF, BPR Records.

57. "Reporting Highway Research," *ENR* 99 (December 15, 1927): 945.

58. "Valuable Discussion," *ENR* 101 (December 20, 1928): 903–4.

59. HRB, *Ideas and Actions*, pp. 57–58; see also ibid., p. 201; *America's Highways*, pp. 208–10; and "Postwar Era," *Concrete* 52 (January 1944): 5.

60. HRB, *Ideas and Actions*, p. 105.

61. *America's Highways*, p. 352; see also National Research Council, HRB, *Organization of Highway Research: Summary and Cases*, Special Report No. 99 (Washington, D.C., 1968); idem, *Getting Research Findings into Practice*, National Cooperative Highway Research Program Synthesis of Highway Practice, No. 23 (Washington, D.C., 1974); idem, *Research Needs in Highway Transportation*, National Cooperative Highway Research Program Report No. 55

(Washington, D.C., 1968); and National Research Council, Transportation Research Board, *America's Highways: Accelerating the Search for Innovation*, Special Report No. 202 (Washington, D.C., 1984).

Chapter Six

1. Crittenden, "Contributions of Government and Private Agencies to National Standards," in Dickson Reck, ed., *National Standards in a Modern Economy* (New York, 1952), p. 36.
2. Quoted in *America's Highways*, p. 178.
3. Hoover, "The Crusade for Standards," in Reck, ed., *National Standards*, p. 4.
4. Sinclair, *Centennial History*, pp. 44–60, 144–57, discusses the role of compromise in setting standards.
5. The development of standards in the late nineteenth century is traced in ibid., pp. 46–60; Bruce Sinclair, "At the Turn of a Screw: William Sellers, the Franklin Institute and a Standard American Screw Thread," *Technology and Culture* 11 (January 1969): 20–34; Steven F. Usselman, "Air Brakes for Freight Trains, Technological Innovation in the American Railroad Industry, 1869–1891," *Business History Review* 58 (Spring 1984): 30–50; Charles B. Dudley, *Memorial Volume Commemorative of the Life and Work of Charles Benjamin Dudley, Ph. D.* (Philadelphia, 1911), pp. 122–61, 276–97; and C. L. Warwick, "The Work in the Field of Standardization of the American Society for Testing Materials," *Annals of the American Academy of Political and Social Science* 137 (May 1928): 49; and Noble, *America by Design*, pp. 69–83.
6. See Thompson, "Intercompany Technical Standardization," pp. 1–20; David Hemenway, *Industrywide Voluntary Product Standards* (Cambridge, Mass., 1975), pp. 13–18; and Sinclair, *Centennial History*, pp. 144–57.
7. Adams, "National Standards Movement—Its Evolution and Future," in Reck, ed., *National Standards*, pp. 21–30; see also National Industrial Conference Board, *Industrial Standardization* (New York, 1929), pp. 11–13; Noble, *America by Design*, pp. 77–81; and U.S., Department of Commerce, The National Bureau of Standards, *Measures for Progress: A History of the National Bureau of Standards*, by Rexmond C. Cochrane (Washington, D.C., 1966), pp. 38–49, 229–37, 253–98. Indications of the interest in standardization in the 1920s abound in the *ENR* and other trade and engineering periodicals.
8. Hoover, "The Crusade for Standards;" see also "Twenty-Four National Standardizing Bodies," *Industrial Standards* 6 (March 1935): 69; "Nearly Half of NRA Codes Have Clauses on Standards," *Industrial Standards* 6 (May 1935): 133–137; C. R. Watts, "Standards in the Federal Supply System—Military Goods," in Reck, ed., *National Standards*, pp. 247–59; Terence R. B. Sanders, ed., *The Aims and Principles of Standardization* (Geneva, Switz., 1972); and Rhonda J. Crane, *The Politics of International Standards: France and the Color T.V. War* (Norwood, N.J., 1979).

9. Herman Oliphant, "How Far Can the Trade Association Go in Cooperative Service to Industry?" *Chemical and Metallurgical Engineering* 30 (February 4, 1924): 190–91; National Industrial Conference Board, *Industrial Standardization*, pp. 256–57; H. P. Baker, "How Can a Trade Association Bring about Standardization in an Industry?" *Paper Trade Journal* 77 (November 22, 1923): 27–28; and P. G. Agnew, "Role of the Trade Association in Industrial Standardization," *Iron Age* 124 (October 31, 1929): 1172.

10. See *America's Highways*, p. 44; "Specifications for Pavements and Road Crusts," *American City* 16 (February 1917): 140–44; "Report of the ARBA Committee on Standards," December 11, 1913, Correspondence File, North Carolina Geological Survey Records; Mississippi Valley Conference, "*Historical Highlights*," pp. i–71, xii-48–xii-49. One state that could act on its own was Illinois. See Clifford Older, "Standard Plans Solve Problems of State Highway Bridge Supervision in Illinois," *ENR* 78 (April 5, 1917): 31–34; and "New Features of Indiana's Concrete Road Specifications," *ENR* 83 (September 25, 1918): 614–15.

11. MacDonald, "Four Years of Building Roads under the Federal-Aid Act," *Public Roads* 3 (June 1920): 11.

12. MacDonald, *Federal Aid*, p. 112.

13. E. W. James to P. St. J. Wilson, October 25, 1919, File 481 General, CCF, BPR Records.

14. MacDonald to Watson, January 4, 1922, File 740.1.1 General, CCF, BPR Records. For further examples, see other correspondence in this file and in File 481 PS & E, CCF, BPR Records.

15. MacDonald to George P. Coleman, December 12, 1916, MacDonald Papers.

16. *America's Highways*, p. 88.

17. Ibid., p. 178. On the AASHO's early standards committee, see Files 481 General and 470 for 1917–18, CCF, BPR Records.

18. Kennedy interview.

19. Minutes of the Executive Committee, June 22–23, 1944, Committees—AASHO, 1943–44, Correspondence, 1943–50, MSHD Records. On the AASHO's committees, see Files 481 Culvert Specifications and 470, CCF, BPR Records; and AASHO, *AASHO: The First Fifty Years*, pp. 343–49, 51.

20. Kennedy interview; see also *BPR Annual Report* (1922–40); Files 440 for Pennsylvania, 440 General, and 481 Correspondence—Michigan, CCF, BPR Records; Johnson, "Committees of AASHO," p. 103; Minutes of the AASHO Executive Committee, June 4, 1923, MacDonald Papers; "Report of Committee on Uniform Highway Accounting," in HRB, *Proceedings of the Annual Meeting*, vol. 17 (1937): 32–37; and AASHO, *Policies on Geometric Highway Design* (Washington, D.C., 1950), preface.

21. Wilson to W. R. Bonsal Company, August 25, 1933, File 470 General, CCF, BPR Records; see also Fairbank to E. D. Buel, September 23, 1931, and other correspondence in ibid.

22. See Correspondence, File 481 PS & E Michigan for 1921–40, CCF, BPR Records.

23. Correspondence, in File 481 PS & E—Michigan, CCF, BPR Records.

24. Memorandum: E. W. James to the Director, June 9, 1916, File 481 General, CCF, BPR Records.

25. P.St.J. Wilson to Voshell, May 1, 1926, File 481 PS & E—Michigan, CCF, BPR Records.

26. Wilson to Mees and Mees, Consulting Engineers, March 9, 1920, File 481 General, CCF, BPR Records; see also MacDonald, "Four Years of Road Building," p. 11. BPR efforts can be traced in File 481 PS & E—Michigan, CCF, BPR Records.

27. Wilson to MacDonald, November 4, 1921, File 481 General, CCF, BPR Records.

28. MacDonald, *Federal Aid*, p. 10; comment about "our standards" from Wilson to Benjamin Arneson, December 14, 1921, File 481 General, CCF, BPR Records. See also MSHD, *Biennial Report* (1921–22—1939–40); and MSHD, *Standard Specifications for Road and Bridge Construction, Adopted September 9, 1934* (Lansing, 1934).

29. See Files 470, 481 PS & E—Michigan, and 481 Tests—Michigan for 1920–40, CCF, BPR Records.

30. "Paternalistic Standardization," *ENR* 83 (October 30–November 6, 1919): 782.

31. Iowa State Highway Commission, *Second Annual Report of the Iowa State Highway Commission* (Ames, 1915), p. 35.

32. MacDonald to Richard L. Humphrey, July 30, 1919, File 481 Culvert Specifications, CCF, BPR Records.

33. See U.S., Bureau of the Census, *Special Report: Mines and Quarries, 1902* (Washington, D.C., 1905), pp. 785, 839; and idem, *Census of Manufactures, 1914*, vol. 2, *Reports for Selected Industries and Detail Statistics for Industries by States* (Washington, D.C., 1919), p. 922.

34. W. K. Hatt to Chairmen, February 21, 1923, File 001.11, CCF, BPR Records; see also George E. Martin to J. S. Crandell, April 3, 1940, Folder 11, J. S. Crandell Papers, Series 11/5/25, University Archives, University of Illinois at Urbana-Champaign.

35. See ASTM, *Yearbook* (Philadelphia, 1920–70).

36. On the BPR-ASTM connection, compare *BPR Annual Report* (1919–38); with ASTM, *Proceedings of the Annual Meeting of the ASTM*, vols. 4–39 (Philadelphia, 1904–39). Also see BPR, *Tentative Standard Methods of Sampling and Testing Highway Materials*, Agriculture Department Bulletin No. 1216 (Washington, D.C., 1924).

37. Scherer, "Standard Sizes of Crushed Stone," *Public Roads* 1 (September 1918): 17; see also numerous articles in *ENR* 78–80 (1917–18); F. H. Jackson and C. W. Mitman, "The Commercial Sizes of Broken Stone Aggregate," *Public Roads* 1 (June 1918): 35–48; Correspondence in File 470 for 1917–23, CCF, BPR Records; *BPR Annual Report* (1920), p. 497; and ASTM, *Proceedings of the Twenty-Third Annual Meeting*, vol. 20, pt. 1 (1920), p. 432.

38. Harmer E. Davis, George E. Troxell, and Clement T. Wiskocil, *The Testing and Inspection of Engineering Materials* (New York, 1941), p. 11.

39. See "Edgar Marburg, 1864–1918," *ENR* 81 (July 4, 1918): 50; HRB, *Proceedings of the Second Annual Meeting of the Advisory Board for Highway Research* (1922); and ASTM, *Proceedings of the Annual Meeting*, vols. 20–40 (1920–40).

40. Hemenway, *Product Standards*, p. 87.

41. "ASCE Tackles Standards," *Civil Engineering* 46 (September 1976): 67.

42. For cases studies involving paving brick, metal and clay culvert pipe, road oils and tars for secondary roads, and other materials, see in Files 481 Culvert Specifications, 481 PS & E—Michigan, and 450.40 General—Correspondence, CCF, BPR Records; and 1932 articles on road oils in *Public Roads* and *ENR*.

43. MacDonald to the Oil Industry, August 1, 1931, File 450.40 General—Correspondence, CCF, BPR Records; other information on this episode in the same file.

44. See Norman F. Harriman, *Standards and Standardization* (New York, 1923), pp. 221–24; and File 470, CCF, BPR Records.

45. See Lacy M. Law, "Asphalt Specification—Past, Present and Future," *Good Roads* 62 (February 22, 1922): 111-13; and Gene Abson, "Asphalt Specifications—Open vs. Closed," *Municipal and County Engineering* 70 (May 1926): 288–90.

46. E. F. Kelley to MacDonald, February 2, 1926, File 001.11, CCF, BPR Records.

47. Victor J. Brown and Carleton N. Conner, *Low Cost Roads and Bridges* (Chicago, 1933), p. viii.

48. MacDonald, paper presented to the Associated General Contractors of America, February 15, 1937, MacDonald Reports.

49. See Joseph A. Pratt, "Creating Coordination in the Modern Petroleum Industry: The Americam Petroleum Institute and the Emergence of Secondary Organizations in Oil," *Research in Economic History* 8 (1983): 179–215; Hawley, ed., *Hoover as Secretary of Commerce*; and David P. Lee, "Herbert Hoover and the Development of Commercial Aviation, 1921–1926," *Business History Review* 58 (Spring 1984): 78–102.

50. See Hawley, ed., *Hoover as Secretary of Commerce*, pp. 43–79; idem, "Three Facets of Hooverian Associationalism; Lumber, Aviation, and Movies, 1921–1930," in Thomas K. McCraw, ed., *Regulation in Perspective: Historical Essays* (Cambridge, Mass., 1981); pp. 95–123; and Fausold and Mazuzan, eds., *Hoover Presidency*.

51. Cuff, *War Industries Board*, p. 225.

Part III

1. Galambos, *America at Middle Age*.

Chapter Seven

1. Chevalier, "What of Tomorrow's Highways—The Need for Coordinated Effort," *ARBA Proceedings* 33 (1936): 258.

2. *Historical Statistics*, pp. 458–59.

3. See *BPR Annual Report* (1934); and "How the $400,000,000 Federal-Aid Highway Appropriation Will Be Spent," *Roads and Streets* 76 (November 1933): 408–12.

4. For example, see U.S., Congress, House, *Hearings on H.R. 11687*, p. 36. In *Toward a Planned Economy: From Roosevelt to Nixon* (New York, 1976), pp. 1–68, Otis L. Graham has shown that although central planning was more a product of World War II, it had begun in the 1930s. See also Galambos, *America at Middle Age.*

5. See "Roads to Politics," *ENR* 116 (January 9, 1936): 62; and H. S. Fairbank, "A Highway Holiday," *ENR* 117 (October 1, 1936): 467.

6. FDR to U.S. Congress, November 29, 1937, Folder 24, Box 8, Cartwright Papers.

7. See Wilburn Cartwright to FDR, December 16, 1937, and other Correspondence, Official File 129 (Roads and Highways, 1937, 1938), FDR Library; articles in *ENR*; Willard T. Chevalier, "What about Federal Aid?" *ARBA Proceedings* 35 (1938): 6–27; File 22, Box 8, and File 3, Box 9, Cartwright Papers; and Cartwright, "Future Roads and Federal Road Legislation," *American Highways* 18 (January 1939): 7–8.

8. Richard N. Chapman, *Contours of Public Policy, 1939–1945* (New York, 1981), pp. 1–59, discusses budget tinkering in 1939–40.

9. Murray D. Van Wagoner, "Superhighways Ahead," *ARBA Proceedings* 36 (1939): 13; see also A. G. Bruce, "Effect of Increased Speed of Vehicles on the Design of Roads," *Public Roads* 10 (March 1929): 10–11. Figures from *Historical Statistics*, p. 462.

10. "Unfit for Modern Motor Traffic," *Fortune* 14 (August 1936): 91–92. On the highway safety issue, see E. Cato, "Speeding to Destruction," *National Safety News* 32 (October 1935): 47–48; and Foster, *From Streetcar to Superhighway*, pp. 111–12.

11. Walter A. Jones, "Toll Road Financing—The Pennsylvania Turnpike," *ARBA Proceedings* 37 (1940): 50.

12. Ben H. Petty, "Highway Demand Is Turning from Main to Local Roads," *ENR* 110 (January 5, 1933): 7.

13. MacDonald, "The Federal Highway Act," *Farm and Fireside*, February 1922, MacDonald Reports.

14. MacDonald, "The Future of Road Building," article for *Curtis 1000*, April 1922, MacDonald Reports.

15. MacDonald, "Federal and State Policies in the Construction of the Federal Aid Highway System," *American Highways* 2 (September 1923): 8; statistics from Holt, *Bureau of Public Roads*, p. 29.

16. MacDonald, "Highways in the Making," *American City* 38 (July 1928): 91.

17. Otto S. Hess, "Report of the Committee Number 1; Legislation, Finance and Administration," *ARBA Proceedings* 27 (1930): 552; see also George W. Farny, "For More Systematic Development of County Highways," *American City* 44 (March 1931): 136. On legislation, see File 740.1.1 for 1928–30, CCF, BPR Records.

18. Roy D. Chapin, "The Relation of the Vehicle Builder to the Highway Maker," *American Highways* 8 (January 1929): 3.

19. W. C. Markham, "Annual Report," *American Highways* 10 (October 1931): 25; see also other articles in that journal and *ENR* for 1930–31. On state secondary road programs, see Samuel Eckels, "Addition of Roads to the State Highway System," *American Highways* 10 (October 1931): 12–13; and Charles Ross, "North Carolina's Adventure in the Field of County Roads," ibid., pp. 14–16.

20. On the BPR's activities, see File 740.1.1, CCF, BPR Records; articles in *ENR* 110–15 (1933–35); and "Secondary Roads," *Roads and Streets* 78 (August 1933): 308–9. On support for a federal rural roads effort, see Charles Davis to Harold Ickes, February 2, 1934, Official File 129 (Roads and Highways, 1934), FDR Library; in File 39, Box 7, Cartwright Papers; and articles in *ARBA Proceedings* 32–33 (1935–36).

21. B. Lascher to James Wilson, May 31, 1897, File 141, CG, BPR Records; see also Julian Hawthorne, "A Transcontinental Roadway," *The Cosmopolitan* 28 (November 1899): 125–28.

22. "Express Roads for High Motor Speeds," *Literary Digest* 64 (January 31, 1925): 46, 48. On prewar cross-country highways, see File 141, GC, BPR Records; Folder 4, Yellowstone Trail Association Records, South Dakota Historical Resource Center, Pierre; and Lincoln Highway Association, *Lincoln Highway*. On post–1920 plans, see File 740.1.1, CCF, BPR Records.

23. Shirley, "Imaginative and Speculative Highways for Tomorrow," *ENR* 81 (December 26, 1918): 1175–77.

24. On superhighway development, see John B. Rae, "Coleman du Pont and His Highway," *Delaware History* 16 (Spring–Summer 1975): 171–83; Spencer Miller, Jr., "History of the Modern Highway in the United States," in Jean Labatut and Wheaton J. Lane, eds., *Highways in Our National Life: A Symposium* (Princeton, 1950), pp. 104–8; R. E. Toms and J. W. Johnson, "Design and Construction of the Mt. Vernon Memorial Highway," *Journal of the American Concrete Institute* 3 (April 1942): 563–84; and articles in *Roads and Streets* and *ENR* in the 1920s.

25. On legislation, see File 740.1.1 for 1926–30, CCF, BPR Records, esp. F. S. Greene to MacDonald, January 24, 1928; Folder 36, Box 7, Cartwright Papers; and "Proposed Federal Superhighway System," *Roads and Streets* 68 (December 1928): 609–10.

26. "Toward New Road Systems," *ENR* 118 (April 1, 1937): 493; see also "Construction of Superhighways," *Canadian Engineering* 59 (December 23, 1930): 738.

27. See Files 740.1.1 and 481 General for 1934–35, CCF, BPR Records; Charles Davis, "What Is the Most Logical Plan for Highway Extension?" *Manufacturers' Record* 103 (January 1934): 18; E. A. Filene, "Railroads, a Superhighway, and the Unemployed," *Survey* 22 (February 1933): 83–85; Fritz Malcher, "Will 1932 Witness the Beginning of a New Epoch in Highway Design and Construction?" *American City* 46 (January 1932): 104–5; and Official File 129 (Roads and Highways, 1935, 1938), FDR Library.

28. See B. Wehner, "Germany Begins Construction of Express Highway System," *ENR* 113 (July 5, 1934): 10–12; A. Rappaport, "New Methods of Highway Control and Development in Germany," *American City* 49 (February 1934): 52; idem, "New Highway Design and Construction in Germany," *American City* 49 (October 1934): 69; and E. Zube, "Germany Builds Express Highways," *Roads and Streets* 78 (June 1935): 187–90. See also J. S. Crandell, "A Touring Engineer's Impressions of the Roads of Italy," *ENR* 115 (November 14, 1935): 669–74.

29. MacDonald, "Remarks to American Automobile Association Luncheon," November 20, 1936, MacDonald Reports; see also idem, "Tomorrow's Roads: Europe's Active Modernization of Its Highways," *ENR* 117 (December 17, 1936): 868–70.

30. MSHD News Release, August 12, 1938, File 481 General for Michigan, CCF, BPR Records. For the reaction of other engineers, see G. Donald Kennedy to Murray D. Van Wagoner, November 21, 1936, Correspondence File, 1935–36, Van Wagoner Papers; and coverage in *ENR* and *Roads and Streets* in 1936–37.

31. *ENR* covered these developments from 1935 through 1938.

32. MacDonald, "The Freedom of the Road," *American Highways* 8 (January 1929): 7–8. On early BPR opposition, see File 71, GC, and MacDonald to Clark Burdick, January 25, 1923, File 740.1.1, CCF, BPR Records; and *BPR Annual Report* (1928). See also F. Lavis, "Letter to the Editor," *ENR* 97 (December 23, 1926): 1050; N. G. Shidle, "Industry against Current Attempt to Establish Toll Roads," *Automotive Industries* 58 (August 7, 1928): 549; and P. K. Schuyler, "Is the Toll Road a Solution?" *Roads and Streets* 70 (April 1930): 129–31.

33. E. E. Duffy, "Toll Highways," *Manufacturers Record* 94 (November 8, 1928): 78; see also other articles in that journal in 1928; "Express Highways and Tolls," *ENR* 97 (December 30, 1926): 1062; "New York-New Haven Toll Road Proposed by Private Corporation," *ENR* 104 (1930): 627; and File 740.1.1, CCF, BPR Records.

34. See U.S., Congress, Senate Post Office and Post Roads Committee, *Hearing on S. J. Res. 106: Transcontinental Highway Commission*, 75th Cong., 1st sess., April 13, 1937; idem, House, Roads Committee, *Hearing on H.R. 7079, H.J. Res. 204, and H. J. Res. 227: Roads*, 75th Cong., 1st sess., May 18, 1937; Charles Davis, "Twenty Year Highway Building" (n.d.), in Folder 31, Box 10, Cartwright Papers; and T. E. Steiner, *The Highways of Tomorrow Made Possible by the Ideal of Today,* 4th ed. (n.p., 1938).

35. McShane, "Use of Urban Space," pp. 291–96.

36. This point is made especially well by Paul Barrett, "Public Policy and Private Choice: Mass Transit and the Automobile in Chicago between the Wars," *Business History Review* 49 (Winter 1975): 473–97; and idem, *Automobile and Urban Transit.* pp. 3–5.

37. See Glen E. Holt, "The Changing Perception of Urban Pathology: An Essay on the Development of Mass Transit in the United States," in Kenneth T. Jackson and Stanley K. Schultz, eds., *Cities in American History* (New York, 1972), pp. 324–43; Stanley Mallach, "The Origins and Decline of Mass Transit in the United States, 1890–1930," *Urban Past and Present* 8 (1979): 1–17.

38. Mark S. Foster, "The Model T, the Hard Sell, and Los Angeles Urban Growth: The Decentralization of Los Angeles during the 1920s," *Pacific Historical Review* 44 (November 1975): 459–84; Blaine Brownell, "A Symbol of Modernity: Attitudes toward the Automobile in Southern Cities," *American Quarterly* 24 (March 1972): 20-44; David Brodsly, *L.A. Freeway: An Appreciative Essay* (Berkeley, 1981), p. 36; and Joel A. Tarr, "From City to Suburb: The 'Moral' Influence of Transportation Technology," in Alexander B. Callow, ed., *American Urban History*, 2nd ed. (New York, 1973), pp. 202–12.

39. Barrett, *Automobile and Urban Transit*, esp. pp. 129–63; and Brownell, "Symbol of Modernity."

40. Barrett, "Public Policy and Private Choice," pp. 482–83. On McClintock, see his *Street Traffic Control* (New York, 1925); his numerous general publications, including *A Better Use of the City Street with Special Reference to Parking On and Off the Street* (New York, 1930); and a lengthy series of reports on individual cities, such as *A Report on the Street Traffic Problem of San Francisco* (San Francisco, 1927); and Harvard University, Albert Erskine Russel Bureau for Traffic Research, *A Report on the Parking and Garage Problem of the Central Business District of Washington, D.C.* (Washington, D.C., 1930). Foster, *From Streetcar to Superhighway*, pp. 160–62, discussed other planners engaged in this work.

41. See Brodsly, *L.A. Freeway*, p. 82; Robert M. Fogelson, *Fragmented Metropolis: Los Angeles, 1850–1930* (Cambridge, Mass., 1967), p. 92; Robert Caro, *The Power Broker: Robert Moses and the Fall of New York* (New York, 1974), p. 329; Paul Barrett, "Mass Transit, The Automobile, and Public Policy in Chicago, 1900–1930" (Ph.D. diss., University of Illinois, Chicago Circle, 1976), pp. 209, 521; and Howard L. Preston, *Automobile Age Atlanta: The Making of a Southern Metropolis, 1900–1935* (Athens, 1979), p. 51.

42. "Traffic Jams Will Be Solved by Engineers," *Nation's Traffic* 1 (May 1927): 25; see also Thomas H. MacDonald, "Parkways for Pleasure and Utility," *American City* 31 (November 1924): 421–22. Urban highway programs can be followed in *American City, ARBA Proceedings, Engineering and Contracting*, and *ENR* during the 1920s.

43. Foster, *From Streetcar to Superhighway*, pp. 80–86, traces Detroit's efforts; see also the Detroit Rapid Transit Commission Papers, Burton Historical Collection, Detroit Public Library; Detroit Rapid Transit Commission, *General Plan of Rapid Transit and Surface Line System for the City of Detroit* (Detroit, 1923); idem, *Proposed Super-Highway Plan for Greater Detroit* (Detroit, 1924); and numerous articles on Detroit's highway program in such technical journals as *American City, ENR*, and *Engineering and Contracting*.

44. See Barrett, *Automobile and Urban Transit*, esp. pp. 67–81, 129–67; Eugene S. Taylor, "Arterial Street Planning," *ARBA Proceedings* 23 (1926): 69–72; and *ENR* 96–105 (1926–30).

45. Brummitt, "Superhighway," *American City* 40 (January 1929): 85–88; see also Foster, "The Model T," pp. 459–84; idem, *From Streetcar to Superhighway*, pp. 86–88; Brodsly, *L. A. Freeway*, pp. 91–97, 151–57; Fogelson, *Fragmented Metropolis*, p. 185; and *ENR* 96–107 (1926–31).

46. See Caro, *Moses*, pp. 328–31; and *ENR* 94–107 (1925–31).

47. See Christopher Tunnard and Boris Pushkarev, *Man-Made America: Chaos or Control?* (New Haven, 1963), pp. 160–65; J. Downer, "Parks and Parkways of Westchester County, New York," *Municipal and County Engineering* 72 (February 1927): 57–58; L. C. Holleran, "Design and Construction of Through Roads for Fast Traffic," *ENR* 105 (August 21, 1930): 286–89; and Sidney Shapiro, "Long Island State Parks and Parkways," *Civil Engineering* 6 (November 1936): 746–50.

48. See Preston, *Automobile Age Atlanta*, pp. 118–44, 45–73; and Brownell, "Symbol of Modernity."

49. Foster, "The Model T," p. 463.

50. Petty, "Ultimate Highway Development," p. 47.

51. Bassett, "The Freeway—A New Kind of Thoroughfare," *American City* 42 (February 1930): 95.

52. "Modern Motor Ways," *Architectural Record* 74 (December 1933): 430–37; see also Robert Whitten, "Facilitating Traffic and Preventing Blight by Spacious Planning of Express Highways," *American City* 45 (August 1931): 80; Barton MacKaye and Lewis Mumford, "Townless Highways for the Motorist," *Harper's Magazine* 163 (August 1931): 347–56; L. C. Squire and Edward M. Bassett, "New Type of Thoroughfare: The Freeway," *American City* 47 (November 1932): 64–66; W. W. Crosby, "Elbow Room: Why Highway Congestion Must Be Relieved," *Roads and Streets* 78 (October 1934): 362–65; J. L. Downer, "Design of Boulevards and Parkways," *Proceedings of the ASCE* 60 (August 1934): 814–15; and Fritz Malcher, *The Steady Flow Traffic System* (Cambridge, Mass., 1935).

53. "Two-Level Traffic," *ENR* 107 (August 27, 1931): 320.

54. See "New Jersey Plans Arterial Road from Holland Tunnel," *ENR* 94 (April 30, 1925): 731–33; "High Viaduct Three Miles Long Will Complete Gap in Holland Tunnel Road," *ENR* 104 (June 12, 1930): 973–74; and F. A. Reddan, "Super-Highway Being Supplemented by Super-Viaduct between Newark and Jersey City," *Scientific American* 147 (August 1932): 83–5.

55. See Miller, "History of the Modern Highway," p. 110; Chicago City Council, Committee on Traffic and Public Safety, *Limited Ways: A Plan for the Greater Chicago Traffic Area* (Chicago, 1932–33); "160-Mile Elevated Superhighway System Proposed for Chicago," *Roads and Streets* 76 (December 1933): 433–37; articles on Chicago and other cities in *ENR* 110–25 (1933–40); and Leroy C. Smith, "Trunk Line Highways in Metropolitan Areas," *Civil Engineering* 7 (October 1937): 685–86; quote from Wayne County, Michigan, Board of Road Commissioners, *Thirty-Third Annual Report* (1939), p. 9.

56. Figures from *Historical Statistics*, p. 461; on urban expressways, see *ENR* 120–25 (1938–40).

57. See "Why States Should Build Pavement through Towns," *Roads and Streets* 68 (November 1928): 523–24; Sidney D. Waldon, "Metropolitan Highways," *American Highways* 10 (January 1931): 3–6, 24–26; and "More Pavements at Lower Tax Rate Is City Problem," *American City* 47 (July 1932): 9.

58. Waldon, "Enlarging Facilities for Distance Travel," *Roads and Streets* 70 (March 1930): 119–22.

59. See *ARBA Proceedings* 28 (1931): 524–25; "Cooperation of Federal,

State, and Municipal Government Urged for Highway Work," *American City* 44 (February 1931): 169; and *American Highways* 12 (October 1933): 31.

60. "Through Road Gaps in Cities Closed," *ENR* 112 (January 18, 1934): 85–86.

61. Sidney Shapiro, "Dictated by Safety," *ENR* 121 (December 15, 1938): 764–65; see also Tunnard and Pushkarev, *Man-Made America*, p. 165; Caro, *Moses*; and articles by Robert Moses in *Civil Engineering* 9, 12 (1939, 1942).

62. "Parkways Go to Town," *ENR* 119 (December 16, 1937): 972.

63. See Caro, *Moses*, p. 465. Caro stressed the importance of federal funds to Moses.

64. See *ARBA Proceedings* 32–34 (1936–37).

65. Robert B. Brooks, "Federal Participation in the Construction of Major Traffic Arteries of Cities," *ARBA Proceedings* 32 (1935): 103; see also "Major Traffic Arteries of Cities Need More Federal Aid," *American City* 50 (March 1935): 56.

66. Quoted in U.S., Congress, House, *Hearing on H.R. 7079*, p. 32.

67. MacDonald, "Broadening the Highway Program," *Public Roads* 16 (March 1935): 1–2, 11.

68. H. C. Whitehurst, "The Highway Situation," *ARBA Proceedings* 32 (1935): 5–15.

69. The highway planning surveys will be discussed in Chapter Eight.

70. Quoted in Taylor, "Arterial Street Planning," p. 72. On the BPR and federal-aid for cities, see P. St. J. Wilson to District Engineers, March 8, 1924, File 481 Vouchers, and Files 740.1.1 and 481 General, CCF, BPR Records; and MacDonald, "Policies in the Highway Field," *Roads and Streets* 67 (November 1927): 498–501.

71. On Voshell and Toms, see File 404 Illinois, 1925–39, CCF, BPR Records; and "Chicago Superhighway Design Wins Prize," *ENR* 98 (March 24, 1927): 476. On the model legislation, see File 740.1.2, 1934–38, CCF, BPR Records; and BPR, *Model State Law for Freeways and Parkways* (Washington, D.C., 1934). After *American City* announced the publication of this booklet, the BPR hastily added that it had only "issued a discussion relative to an act"; see H. S. Fairbank to A. S. Robinson, April 21, 1936, File 740.1.2, CCF, BPR Records.

72. See L. I. Hewes, "Highway Transportation since 1928," *Civil Engineering* 8 (September 1938): 587.

73. F. T. Sheets to MacDonald, November 13, 1936, File 481 General, CCF, BPR Records.

74. See *BPR Annual Report* (1936); Bert Lord to Cartwright, May 22, 1937, Folder 21, Box 8, and Karl Stefan to Cartwright, May 25, 1938, Folder 11, Box 9, Cartwright Papers.

75. *BPR Annual Report* (1937), p. 2.

76. *BPR Annual Report* (1938), p. 4; see also C. N. Connor, "Secondary or Feeder Roads," *ENR* 117 (December 24, 1936): 902–4; and "Regulations Issued for Federal Aid," *ENR* 120 (February 18, 1937): 278, 282.

77. See G. D. Tramp, "Report of the County Committee on Legislation, Finance and Administration," *ARBA Proceedings* 33 (1936): 665-68; and H. G. Sours, "Federal Funds for Local Road Construction," ibid., pp. 669–73.

78. *American Highways* 17 (April 1938): statement on back cover.

79. Willard T. Chevalier to MacDonald, April 5, 1937, File 481 Correspondence—Michigan, CCF, BPR Records.

80. Hess to Charles M. Upham, March 22, 1937, File 481 General—Michigan, CCF, BPR Records; see also *ARBA Proceedings* 34, 36 (1937, 1939); Ross Buchanan to Cartwright, February 23, 1938, File 481 General, and MacDonald to Clarence Cannon, June 1, 1937, File 740.1.1, CCF, BPR Records.

81. Sheets, "The Highway of Tomorrow," April 14, 1937, Folder 2, Box 9, Cartwright Papers; see also idem, "A Master Plan for Highways," *ENR* 118 (June 17, 1937): 909–11.

82. George P. Woodleton, "Great Benefits from a National Highway," *Civil Engineering* 8 (April 1938): 274.

83. Quoted in U.S., Congress, House, Roads Committee, *Hearings on H.R. 11687: Roads*, 74th Cong., 2d sess., 1936, p. 40.

84. William S. Bennet to FDR, March 2, 1937, File 740.1.1, CCF, BPR Records.

85. MacDonald, "The Trend of Modern Highways," paper presented to the 67th Convention of the ASCE, July 21, 1937, MacDonald Reports; see also idem, "Tomorrow's Roads," *ENR* 117 (December 17, 1936): 868–70.

86. "Express Highways," *ENR* 115 (November 14, 1935): 691; and "Toward New Road Systems," *ENR* 118 (April 1, 1937): 493.

87. Brooks, "Place of Modern Express Highway," *Civil Engineering* 8 (January 1938): 39; and Worley, "Express Highways," *ENR* 117 (December 10, 1936): 830–31.

88. See Noble, "The Modern Express Highway," *Transactions of the ASCE* 107 (October 1937): 1068–1172.

89. Thomas P. Henry, "Toll Superhighways Are Incompatible with Sound Road Policy," *Better Roads* 8 (May 1938): 13; see also transcript of radio interview with Hubert Holloway on WOL, March 1, 1938, Legislative Correspondence, 1938, McKellar Papers; and material from the National Highway User Conference, Folders 8 and 14, Box 9, Cartwright Papers.

90. T. E. Steiner to Jennings Randolph, May 24, 1938, File 740.1.1, CCF, BPR Records; see also other material in the file.

91. Quotation from "Transcontinental Superhighways: Traffic Needs or Fairy Tale?" *Better Roads* 8 (March 1938): 16. On the Bulkey Bill, see File 740.1.1 for 1938, CCF, BPR Records; and "Transcontinental Highways," *ENR* 120 (February 17, 1938): 253.

92. Frederick S. Greene to MacDonald, February 15, 1938, File 740.1.1, CCF, BPR Records.

93. See "Designing the Roadway: Pennsylvania's Turnpike," *Roads and Streets* 82 (October 1939): 40–48, 67–71; and J. K. Kroerle, "Construction Progress on the Pennsylvania Turnpike," *Civil Engineering* 10 (March 1940): 137–41.

Chapter Eight

1. Fossnight, "Letter to Editor," *ENR* 123 (August 17, 1939): 202–3.

2. See Graham, *Toward a Planned Economy*, pp. 1–68; and Galambos, *America at Middle Age.*

3. *BPR Annual Report* (1925), p. 34.

4. Summaries of the BPR's studies, which were published separately, can be found in *BPR Annual Report* (1922–30); see also MSHD, *Thirteenth Biennial Report* (1929–30), pp. 85–89; and File 710.6, CCF, BPR Records.

5. See Bruce D. Greenshields, "Traffic Surveys to Solve Highway Problems," *Civil Engineering* 2 (January 1932): 21–24; Frank T. Sheets, "Rational Planning of a Public Highway Program," *Roads and Streets* 77 (January 1934): 37–40; and "Public Works—or Raking Leaves?" *American City* 50 (March 1935): 83.

6. MacDonald to Hayden, March 7, 1934, File 740.1.1, CCF, BPR Records.

7. Kennedy interview; see also *America's Highways*, pp. 181–82, 268–71.

8. Following information from Files 481 Correspondence—FAS Michigan, 481 PS & E—Michigan, and 710.6 Michigan, 1931–38, CCF, BPR Records; MSHD, *Biennial Report* (1927–38); and *BPR Annual Report* 1935–40). The BPR also made a movie about Michigan's survey to train other state officials.

9. See File 481 Correspondence—FAS Michigan, and G. Donald Kennedy, "Progress Report on Federal-Aid Secondary System," 1941, File 710.6 Michigan, CCF, BPR Records; and BPR, *Rules, Regulations, and Recommendations for Carrying Out Federal Aid Highway Programs* (Washington, D.C., 1939).

10. Fairbank, "State-Wide Highway Planning Surveys," *Civil Engineering* 7 (March 1937): 178–81.

11. M. L. Wilson to McKellar, February 25, 1938, Legislative Correspondence, 1938, McKellar Papers; see also "Developments in Highway Transportation," *Civil Engineering* 7 (October 1937): 683–86; and L. I. Hewes, "Highway Transportation since 1928," *Civil Engineering* 8 (September 1938): 585–88.

12. MacDonald to James Roosevelt, April 16, 1938, Official File 129 (Roads and Highways, 1938).

13. Memo: FDR to MacDonald to February 16, 1938, Official File 1E (Bureau of Public Roads, 1938), FDR Library; *America's Highways*, p. 136; MacDonald to C. R. Thomas, August 20, 1928, File 481 Right-of-Way, Maintenance and Inspection, H. S. Fairbank to J. T. Voshell, December 27, 1937, File 710.8 District 7, and a 1937 traffic survey map in File 481 Correspondence—Michigan, CCF, BPR Records.

14. See BPR, *Toll Roads and Free Roads: Message from the President of the United States* . . . , 76th Cong., 1st sess., 1939, H. Doc. 272; and "Toll Superhighway System Rejected," *ENR* 122 (May 11, 1939); 655–58.

15. Mary Huss [Wallace's Secretary] to Henry Kannee, March 29, 1939, Official File 129 (Roads and Highways, 1939), FDR Library. On FDR's interest in highways, see Caro, *Moses*, pp. 288–89; File 740.1.1, February 1938, CCF, BPR Records; and Files 96 and 455, President's Personal File, FDR Library.

16. Quoted in BPR, *Toll Roads and Free Roads*, p. vii.

17. Ibid., p. 93; see also pp. 89–102. For summary, see H. S. Fairbank, "Interregional Highways: Indicated by State-Wide Highway Planning Surveys," *Roads and Streets* 83 (January 1940): 37–44.

18. See BPR, *Tolls Roads and Free Roads*, pp. 90–98; see also Fairbank, "Interregional Highways," pp. 43–44.

19. *BPR Annual Report* (1939), p. 2.

20. Fairbank, "Interregional Highways," pp. 38–39.

21. Charles M. Upham, "A Proposed National Highway System," *ARBA Proceedings* 36 (1939): 101–5; also p. 590.

22. "Express Roads," *ENR* 122 (May 11, 1939): p. 644; see also "Toll Superhighway System Rejected," ibid., p. 658.

23. Van Wagoner, "President's Page," *Road Builders' News* (June 1939); 3.

24. Hoffman, "Interregional Highways: America's Highways—New Frontiers," *Roads and Streets* (January 1940): 35–36. See also File 740.1.2, CCF, BPR Records.

25. Van Wagoner, "Superhighway Ahead," *ARBA Proceedings* 36 (1939): 16–28, quotation on p. 19.

26. See "Resolutions," *ARBA Proceedings* 37 (1940): 848–49; see also articles in *ENR, Civil Engineering*, and *Roads and Streets* for 1939–41.

27. MacDonald, "Accord or Discord?" *ARBA Proceedings* 37 (1940): 20–27; see also idem, "National Road Building through the Cooperation of the State and Federal Government," *Roads and Streets* 82 (December 1939): 50, 52, 54–55.

28. See *ENR*, 122–25 (1939–40); and File 740.1.1, CCF, BPR Records.

29. Richard Polenberg, "The Decline of the New Deal," in John Braeman et al., eds., *The New Deal: The National Level* (Columbus, 1975), pp. 246–66, discusses the general resistance to FDR's social program in the late 1930s.

30. The bill's progress can be followed in File 740.1.1 for 1940, CCF, BPR Records; *ENR* 123–24 (1940); and U.S., Congress, House, Roads Committee, *Report on H.R. 9575: The Federal Highway Act of 1940*, 76th Cong., 3d sess., 1940, H. Rpt. 2041.

31. O. S. Hess, "Secondary Federal-Aid from the County Viewpoint," *ARBA Proceedings* 37 (1940): 621–26; J. E. Pennybacker, "Local Roads Most Needed," *National Petroleum News* 31 (March 8, 1939): 15; J. S. Bursh, "Rural Mail Routes and Road Problems," *Roads and Streets* 82 (August 1939): 36–38: and File 740.1.1 for 1939–40, CCF, BPR Records. On toll roads, see *ENR* (122–27 (1939–41).

32. Rose, *Interstate*, pp. 1–2.

33. Sheets, "The Development of Primary Roads during the Next Quarter Century," *American Highways* 18 (October 1939): 34, also pp. 31, 35-9; see also Norman Bel Geddes, *Magic Motorways* (New York, 1940); Jeffrey Meikle, *Twentieth Century Limited: Industrial Design in America, 1925–1939* (Philadelphia, 1981), pp. 200–209.

34. J. W. Beretta, "Letter to the Editor," *ENR* 123 (December 7, 1939): 742. On the Merritt and Hutchinson River Parkways, see numerous articles in *ENR* 122–24 (1939–40); and *America's Highways*, p. 137.

35. The Pennsylvania Turnpike was widely covered in contemporary technical and popular magazines, while *ENR*, 124–27 (1940–41), and File 740.1.1 for 1941, CCF, BPR Records, covered other toll plans.

36. See File 481 General for 1919, CCF, BPR Records; General George V. Strong, "War Department's Interest in Highway Development," *American Highways* 18 (October 1939): 17–19; and H. S. Fairbank, "War and the Interregional Highway Plan," *ENR* 124 (January 18, 1940): 87–90, 91–93.

37. The BPR's report, "Highways for the National Defense," appeared in U.S., Congress, House, Roads Committee, *Hearings on H.R. 4935: Defense Highway Act of 1941*, 77th Cong., 1st sess., June 5–11, 1941, pp. 219–60. The struggle between FDR and Congress can be traced in numerous articles in *ENR*, 124–27 (1940–41); and File 740.1.1, CCF, BPR Records. See also *PRA Annual Report*, (1941–42): U.S., Congress, Senate, Post Office and Post Roads Committee, *Expansion of Highway Facilities for National Defense*, 77th Cong., 1st sess., June 2, 1941, S. Doc. 62; and idem, *Defense Highway Act of 1941: Veto Message*, 77th Cong., 1st sess., August 4, 1941, S. Doc. 103.

38. *America's Highways*, p. 144; see also "State Road Officials Discuss Highway Problems," *ENR* 125 (September 26, 1940): 385, 391; and "PRA Demands Defense Priority for Federal Aid Work," *ENR* 125 (September 5, 1940): 307.

39. Samuel C. Hadden, "Presidential Address," *American Highways* 25 (January 1945): 17, 20.

40. On wartime construction, see "Programming Federal Aid Projects Essential to the National Defense," *Roads and Streets* 85 (February 1942): 42–43; M. J. Hoffmann, "Effects of War on the State Highway Departments and Their Programs," *American Highways* 22 (January 1943): 15–18; *ENR* 128–29 (1942); File 740.1.1 General, CCF, BPR Records; and U.S., Congress, House, Roads Committee, *Hearings on H.R. 6908; Additional Authorization for Access Roads*, 77th Cong., 2d sess., 1942, pp. 87–103.

41. See Joseph Barnett, "New War Department Building Road Network," *Civil Engineering* 13 (March 1943): 127–30; File 481 Correspondence—Michigan, CCF, BPR Records; and *Roads and Streets* 87–88 (1944–45).

42. On postwar planning, see File 740.1.1 for 1941–44, CCF, BPR Records; and *ENR* 126–27, 130–33 (1941, 1943-44).

43. "Report of the Special Committee on Post-War Economic Policy and Planning," c. October 1943, File 740.1.1 General, CCF, BPR Records; see also "Legislative Program of the American Association of State Highway Officials," *American Highways* 22 (April 1943): 3–9; *ENR* 130–31 (1943); and U.S., Congress, House, *Interregional Highways: Message from the President of the United States Transmitting a Report of the National Interregional Highways Committee*, 78th Cong., 2d sess., January 12, 1944, H. Doc. 379, pp. 110–12.

44. This and the following information from Minutes of the Initial Meeting of the National Interregional Highway Committee, June 24, 1941, Box 9, Tugwell papers.

45. Minutes of the Third Meeting of the National Interregional Highway Committee, October 2, 1941, File 790 for Michigan, CCF, BPR Records.

46. See C. S. Hill, "An Interregional Highway System Proposed for Postwar Construction," *ENR* 132 (January 27, 1944): 144–46; and Thomas H. MacDonald and H. S. Fairbank, "American Interregional Highway System," *Civil Engineering* 14 (May 1944): 187–90, (June 1944): 229–32.

47. Compare the presentation of information on p. 6 and pp. 51–52 in U.S., Congress, House, *Interregional Highways*.

48. For MacDonald's views on planning, see "Preparing for the Postwar Highway Program," February 7, 1944, MacDonald Reports. Tugwell's acceptance of this philosophy can be seen in his correspondence with MacDonald in NIHC Folder, Box 9, Tugwell Papers. However, *America's Highways*, p. 274, argued that the report actually demonstrated a broader view of highway problems, a position supported by Rose, *Interstate*, pp. 19–20.

49. See PRA, *Public Control of Highway Access and Roadside Development* (Washington, D.C., 1943); MacDonald, "The City's Place in Post-War Highway Planning," *American City* 58 (February 1943): 42–44; and Fairbank, "Declares Short-Distance Superhighways Most Urgent Post-War Necessity," *Concrete* 51 (April 1943): 35–36.

50. See "Statement of G. Donald Kennedy before the Board of Investigation and Research on Public Aids to Transportation," June 29, 1942, in Hearings on Public Aid to Transportation File, Administrative and Planning Correspondence of Planning and Traffic Division, 1942–49, RG 58-5, MSHD Records; and articles by Kennedy in *Civil Engineering, ENR, American Highways, Concrete, Roads and Streets,* and other technical periodicals from 1942–44; and Memo: Kennedy to Carl Hayden, "Role of the Federal Government in Highway Development: Post War Highway Needs," 1943, File 740.1.1 General, CCF, BPR Records.

51. See "Concentrate Postwar Planning on Urban Highways," *ENR* 129 (September 24, 1942): 401.

52. See "Legislative Program of the AASHO," pp. 3–9; U.S. Congress, Senate, "Amendment of Federal Highway Act of July 11, 1916," *Congressional Record*, vol. 89, pt. 4 (June 8, 1943): 5496–5504; and idem, "Amendment of Federal Highway Act," *Congressional Record*, vol. 89, pt. 5 (June 23, 1943): 6309–17.

53. See "Interest in Postwar Work Renewed in Congress," *ENR* 130 (March 11, 1943): 333, 338; J. W. Robinson, "Post-War Highway Legislation," *American Highways* 22 (October 1943): 3–4; and Memo: Kennedy to Automotive Safety Foundation, March 6, 1943, Brady Gentry Correspondence, Box 2, Kennedy Papers.

54. Memo: Kennedy to Operating Committee, InterIndustry Committee, and Board of Trustees of the Automotive Safety Foundation, March 6, 1943, Box 2, Kennedy Papers; see also Brady Gentry Correspondence for 1943–44, ibid.; and MacDonald's correspondence in File 740.1.1 General, CCF, BPR Records.

55. The ARBA's postwar proposal, also released in March, called for $500 million in annual federal-aid grants to be supplemented by $500 million in PWA-type grants for urban arterial roads. See Charles M. Upham, "Highway Plans for the Post-War Period Are Urgently Needed," *Roads and Bridges* 81 (April 1943): 70; and numerous articles in *ENR*, 130–31 (1943).

56. MacDonald, "The Future Development of Highway Transportation," December 4, 1941, MacDonald Reports.

57. See File 740.1.1 General, CCF, BPR Records; and "New York to Reduce Work on Projected Parkways," *ENR* 128 (January 22, 1942): 156.

58. See numerous articles in *ARBA Proceedings* 38 (1941); "New Local Highway Agency Proposed in Congress," *ENR* 131 (November 18, 1943): 727; "Needless Duplication," *ENR* 131 (December 2, 1943): 801; and F. E. Sargent to MacDonald, December 22, 1943, File 740.1.1 General, CCF, BPR Records.

59. Johnstone to Fairbank, November 3, 1943, File 740.1.1 General, CCF, BPR Records.

60. See Otto Hess, "State-County Cooperation on Federal-Aid Secondary Roads," *American Highways* 24 (October 1944): 7–8, 21.

61. Cox to Moses, December 14, 1943, and reply, December 18, 1943, File 740.1.1 General, CCF, BPR Records. See also other material in this file; Brady Gentry Correspondence, Box 2, Kennedy Papers; and Legislative Correspondence—AASHO, 1943–49, RG 58–10, MSHD Records.

62. Hadden to State Highway Departments, June 17, 1943, Correspondence of the Commissioners, Congressional Correspondence, 1943–49, RG 64–16, MSHD Records. See also William J. Cox, "Highway Program Fosters Waste, "*Railway Age* 115 (December 25, 1943): 1002–5; and Brady Gentry, "Annual Convention Address," *American Highways* 23 (January 1944): 3–4; and "Bases for Federal Aid," *Better Roads* 14 (January 1944): 3.

63. Coykendall, "Federal Highway Projects," *American Highways* 20 (October 1941): 4; see also John U. Shroyer, "Federal-Aid and Its Relationship with the States," *Roads and Streets* 86 (December 1943): 57–61.

64. H. D. Fallon to Bruce, December 29, 1943, File 740.1.1 General, CCF, BPR Records.

65. National Highways Users Conference, *Save the Federal Aid Highway Principle: The Story of What It Is; How It Operates and the Attacks against It* (Washington, D.C., 1942).

66. Moses to J. W. Robinson, March 29, 1943, File 740.1.1 General, CCF, BPR Records.

67. Gentry to Hadden, October 15, 1943, File 740.1.1 General, CCF, BPR Records.

68. "The Interregional Report," *Better Roads* 14 (February 1944): 3.

69. Reid to MacDonald, March 13, 1944, File 481 Correspondence—Michigan, CCF, BPR Records; see also "Interregional Roads," *ENR* 132 (January 27, 1944): 94.

70. U.S., Congress, House, Roads Committee, *Hearings on H.R. 2426: Federal-Aid for Post-War Highway Construction*, 2 vols., 78th Cong., 2d sess., 1944. The following discussion is drawn from this document.

71. Quoted in ibid., p. 963; see also pp. 327, 595, 855.

72. Ibid., pp. 160–61, 772, 749–51.

73. See U.S., Congress, House, Roads Committee, *Post-War Federal-Aid Highway Act of 1944*, 78th Cong., 2d sess., 1944, H. Rpt. 1597; also articles in

ENR 132 (June 1944): and Wilburn C. Cartwright, "Federal Cooperation in Road Building," *ARBA Proceedings* 37 (1940): 19.

74. See U.S., Congress, Senate, Post Office and Post Roads Committee, *Post-War Federal-Aid Highway Act of 1944: Report to Accompany S. 2105*, 78th Cong., 2d sess., 1944; and "Highway Bill Approved by Senate Committee," *ENR* 133 (August 24, 1944): 208.

75. "Get Together and Push," *Roads and Streets* 87 (September 1944): 70. The debates are in U.S., Congress, *Congressional Record*, 78th Cong., 2d sess., vol. 90, pt. 6 (September 12 and 15, 1944), and (November 22, 24, 28, and 29, 1944). See also "Comparison of Apportionment of Post-War Highway Funds between S. 2105 . . . and H.R. 4915," *American Highways* 23 (October 1944): 14.

76. See coverage in *ENR* 132–33 (1944).

77. Quoted in U.S., Congress, House, *Hearings on H.R. 2426*, p. 956.

78. Hale, "Highways—Major Postwar Public Works," *Civil Engineering* 15 (April 1945): 161–62.

Chapter Nine

1. Kennedy interview.

2. Mark Rose's fine study, *Interstate*, showed that highway policy making from 1944 through 1956 resulted from a pluralistic political process that accepted a narrow goal of moving traffic that had been proposed by, among others, highway engineers. This chapter follows Rose's basic argument, but ascribes a larger role to the BPR.

3. See G. Donald Kennedy, "Postwar Highways in the Cities," File F—Speeches, 1945, Box 3, Kennedy Papers; and "Roadbuilders Study Ways to Implement Federal-Aid Highway Act," *ENR* 134 (January 25, 1945): 103.

4. *BPR Annual Report* (1951), p. 1; see also ibid. (1952); *Historical Statistics*, p. 462; H. A. MacDonald, "Urban Traffic Congestion Problem," *Roads and Bridges* 84 (April 1946): 76–79; and W. S. Chapin, "What Can Be Done about Traffic Congestion? Urban Parkways and Expressways," *Civil Engineering* 16 (April 1946): 151–53.

5. On urban highways, see *ENR* and numerous articles in *Civil Engineering* for 1944–45; an annual summary by Harold McKeever in *Roads and Streets* from 1947–55 also was useful. On Los Angeles, see Brodsly, *L.A. Freeway*, pp. 105–31.

6. MacDonald quoted in "Engineers, Businessmen Cooperate on Solution of Urban Traffic Problems," *Civil Engineering* 17 (October 1947): 640; see also MacDonald "The Case for Urban Expressways: Long-Range Planning of Adequate Highway Facilities Will Save Many Cities from Stagnation," *American City* 62 (June 1947): 92–93; and idem, "The Interstate System in Urban Areas," *American Planning and Civic Annual* (1950): 114–19. BPR engineers were especially active in Detroit; see Files 481 Correspondence—Michigan and 481 PS & E—Michigan, 1943–50, CCF, BPR Records; Administrative Correspondence, Planning and Traffic Division, 1942–49, Detroit Crosstown Expressway—1944, RG 58-5, and File John C. Lodge and Edsel Ford Expressways, 1944–47,

Correspondence of the Commissioner, RG 64-16, MSHD Records; *America's Highways*, pp. 157, 188–90; and *PRA Annual Report* (1945–49).

7. See *PRA Annual Report* (1947). Designation of the Interstate system can be followed in ibid (1946–47); and "Ask States to Designate National Roads," *Concrete* 53 (April 1945): 4. The BPR's guidance of the selection process can be seen in Administrative Correspondence, Planning and Traffic Division, 1942–49, and County Road Association of Michigan, General Correspondence, Planning and Traffic Division Relating to Associations, 1942–49, RG 58-5, MSHD Records. On construction progress, see "Federal-Aid System Provides Network of Arterial Routes in Urban Areas," *Civil Engineering* 17 (November 1947): 651, 718; and R. E. Jorgenson, "Urban Highways and the Interstate System," *Roads and Streets* 89 (October 1946): 84–86.

8. MacDonald, "Interstate or Super Highways," *American Highways* 26 (April 1947): 5–6, 15-6.

9. See *PRA Annual Report* (1947). The routes of the additional mileage were designated in 1955; see *BPR Annual Report* (1956).

10. See Rose, *Interstate*, pp. 60–63; Thomas H. MacDonald, "Housing and Express Highways" (Testimony before the Subcommittee on Housing in the Post-War Period), November 1, 1945, MacDonald Reports; and *PBA Annual Report* (1948), p. 4.

11. "Better Roads Are Years Off," *Business Week* (July 31, 1948), pp. 22–23. Figures from *Historical Statistics*, p. 458.

12. Leslie Williams, "The Problem We Face," *Civil Engineering* 16 (April 1946): 149–51. On integrated transit, see Charles Gordon, "Transportation in the Future City," *Better Roads* 14 (November 1944): 18–20, 36; C. G. Bradfield, "Relation of Urban Expressways to Mass Transportation Facilities," *Proceedings of the ASCE* 74 (December 1948): 1597–1604; and H. C. Coons, "Relationship of City Planning, Traffic, and Mass Transit in Construction of Detroit's Highways," *Midwest Engineer* 1 (March 1949): 9–11. Statistics from *PRA Annual Report* (1947–49).

13. See *PRA Annual Report* (1945–49); and articles in *ENR* and *American Highways* for 1946–49.

14. On roads and economic policy, see Rose, *Interstate*, pp. 35, 61; Hal H. Hale, "Highways—Major Postwar Public Works," *Civil Engineering* 15 (April 1945): 161–62; Charles M. Upham, "Highway Construction and Our National Economy," *Civil Engineering* 16 (August 1946): 340–41; and Thomas H. MacDonald, "The Highway Improvement Program in Relation to the Conventional Public Works Concept," *American Highways* 26 (January 1947): 8–10, 28.

15. See "Federal-Aid Highway Act of 1948," *Public Roads* 25 (September 1948): 107–8; and "Congress Gave Us All a Scare," *Roads and Streets* 91 (August 1, 1948): 68.

16. See MacDonald, "Girding for the Big Postwar Task," *Roads and Streets* 89 (February 1946): 88–91; idem, "A National Program for Secondary Road Development," *Roads and Streets* 89 (August 1946): 84–85, 99–101, 124; and *PRA Annual Report* (1948–49). See also H. G. Sours, "State-County-City Rela-

tionships in the Postwar Highway Program," *Roads and Streets* 88 (February 1945): 63; and Charles M. Upham, "Secondary Roads Need Action Now," *Roads and Streets* 88 (May 1945): 79, 91.

17. Quoted in U.S., Congress, Senate, Public Works Committee, *Hearings on H.R. 5888: Federal Aid for Highways*, 80th Cong., 2d sess. 1948, p. 300; see also pp. 135, 176–81, and 201; MacDonald, "State-Federal Relations in Highway Development," *American Highways* 27 (October 1948): 8–9, 25–27, 29.

18. Gomez, *Intergovernmental Relations*, p. 29.

19. Quoted in U.S., Congress, Senate, Public Works Committee, *Hearings on S. 244 and S. 1471: Rural Local Roads*, 81st Cong., 1st sess., 1949, p. 256; see also pp. 237–77; and "Federal Aid to Local Roads Opposed at Senate Hearings," *ENR* 142 (May 12, 1949): 51. Rose, *Interstate*, p. 36, suggested that the congressmen intended these bills to soften MacDonald's stance on rural roads, but Kerr's attitude was actually less subtle—he seemed intent on discrediting MacDonald.

20. See U.S. Congress, Senate, *Hearings on S. 244 and S. 1471*, esp. pp. 57–58, 144–49, 180–85, 227–37, 315–49.

21. Roy E. Jorgenson, "Free Roads vs. Toll Roads," *Civil Engineering* 20 (September 1950): 590–94; see also Carl Brown, "Annual Address," *American Highways* 28 (October 1949): 3–5; and Hal H. Hale, "Highways at the National, State, and Local Level," *American Highways* 28 (January 1949): 12–13, 22.

22. See BPR, *The Rural Local Road Problem: Report to the Subcommittee on Roads of the Committee on Public Works, U.S. Senate* (Washington, D.C., 1950); see also "BPR Proposes Taking over Additional Secondary Roads," *ENR* 144 (February 23, 1950): 23–24; "The Local Road Situation," *Roads and Streets* 93 (March 1950): 56–58; and U.S., Congress, Senate, *Hearings on S. 244 and S. 1471*, p. 253.

23. Quoted in U.S., Congress, House Public Works Committee, *Hearings on H.R. 7398 and H.R. 7491, The Federal-Aid Highway Act of 1950*, 81st Cong., 2d sess., 1950, pp. 356-57.

24. Quoted in U.S., Congress, Senate, Committee on Public Works, *Hearings on S. 3424 and H.R. 7491; The Federal-Aid Highway Act of 1950*, 81st Cong., 2d sess., 1950, p. 547; see also pp. 319–21.

25. Quoted in ibid., p. 553. Statistics from *BPR Annual Report* (1950), p. 15.

26. C. W. Phillips, "Annual Address," *American Highways* 26 (October 1947): 3–4, 27–29; see also Ben H. Petty, "Co-Ordination of Activities of Highway Organizations," *American Highways* 27 (July 1948): 7, 28–29.

27. Hal H. Hale, "Highway Engineers Must Gear up Public Relations," *Roads and Streets* 90 (September 1947): 85, 87, 123; several articles in *ENR*, 142–43 1949); G. Donald Kennedy, "Statewide Highway Studies as a Public Relations Tool," Address to the 34th Annual Meeting of AASHO, New York, September 25, 1947, Kennedy Papers; and Kennedy interview.

28. Robinson, "Your Highway Program," *American Highways* 26 (January 1947): 11, 22–3.

29. "State Highway Leaders Seek Doubled U.S. Program," *ENR* 143 (December 1, 1949): 13.

30. Quoted in U.S., Congress, House, *Hearings on H.R. 7398 and H.R. 7491*, p. 13–7; see also pp. 130–37; 591, 181, and testimony of Samuel C. Hadden in U.S., Congress, Senate, *Hearings on S. 3424 and H.R. 7941*, p. 198.

31. See U.S., Congress, House, *Hearings on H.R. 7398 and H.R. 7941*, pp. 242, 494, 294, 500; detailed coverage in *ENR* 142–45 (1949–50); and "New Federal Aid Highway Act to Spur State-County Cooperation," *Roads and Streets* 93 (October 1950): 63–64.

32. See U.S., Congress, House, Public Works Committee, *Hearings on H.R. 5888; Federal-Aid for Highways*, 80th Cong., 2d sess., 1948, pp. 22–23; "Bill Asks Indefinite Federal Road Aid," *ENR* 140 (January 22, 1948): 96; PRA, *Highway Needs of the National Defense*, 81st Cong., 1st sess., 1949, H. Doc. 249; and *PRA Annual Report* (1949).

33. See *ENR* 142–45 (1949–50); and U.S., Congress, House, *Hearings on H.R. 7398 and H.R. 7941.*

34. Quoted in U.S., Congress, Senate, *Hearings on S. 3424 and H.R. 7941*, p. 185.

35. "Stepping up the Road Program," *ENR* 143 (December 8, 1949): 23.

36. On the changing reaction to rural roads, see U.S., Congress, House, *Hearings on H.R. 7398 and H.R. 7941*, pp. 11–15, 97; idem, *Hearings on H.R. 6094, H.R. 6180, and H.R. 6390: Federal-Aid Highway Act of 1952*, 82nd Cong., 2d Sess., 1952, esp. pp. 45, 121–22; and idem, *Hearings: National Highway Study*, 2 vols., 83d Cong., 1st sess., 1953, p. 92.

37. See "All Road Building Records Fall; Top $1 Billion," *ENR* 145 (October 26, 1950): 27; L. D. Jackson, "Cities at a Disadvantage in Road Improvement," *Roads and Engineering Construction* 89 (March 1951): 114; and *Historical Statistics*, p. 462.

38. MacDonald, "Critical Aspects of the Current Highway Program," *American Highways* 31 (January 1952): 7. Mileage figures from *PRA Annual Report* (1947–49); *BPR Annual Report* (1950–52); and numerous articles in *Civil Engineering* and *ENR* in 1950. One call for integrated planning was George F. Emery, "Urban Expressways," *American Planning and Civic Annual* (1946–47): 123–29.

39. See "Highways Run Losing Race with Traffic," *Business Week* (December 8, 1951): 108–10, 112; *BPR Annual Report* (1952–53); and *American Highways* 31 (1952).

40. Quoted in PRA, *Highway Needs of the National Defense*, pp. v–vi. On roads and the military, see James K. Knudson, "Defense Transportation and Highway Use," *American Highways* 30 (January 1951): 14–15, 30–31; and "Wholly Inadequate Highways," *ENR* 146 (March 15, 1951): 21. See also U.S., Congress, Senate, Public Works Committee, *Hearings on S. 2035 and S. 1956: Emergency Highway Repairs*, 82d Cong., 1st sess., 1951.

41. Hal Hale, "Let's Face It," *American Highways* 29 (October 1950): 4.

42. See C. W. Gilchrist, "Roads Are Everybody's Business," *Roads and Engineering Construction* 89 (March 1951): 117; Lacey V. Murrow, "How to Plan and Pay for Safe and Adequate Highways We Need," *Civil Engineering* 23

(August 1953): 524–27; and articles on highway publicity in *ENR* 146–51 (1951–53).

43. See coverage in *Bus Transportation* 31 (1952); "National Good Roads Association Formed," *Road and Streets* 95 (August 1952: 62–63; and "Better Roads Movement Gains," *ENR* 149 (September 11, 1952): 96. Rose, *Interstate*, pp. 41–49, noted that truckers formed the core of Project Adequate Roads, with the goal of eliminating the federal gas tax, which the AASHO and the BPR did not support. Still, the effort to teach the public about the need for better roads was advanced by the project, regardless of its specific political intentions.

44. Quoted in U.S., Congress, House, *Hearings on H.R. 6094, H.R. 6180, and H.R. 6390*, pp. 37–38.

45. "Builders Fight Cut in Highway Funds," *ENR* 148 (February 14, 1952): 43–44.

46. Quoted in "AASHO Leaders Scan $45 Billion Job," *Roads and Streets* 96 (January 1953): 68–69; see also "Lawmakers Heed Road Men a Little," *ENR* 148 (April 10, 1952): 19.

47. Rose, *Interstate*, pp. 45–49.

48. Quotation from "Q. & A. Toll Roads Aren't the Answer!" n.d., File F, 1943–45, Ziegler Papers. On New York, see Bertram D. Tallamy, "New York State's Urban Program: Arterial Routes at State Expense," *Roads and Streets* 89 (August 1946): 94, 96, 98; and "Toll Highways Again," *ENR* 134 (February 22, 1945): 249.

49. On postwar toll roads see Rae, *The Road and the Car*, pp. 173–83; Charles L. Dearing and Wilfred B. Owen, *Toll Roads and the Problem of Toll Road Modernization* (Washington, D.C., 1951); and *America's Highways*, pp. 166–70, 190–92.

50. See MacDonald, "Why Toll Roads Aren't the Answer," *Roads and Streets* 90 (September 1947): 91–92, 94; idem, "The Case against Toll Roads," *American City* 62 (July 1947): 153; and Joseph Barnett, "Few Locations Are OK for Toll Projects," *Roads and Streets* 90 (September 1947): 94, 96, 98. See also MacDonald's testimony in U.S., Congress, House, Public Works Committee, *Hearings on H.R. 6572: Roads and Bridges Legislation*, 80th Cong., 2d sess., 1948, p. 124.

51. "More Money for Roads," *ENR* 141 (November 11, 1948): 79.

52. The changing attitude can be seen in *ENR* 144–49 (1950–52); Jorgenson, "Free Roads vs. Toll Roads," pp. 590–94; "Toll Road Proposals 'Bustin' Out All Over,'" *Roads and Streets* 96 (April 1953): 62–63; and Harold J. McKeever, "Year-End Look at Roadbuilding," *Roads and Streets* 97 (January 1954): 30–34, 62–65. See also Enoch R. Needles, "Turnpikes Are Economically Justified," *Civil Engineering* 21 (August 1951): 442–44.

53. The AASHO report was discussed in "Changes in Toll-Road Policy Aired," *ENR* 149 (August 7, 1952): 28; and "Crowded Roads Result in Jumbled Network of Toll Roads," *Business Week* (September 13, 1952): 82–86, 88, 90; see also "Integrate Toll Roads." *ENR* 150 (January 1, 1953): 80.

54. See MacDonald's testimony in U.S., Congress, and Senate, *Hearings on*

S. 3424 and H.R. 7941, p. 9; and L. I. Hewes, "Are Toll Roads the Only Answer? Plan for Financing Badly Needed Highways," *ENR* 144 (April 13, 1950): 42–43.

55. See Rae, *The Road and the Car*, pp. 175, 181; *ENR* 152–53 (1954); and McKeever, "Year-End Look at Roadbuilding," p. 65.

56. U.S., Congress, Senate, Public Works Committee, *Hearings on S. 796: Tolls on Federal-Aid Highways*, 83d Cong., 1st sess., 1953, esp. pp. 95–98; and idem, *Permitting the Charging of Tolls on Certain Highways Constructed with Federal Aid: Minority Views Concerning S. 796*, 83d Cong., 1st sess., 1953, S. Rpt. 715.

57. *BPR Annual Report* (1952), p. 7; statistics from *ibid.* (1952–53).

58. See M. W. Torkelson, "Laying the Foundation for Future Highway Developments," *ARBA Proceedings* 33 (1936): 872–81; and Caro, *Moses*, pp. 515–18, 897–98, 911–12, 930.

59. Quoted in U.S., Congress, House, Public Works Committee, *Hearings: National Highway Study*, 2 pts., 83d Cong., 1st sess., 1953, p. 10. The questions appear in ibid., p. 2; see also "Congress Launches Far-Reaching Review of Federal Highway Policy," *Roads and Streets* 96 (May 1953): 57.

60. Quoted in U.S., Congress, House, *National Highway Study*, p. 369.

61. Quotation from ibid., pp. 322; see also pp. 360–68; and coverage in *ENR* 150–51 (1953).

62. See U.S., Congress, House, *National Highway Study*, pp. 134, 139, 196–250, 314; see also "State Highway Officials Urge Raising Federal Aid to Full Gas Tax Income," *ENR* 150 (May 21, 1953): 24.

63. Quoted in U.S., Congress, House, *National Highway Study*, pp. 134, 139.

64. See MacDonald, "The Future of Road Building"; "A Gasoline Tax on Motor Vehicles," *ENR* 82 (April 17, 1919): 750–51; Burnham, "The Gasoline Tax" and File 740.1.2 General for the 1920s, CCF, BPR Records.

65. See St. Clair, "Entrepreneurship and the American Automobile Industry," p. 206; Burnham, "The Gasoline Tax;" Mississippi Valley Conference, "Historical Highlights," pp. ix-8–ix-9; Frederic E. Everett, "Diversion Dangers and Gasoline Tax Evasions," *American Highways* 11 (January 1932): 1–2; articles and statistical compilations by engineers in *Public Roads* in the 1930s; and BPR, *The Taxation of Motor Vehicles in 1932* (Washington, D.C., 1934). On state legislation, see Dearing, *American Highway Policy*, pp. 106–7, 180–81; material from the National Highway Users Conference in File 740.1.1 General for 1935–45, CCF, BPR Records; and Rose, *Interstate*, pp. 32–33. Statistics from *Historical Statistics*, p. 459.

66. See Rose, *Interstate*, pp. 11, 20; Dearing and Owen, *Toll Roads*, pp. 156–58; "Stepping up the Road Program," *ENR* 143 (December 8, 1949): 23; and U.S., Congress, Senate, *Hearings on S. 3424 and H.R. 7941*, p. 120.

67. See Rose, *Interstate*, pp. 46–47, 49; on the AASHO's position, see *American Highways* 28–32 (1949–53); *ENR* (1952–53); and Senator Frances Case, "Six Suggestions for Safer Highways," *American Highways* 33 (January 1954): 8, 31–32.

68. "Federal Highway Aid Continues to Hold Congressional Support," *ENR*

151 (November 26, 1953): 21; see also "Highway Officials Call Construction Funds Most Pressing Problem for Nation's Roads," *ENR* 151 (November 19, 1953): 25–26; and *American Highways* 33 (1954).

69. A. E. Johnson, "Adequate Highways for a Prosperous America," *American Highways* 33 (April 1954): 3, 17–19. On legislative plans, see U.S., Congress, House, Public Works Committee, *Hearings on H.R. 7678, H.R. 7818, H.R. 7841, H.R. 7124, H.R. 7207, H.R. 14, H.R. 1407, H.R. 3258, and H.R. 3529: Federal-Aid Highway Act of 1954*, 83d Cong., 2d sess., 1954.

70. See U.S., Congress, House, *Hearings on H.R. 7678*; and U.S., Congress, Senate, Public Works Committee, *Hearings on S. 2859, S. 2982, S. 3069, and S. 3184: Federal-Aid Highway Act of 1954*, 83d Cong., 2d sess., February 19, 26 and March 1, 3, 4, 11, 1954.

71. See U.S., Congress, House, *Hearings on H.R. 7678*; U.S., Congress, Senate, *Hearings on S. 2859*; and Johnson, "Adequate Highways," pp. 17–18.

72. Duane L. Cronk, "1954—Turning Point Year," *Roads and Streets* 98 (January 1955): 44–45; see also *ENR* 152–53 (1954); and Rose, *Interstate*, pp. 52–53.

73. "Congress Fulfills Highway Chiefs' Hopes," *ENR* 152 (April 22, 1954): 26; see also "Cause for Optimism," ibid., p. 112; and *BPR Annual Report* (1954).

74. Examples include Rose, *Interstate*, pp. 69–94; Rae, *The Road and the Car*, pp. 187–94; Richard O. Davies, *The Age of Asphalt: The Automobile, the Freeway, and the Condition of Metropolitan America* (Philadelphia, 1975), pp. 1–27; Helen Leavitt, *Superhighway—Superhoax* (Garden City, N.Y., 1971), pp. 20–51; A. Q. Mowbray, *Road to Ruin* (Philadelphia, 1968); John R. Meyer and Jose A. Gomez-Ibanez, *Auto Transit and Cities* (Cambridge, Mass., 1981); and George M. Smerk, *Urban Mass Transportation: A Dozen Years of Federal Policy* (Bloomington, Ind., 1974).

75. "MacDonald Retires as Commissioner of Public Roads," *ENR* 150 (April 2, 1953): 59.

76. The following discussion, unless otherwise noted, is from Rose *Interstate*, pp. 69–94; and extensive coverage in *ENR* and *Roads and Streets* for 1954–56.

77. See U.S., The President's Advisory Committee on a National Highway Program, *A Ten-Year National Highway Program: A Report to the President* (Washington, D.C., 1955); see also "A National Highway Program," *Civil Engineering* 25 (February 1955): 82–84.

78. "Rough Road Ahead," *ENR* 154 (March 17, 1955): 23.

79. To follow these events, see U.S., Congress, Senate, Public Works Committee, *Hearings on S. 1048, S. 1072, S. 1160, and S. 1573; National Highway Program*, 84th Cong., 1st sess., 1955; U.S., Congress, House, Public Works Committee, *Hearings on H.R. 4260: National Highway Program*, 84th Cong., 1st sess., 1955; U.S., Congress, Senate, *Federal-Aid Highway Act of 1955*, 84th Cong., 1st sess., 1955, S. Rpt. 350, quotation from p. 33; U.S., Congress, House, *To Authorize Appropriations for Continuing the Construction of Highways*, 84th Cong., 1st sess., 1955, H. Rpt. 1336; see also comment by Senator Dennis Chavez in "ARBA Appraises Road-Building Capacity of U.S. Industry," *Civil Engineer-*

ing 25 (February 1955): 82–84; detailed coverage in *ENR* 154–55 (1955); and Rose, *Interstate*, pp. 76–84.

80. Quoted in "Nation's Press Voiced Its Feelings on Highway Bill," *Roads and Streets* 98 (September 1955): 43.

81. Cronk, "How Did We Do in 1955?" *Roads and Streets* 99 (January 1956): 58–59.

82. "Pressure for Big Road Program Will Mount," *ENR* 155 (August 14, 1955): 21–24; see also Duane L. Cronk, "Washington Newsletter," *Roads and Streets* 99 (January 1956): 19–20; and Dero A. Saunders, "Those Expensive Highways," *Fortune* 51 (May 1955): 116–20, 154, 156, 158.

83. *BPR Annual Report* (1955), pp. 1–2.

84. Cronk, "Washington Newsletter," *Roads and Streets* 99 (February 1956): 19–20.

85. Cronk, "Washington Newsletter, *Roads and Streets* 99 (June 1956): 19–20; see also "The Grand Plan Becomes a Big Job," *ENR* 157 (July 12, 1956): 128.

86. Rose, *Interstate*, pp. 88–91.

87. See U.S., Congress, House, Public Works Committee, *Hearings on H.R. 12110 and H.R. 11990: Pentagon Road Bill and Federal Highway Administrator*, 84th Cong., 2d sess., 1956, pp. 1–14.

88. du Pont, "What's Going on in Public Roads?" *American Highways* 33 (January 1954): 11. Edwin Layton, in *Revolt of the Engineers*, discussed this attitude toward politics at length.

89. du Pont, "What's Going on;" and idem, "Commissioner's Annual Address," *American Highways* 34 (January 1955): 5, 26.

90. See U.S., Congress, Senate, *Hearings on S. 796*, p. 29.

91. See Rose, *Interstate*, pp. 69–75; on Murray's views, see "Toll Roads Getting Federal Attention," *ENR* 152 (February 14, 1954): 21–22.

92. See BPR, *Report on Needs of Highway Systems, 1955–84: Cost of Construction and Statement of Financing*, 84th Cong., 1st sess., March 28, 1955, H. Doc. 120; and *BPR Annual Report* (1955).

93. V. T. Boughton, "Toll Highway Construction to Reach Peak in 1954," *ENR* 152 (January 17, 1954); 35.

94. "A Five-Year Boom Draws to a Close," *ENR* 156 (April 26, 1956): 22; also see BPR, *Progress and Feasibility of Toll Roads and Their Relation to the Federal-Aid Program*, 84th Cong., 1st sess., April 14, 1955, H. Doc. 139; "Toll Road Study," *ENR* 154 (June 2, 1955); 24; and *BPR Annual Report* (1955). On toll road problems in 1955, see "Skeleton at the Feast; West Virginia Turnpike a Warning to Investors," *Barrons's* 25 (December 1, 1955): 1; "Turnpike Troubles," *ENR* 154 (March 10, 1955): 23; and "Toll Roads: Is It the End of the Boom?" *Business Week* (February 11, 1956), pp. 43–44, 46, 48.

95. *BPR Annual Report* (1955), p. 1.

96. Cronk, "1954—Turning Point Year," pp. 44–45.

97. "Sound Start on Road Planning," *ENR* 153 (August 5, 1954): 96; see also "A National Highway Program," *Civil Engineering* 25 (February 1955): 82–84; and "Highways on a Grand Scale," *ENR* 153 (November 18, 1954): 21.

98. *BPR Annual Report* (1955), p. 1.

99. See BPR, *Public Utility Relocation Incident to Highway Improvement*, 84th Cong., 1st sess., 1955, H. Doc. 127; and *BPR Annual Report* (1955).

100. Duane L. Cronk, "Washington Newsletter," *Roads and Streets* 99 (August 1956): 19–20; see also U.S., Commission on Intergovernmental Relations, Study Committee on Federal Aid to Highways, *Federal Aid to Highways* (Washington, D.C., 1955).

101. Quoted in U.S., Congress, House, Public Works Committee, *Hearings on H.R. 234, H.R. 235, H.R. 2127: Restatement of Highway Laws*, 84th Cong., 1st sess., 1955, p. 29. This report was the basis for legislation passed in 1958. See *BPR Annual Report* (1958), pp. 6–7.

102. See Cronk, "Washington Newsletter," August 1956, pp. 19–20; see also "Road Bill Ready," *ENR* 156 (June 28, 1956): 25.

103. Kennedy interview.

104. See, for example, Owen and Dearing, *Toll Roads*, pp. 156–60.

105. Davies, *Age of Asphalt*, p. 4.

Conclusion

1. See F. W. Sargent, "Hidden Subsidies," *World's Work* 60 (July 1931): 25–27; and Thomas H. MacDonald, "Highways and Railroads," *Scientific American* 146 (April 1932): 222–25.

2. U.S., Commission on Intergovernmental Relations, *Federal Aid to Highways*, p. 14.

3. See Gunnel, "Technocratic Image," pp. 392–416.

4. Quoted in U.S., Congress, Senate, *Hearings on S. 3424 and H.R. 7941*, p. 185.

5. MacDonald, "A Choice of Guides," *American Highways* 32 (January 1953): 6.

6. "More Roads, Less Politics," *Business Week* (March 15, 1952), pp. 95–98.

7. Duane L. Cronk, "Washington Newsletter," *Roads and Streets* 99 (May 1956): 19-20.

8. "Normal Politics Won't Work," *ENR* 158 (March 3, 1955): 10.

9. Jerome Allan Cohn, "Urban Background to the Interstate Highway Program: The Planning and Politics of Highways in Syracuse, New York, 1944–60." (Ph.D. diss., Syracuse University, 1978); see also Leavitt, *Superhighway—Superhoax*; and Caro, *Moses*, pp. 850–958. For a positive assessment, see Rae, *The Road and the Car*, pp. 187–94, 306–31. The following information on the BPR is from *America's Highways*, pp. 213–37, 318–19.

10. See HRB, *The WASHO Road Test*, pt. 1, Special Report No. 18 (Washington, D.C., 1954), pt. 2, Special Report No. 22 (Washington, D.C., 1955); and idem, *The AASHO Road Test: History and Description of the Project*, Special Report No. 61A (Washington, D.C., 1961).

11. HRB, *Ideas and Actions*, pp. 136, 138–39.

12. "The 'Chief' Retires," *ENR* 150 (April 9, 1953): 128; "He Revised Our

Roads Maps," *Compressed Air Magazine* 58 (August 1953): 228; and "Mr. Highways; Retires from Bureau of Public Roads," *Roads and Streets* 96 (May 1953): 24. Personnel changes were reported in *American Highways* 29–35 (1950–56); see also *America's Highways*, pp. 176-84, 490–91, 526–27.

13. du Pont, "What's Going on in Public Roads?" p. 11; see also *BPR Annual Report* (1953), pp. 4–5.

14. See "Changes in Federal Highway Policies to Await Administration Studies," *ENR* 15 (April 23, 1953): 21–22; and "Important Bureau of Public Roads Division Merged," *American Highways* 32 (October 1953): 15.

15. See Duane L. Cronk, "Preview of the Bureau's Big New Responsibility," *Roads and Streets* 99 (July 1956): 96–98, 107; idem, "Washington Newsletter," August 1956, pp. 19–20; and John A. Volpe, "Administrator's Annual Address," *American Highways* 26 (January 1957): 4, 25–28.

16. See *BPR Annual Report* (1957), p. 4; (1958), pp. 35–36.

17. See "MacDonald Retires as Commissioner of Public Roads." Fred Benson, Dean-Emeritus of Engineering, Texas A & M University, College Station, who worked with the chief when he arrived at the Texas Transportation Institute at the university, confirmed this point. According to Benson, MacDonald saw the 90 to 10 formula as the greatest mistake in highway development, one that was guaranteed to destroy the harmony between federal and state authorities.

18. Information from Dr. Chris Hendrickson, Associate Professor of Civil Engineering, Carnegie-Mellon University, Pittsburgh.

19. Wright, "Jonah and the Whale," *American Highways* 42 (October 1963): 3, 8–9.

20. Interview with David Cook, Staff Archaeologist, Environmental Section Staff, Iowa Department of Transportation, Ames, November 17, 1980.

21. *America's Highways*, p. 297.

Note on Sources

In spite of the importance of highways in contemporary American society, there have been few efforts to examine the development of either roads or the policies that guided their construction; in fact, no general history of American highways exists. Highway policy development before 1940 has been examined by Charles L. Dearing in his brief study for the Brookings Institution, *American Highway Policy* (Washington, D.C., 1942), and the U.S., Department of Transportation, Federal Highway Administration, *America's Highways, 1776–1976: A History of the Federal-Aid Program* (Washington, D.C., 1977), provides an uncritical overview of events related to roads and the BPR. A better starting point is John B. Rae, *The Road and the Car in American Life* (Cambridge, Mass., 1971), whose optimistic evaluation of the automobile's impact on American society also discusses the emergence of the highway system. But Rae, like other writers on America's adoption of the car, focuses more on the car than the highway, thus implicitly presenting the road as an adjunct to the automobile.

Road building, however, involved a far larger cast of characters than just the automobile, and two main sources provided the foundation for this attempt to delineate the BPR's role in those complex proceedings. The first are technical and engineering periodicals, which published numerous articles on new machinery, engineering details, and other innovations as well as discussions of individual projects of special note; they are invaluable for understanding the basic technical factors of road building. But these journals also provide clear and detailed coverage of political developments related to highways. The most important are the *Engineering News-Record* and its two predecessors, *Engineering News* and *Engineering Record*; they were an informal spokesperson for the engineering profession. Equally important, the engineering image of apolitical expertise emerges clearly from the journal's pages, especially through the editorials. Other periodicals that were similarly essential to this study were *American Highways* (begun by the American Association of State Highway Officials in 1921), *Public Roads* (the research journal launched by the BPR in 1918), *Good Roads* (1901–31), *Roads and Streets* (1926–56), *Proceedings of the American Road Builders' Association* (1911–41), *Better Roads*, *Municipal and County Engineering* (1896–1928), *Engineering and Contracting* (1905–25), and *Civil Engineering* (1930–56). An important means of access to these journals are the annual volumes of the *The Engineering Index*, later known as *The Industrial Arts Index*.

Another important source of material was the Records of the Bureau of Public Roads, Record Group 30, National Archives, Washington National Records Center, Suitland, Maryland. Consisting of some 5,100 manuscript boxes, these records document all aspects of the bureau's activities, although the quality of data declined in the 1940s, and post-1950 material is available only by permission. Additional information can be found in the Records of the Secretary of Agriculture, Record Group 16, National Archives, Washington, D.C. As helpful as the manuscript material were the bureau's annual report, released from 1893 through 1938 by the Department of Agriculture, from 1939 to 1949 by the Federal Works Administration, from 1950 to 1968 by the Department of Commerce, and thereafter by the Department of Transportation; one must compensate for the changing name of the agency itself (ORI, 1893–98; OPRI, 1899–1904; OPR, 1905–17; BPR, 1918–38; PRA, 1939–49; BPR, 1950–68; Federal Highway Administration, 1968–date). These reports document not only the BPR's activities but also its opinions on all important issues. Their central theme was the ideology of apolitical expertise, an argument repeated in the large array of technical bulletins and circulars released by the bureau from 1894 to 1916 and in the technical studies published less frequently after 1920. A brief but useful overview of the BPR's early history is W. Stull Holt, *The Bureau of Public Roads: Its History, Activities and Organization* (Baltimore, 1923); and one should also see A. N. Johnson, "Aims and Purposes of the Office of Public Roads, United States Department of Agriculture," *Engineering Record* 52 (September 16, 1905): 326; "The Organization and Equipment of the Office of Public Roads, United States Department of Agriculture," *Engineering Record* 54 (August 25, 1906): 205–7; "The United States Office of Public Roads," *Engineering Record* 62 (December 17, 1910): 709–11; and Thomas H. MacDonald, "The Bureau of Public Roads," *Civil Engineering* 1 (March 1931): 494.

Little personal information exists about Logan Page, head of the agency from 1905 to 1918, and Thomas MacDonald, chief of the BPR from 1919 to 1953. MacDonald especially is portrayed only as the stereotypical engineer absorbed in his work in William E. Lind, "Thomas H. MacDonald: A Study of the Career of an Engineer and His Influence on Public Roads in the United States, 1919–1953" (M.A. thesis, The American University, 1965), and the uncatalogued Papers of Thomas H. MacDonald, University Archives, Texas A & M University, College Station, add little to this image. The Reports of Thomas H. MacDonald, Bentley Historical Library, University of Michigan, Ann Arbor, contain only copies of the articles and addresses prepared by "the Chief" during his long career.

Although the secondary literature on the general developments in the history of American highways is limited, a number of important studies have treated various aspects of the larger story. For the period before 1900, Phillip Mason, "The League of American Wheelmen and the Good Roads Movement" (Ph.D. diss., University of Michigan, 1957), provides an important examination of the good roads movement that stresses the bicycle's role. One should also see Peter Hugill, "Good Roads and the Automobile in the United States, 1880–1929," *Geographical Review* 72 (July 1982): 332–40. Roy V. Scott, *Railroad Development Programs in the Twentieth Century* (Ames, Iowa, 1985), contains a short section on railroad involvement in the good roads movement, an issue explored more fully in Bruce E. Seely, "Railroads, Good Roads, and Motor Vehicles," *Railroad History* No. 155 (Autumn 1986): 35–63. As the push for good roads was largely local until 1910, developments in individual states should be followed in such studies as Harry Wilson McKnown, "Roads and Reform: The Good Roads Movement in North Carolina, 1885–1921" (M.A. thesis, University of North

Carolina at Chapel Hill, 1972); Arthur W. Dean, "Massachusetts Highways," *Proceedings of the Boston Society of Civil Engineers* 16 (December 1929): 493–510; Hugh Myron Hoyt, Jr., "The Good Roads Movement in Oregon, 1900–1920" (Ph.D. diss., University of Oregon, 1966); and Kenneth Earl Peters, "The Good Roads Movement and the Michigan State Highway Department, 1905–1917" (Ph.D. diss., University of Michigan, 1972). Wayne E. Fuller provides vital information on the connections between mail service and the campaign for good roads in *RFD: The Changing Face of Rural America* (Bloomington, Ind., 1964).

After 1900, road-building efforts accelerated as highway improvement became a typical reform of the Progressive era. The general studies of this period that this work drew upon are Robert Wiebe, *The Search for Order, 1877–1920* (New York, 1967); Samuel P. Hays, *Conservation and the Gospel of Efficiency: The Progressive Conservation Movement, 1840–1920* (Cambridge, Mass., 1959), both of which discuss the role of experts in reform. John C. Burnham's "Essay" in John D. Buenker, John C. Burnham, and Robert M. Crunden, eds., *Progressivism* (Cambridge, Mass., 1977), pp. 3–29, provided an essential understanding of the Progressives as a whole; see also Daniel T. Rogers, "In Search of Progressivism," *Reviews in American History* 10 (December 1982): 113–32. The Progressive embrace of expertise as a solution to technical and social problems has been shown in such works as Martin V. Melosi, *Garbage in the Cities: Refuse, Reform, and the Environment, 1880–1980* (College Station, Tex., 1981); and Stanley K. Schultz and Clay McShane, "To Engineer the Metropolis: Sewers, Sanitation, and City-Planning in Late Nineteenth Century America," *Journal of American History* 65 (September 1978): 389–411.

The actual construction of roads after 1900 is best observed in the technical periodicals, which consistently echoed the idea that highways should be a matter for experts, not politicians. A publication by the OPRI, *Public Road Mileage, Revenues and Expenditures in the United States in 1904*, Bulletin No. 32 (Washington, D.C., 1907), provided the first statistical breakdown of American highways; this was updated by U.S., Department of Agriculture, Office of Public Roads, *Mileage and Cost of Public Roads in the United States in 1909*, Bulletin No. 41 (Washington, D.C., 1912); and idem, *Public-Road Mileage, Revenues, and Expenditures in the United States, 1914—A Summary*, Department Bulletin No. 390 (Washington, D.C., 1917). One reason for the improvements that became apparent in successive reports was the creation of state highway departments. The activities of two of these agencies can be followed in Iowa State Highway Commission, *Annual Report* (Ames, Iowa, 1905–16); and Michigan State Highway Department, *Biennial Report of the Michigan State Highway Department* (Lansing, 1905/6–1915/16). The effort to pass federal legislation to fund road building from 1910 to 1921 can be traced in the BPR's records, *Engineering News-Record*, the debates in the *Congressional Record*, congressional hearings, and other documents. The most important such document is U.S., Congress, *Report of the Joint Congressional Committee on Federal-Aid in the Construction of Post Roads*, 63d Cong., 3d sess., 1915, H. Doc. 1510. Crucial for following the legislative struggle of 1918–21 are the Papers of Roy D. Chapin, president of the Hudson

Motor Car Company and head of the National Automobile Chamber of Commerce's Highway Committee, in the Michigan Historical Collection, Bentley Historical Library, University of Michigan, Ann Arbor. Chapin's daily correspondence with the chamber's lobbyist, Pyke Johnson, details the abortive effort to create a National Highway Commission.

Road-building developments in the 1920s and 1930s conformed to the associative ideal of cooperation that dominated thinking about business-government relations after World War I. The best discussions of this philosophy are Ellis Hawley, "Herbert Hoover, the Commerce Secretariat, and the Vision of an 'Associative State,'" *Journal of American History* 61 (June 1974): 116–40; and Ellis Hawley, ed., *Herbert Hoover as Secretary of Commerce: Studies in New Era Thought and Practice* (Iowa City, 1981). One should also see Louis Galambos, "The Emerging Organizational Synthesis in Modern American History," *Business History Review* 44 (Summer 1970): 279–90. An analysis of the federal-aid system of road building can be found in Austin F. MacDonald, *Federal Aid: A Study of the American Subsidy System* (New York, 1928). Another development central to the history of highway financing is documented by John C. Burnham, "The Gasoline Tax and the Automobile Revolution," *The Mississippi Valley Historical Review* 48 (December 1961): 435–59. Despite the importance of road building as a relief measure during the Depression, the origin and development of these programs must be pieced together from the traditional sources on FDR and the New Deal. These include Harry Hopkins, *Spending to Save: The Complete Story of Relief* (New York, 1936); Harold L. Ickes, *Back to Work: The Story of the PWA* (New York, 1935); Rexford G. Tugwell, *Roosevelt's Revolution: The First Year—A Personal Perspective* (New York, 1977); Elliott Rosen, *Hoover, Roosevelt, and the Brain Trust: From Depression to New Deal* (New York, 1977); and Robert Sherwood, *Roosevelt and Hopkins: An Intimate History*, rev. ed. (New York, 1950). The weight of the evidence confirms the suggestion of recent scholarship that continuity existed between Hoover and the New Deal, as shown by Martin A. Fausold and George T. Mazuzan, eds., *The Hoover Presidency: A Reappraisal* (Albany, 1974).

The BPR was, of course, active in industrial research, a topic that has recently attracted a great deal of attention by historians of technology. Most of the resulting studies have focused on the development of industrial research laboratories in large corporations, with the most important being Leonard S. Reich, *The Making of American Industrial Research: Science and Business at GE and Bell, 1876–1926* (New York, 1985); George Wise, *Willis R. Whitney, General Electric, and the Origins of U.S. Industrial Research* (New York, 1985); David F. Noble, *America by Design: Science, Technology, and the Rise of Corporate Capitalism* (New York, 1977); and Stuart W. Leslie, *Boss Kettering: Wizard of General Motors* (New York, 1983). Other, older works have shown the role of the federal government in developing research programs, with A. Hunter Dupree, *Science in the Federal Government: A History of Policies and Activities to 1940* (Cambridge, Mass., 1957), and T. Swann Harding, *Two Blades of Grass: A History of Scientific Development in the U.S. Department of Agriculture* (Norman, Okla., 1947), being the most significant. Dupree and Harding indicate that the BPR's research program was by no means unusual, for the research motives visible in large corporations also applied to research conducted in the associative style by the BPR. Highway research efforts can be followed in numerous articles in engineering periodicals as well as in National Academy of Sciences, National Research Council, HRB, *Ideas and Actions: A History of the Highway Research Board, 1920–1970* (Washington, D.C., c. 1970); and idem, *Proceedings of the Annual Meeting of the Highway Research Board*, vols. 1–25 (Washington, D.C., 1920–45). In Bruce E. Seely, "The Scientific Mystique in Engineering: Highway Research at the Bureau of Public Roads, 1918–1940," *Technology and Culture* 25 (October 1984): 798–831, the BPR's research program is considered in greater detail than in this book.

The establishment of industrial standards has received far less attention from historians than research, but recent studies have stressed the dominance of corporations in this area. The BPR's experience, however, suggests that the desire for the efficiency of standards was broadly based in the engineering profession by 1900. The best discussions of standardization are Bruce Sinclair, *Centennial History of the American Society of Mechanical Engineers, 1880–1980* (Toronto, 1980); Noble, *America by Design*; Dickson Reck, ed., *National Standards in a Modern Economy* (New York, 1952); and George V. Thompson, "Intercompany Technical Standardization in the Early American Automobile Industry," *Journal of Economic History* 14 (Winter 1954): 1–20. A brief introduction to the ASTM is in C. L. Warwick, "The Work in the Field of Standardization of the American Society for Testing Materials," *Annals of the American Academy of Political and Social Science* 137 (May 1928): 49; see also ASTM, *Fifty Year Index of ASTM Technical Papers and Reports, 1898–1950* (Philadelphia, 1952); ASTM, *Yearbook* (Philadelphia, 1905–15); and ASTM, *Proceedings of the Annual Meeting of the American Society for Testing Materials*, vols. 4–39 (Philadelphia, 1904–39). Bruce E. Seely, "Engineers and Government-Business Cooperation: Highway Standards and the Bureau of Public Roads, 1900–1940," *Business History Review* 58 (Spring 1984): 53–60, explains the bureau's work in establishing highway standards, stressing the central role of engineers in standardization regardless of their institutional affiliation.

Several aspects of road building after 1935 have attracted historical attention. This is especially true of urban highways. One can start with Clay McShane, "Transforming the Use of Urban Space: A Look at the Revolution in Street Pavements," *Journal of Urban History* 5 (May 1979): 289–307, and then proceed to Mark S. Foster, *From Streetcar to Superhighway: American City Planners and Urban Transit, 1900–1940* (Philadelphia, 1981); and Blaine Brownell, *The Urban Ethos in the South, 1920–1930* (Baton Rouge, 1975), for discussions of early urban road building. Several excellent case studies demonstrate how cities came to rely on automotive transit and the ramifications of that decision. On Chicago, see Paul Barrett, *The Automobile and Urban Transit: The Formation of Public Policy in Chicago, 1900–1930* (Philadelphia, 1983); on Los Angeles, see Robert M. Fogelson, *Fragmented Metropolis: Los Angeles, 1850–1930* (Cambridge, Mass., 1967);

Mark S. Foster, "The Model T, the Hard Sell, and Los Angeles Urban Growth: The Decentralization of Los Angeles during the 1920s," *Pacific Historical Review* 44 (November 1975):459–84; and David Brodsly, *L.A. Freeway: An Appreciative Essay* (Berkeley, 1981); on New York, see Robert Caro, *The Power Broker: Robert Moses and the Fall of New York* (New York, 1974); and on Atlanta, see Howard L. Preston, *Automobile Age Atlanta: The Making of a Southern Metropolis, 1900–1935* (Athens, 1979). The evolution of expressways has been less well documented, but information can be found in Norman Bel Geddes, *Magic Motorways* (New York, 1940); and Jean Labatut and Wheaton J. Lane, eds., *Highways in Our National Life: A Symposium* (Princeton, 1950); the only historical treatment is in Mark Rose's excellent volume, *Interstate: Express Highway Politics, 1914–1956* (Lawrence, Kans., 1979).

The sources of policy development after the mid-1930s consist, once again, of largely primary material. Two studies, however, do provide a general background: Louis Galambos's delightful analysis of the changing role of government in American society during the twentieth century, *America at Middle Age: A New History of the United States in the 20th Century* (New York, 1983); and Otis L. Graham's examination of government planning, *Toward a Planned Economy: From Roosevelt to Nixon* (New York, 1976). Mark Rose's *Interstate* offers the only in-depth analysis of the chaotic attempts to fashion a postwar road policy; this account generally conforms to Rose's argument while suggesting that engineers held a more central position in that struggle than Rose ascribed to them.

To trace the events of this period, however, one must rely on the BPR's records, the coverage in the *Engineering News-Record* and *Roads and Streets*, and the numerous congressional hearings and reports on highway legislation. The BPR's close relationship with the state highway officials during these and earlier years emerges in the AASHO, *AASHO—The First Fifty Years, 1914–1964* (Washington, D.C., 1965), and in the Papers of G. Donald Kennedy, Michigan Historical Collections, Bentley Historical Library, University of Michigan, Ann Arbor. The most important congressional documents include the BPR's *Toll Roads and Free Roads: Message from the President of the United States . . .*, 76th Cong., 1st sess., 1939, H. Doc. 272, which laid out the BPR's initial plan for an enlarged highway system. One should also see the U.S., Congress, House, *Interregional Highways: Message from the President of the United States Transmitting a Report of the National Interregional Highways Committee*, 78th Cong., 2d sess., 1944, H. Doc. 379, the updated version of the 1939 report that provided the basis for a postwar highway policy; U.S., Congress, House, Roads Committee, *Hearings on H.R. 2426: Federal-Aid for Post-War Highway Construction*, 2 vols., 78th Cong., 2d sess., 1944, the in-depth examination of roads that preceded the passage of postwar legislation; and U.S., Congress, House, Public Works Committee, *Hearings: National Highway Study*, 2 vols., 83d Cong., 1st sess., April 15–May 13, 1953, another extensive review of highway policy prior to the passage of the Interstate legislation in 1956.

Finally, on the role of experts in American society in general, one should consult the works on engineers in the Progressive period noted above as well as

Donald T. Critchlow, *The Brookings Institution, 1916–1952: Expertise and the Public Interest in a Democratic Society* (DeKalb, Ill., 1985); Edwin T. Layton, Jr., *The Revolt of the Engineers: Social Responsibility and the American Engineering Profession* (Cleveland, 1971); and John G. Gunnel, "The Technocratic Image and the Theory of Technocracy," *Technology and Culture* 23 (July 1982): 392–416.

Index